Nucleic Acids and Molecular Biology

Volume 27

Series Editor
Janusz M. Bujnicki
International Institute of Molecular
and Cell Biology
Laboratory of Bioinformatics and
Protein Engineering
Trojdena 4
02-109 Warsaw
Poland

For further volumes:
http://www.springer.com/series/881

Neocles Leontis • Eric Westhof
Editors

RNA 3D Structure Analysis and Prediction

Springer

Editors
Neocles Leontis
Bowling Green State University
Dept. Chemistry
Bowling Green Ohio
USA

Eric Westhof
Université de Strasbourg
Institut de biologie moléculaire
et cellulaire du CNRS,
Strasbourg France

ISSN 0933-1891
ISSN 1869-2486 (electronic)
ISBN 978-3-642-25739-1
ISBN 978-3-642-25740-7 (eBook)
DOI 10.1007/978-3-642-25740-7
Springer Heidelberg New York Dordrecht London

Library of Congress Control Number: 2012937843

Printed on acid-free paper

Springer is part of Springer Science+Business Media (www.springer.com)

Contents

1 **Introduction** 1
Michael Levitt

2 **Modeling RNA Molecules** 5
Neocles Leontis and Eric Westhof

3 **Methods for Predicting RNA Secondary Structure** 19
Kornelia Aigner, Fabian Dreßen, and Gerhard Steger

4 **Why Can't We Predict RNA Structure At Atomic Resolution?** . . . 43
Parin Sripakdeevong, Kyle Beauchamp, and Rhiju Das

5 **Template-Based and Template-Free Modeling of RNA 3D Structure: Inspirations from Protein Structure Modeling** 67
Kristian Rother, Magdalena Rother, Michał Boniecki, Tomasz Puton, Konrad Tomala, Paweł Łukasz, and Janusz M. Bujnicki

6 **The RNA Folding Problems: Different Levels of sRNA Structure Prediction** 91
Fredrick Sijenyi, Pirro Saro, Zheng Ouyang, Kelly Damm-Ganamet, Marcus Wood, Jun Jiang, and John SantaLucia Jr.

7 **Computational Prediction and Modeling Aid in the Discovery of a Conformational Switch Controlling Replication and Translation in a Plus-Strand RNA Virus** 119
Wojciech K. Kasprzak and Bruce A. Shapiro

8 **Methods for Building and Refining 3D Models of RNA** 143
Samuel C. Flores, Magdalena Jonikas, Christopher Bruns, Joy P. Ku, Jeanette Schmidt, and Russ B. Altman

9 **Multiscale Modeling of RNA Structure and Dynamics** 167
Feng Ding and Nikolay V. Dokholyan

10 Statistical Mechanical Modeling of RNA Folding: From Free Energy Landscape to Tertiary Structural Prediction 185
Song Cao and Shi-Jie Chen

11 Simulating Dynamics in RNA–Protein Complexes 213
John Eargle and Zaida Luthey-Schulten

12 Quantum Chemical Studies of Recurrent Interactions in RNA 3D Motifs 239
Jiří Šponer, Judit E. Šponer, and Neocles B. Leontis

13 Nonredundant 3D Structure Datasets for RNA Knowledge Extraction and Benchmarking 281
Neocles B. Leontis and Craig L. Zirbel

14 Ions in Molecular Dynamics Simulations of RNA Systems 299
Pascal Auffinger

15 Modeling RNA Folding Pathways and Intermediates Using Time-Resolved Hydroxyl Radical Footprinting Data 319
Joshua S. Martin, Paul Mitiguy, and Alain Laederach

16 A Top-Down Approach to Determining Global RNA Structures in Solution Using NMR and Small-Angle X-ray Scattering Measurements 335
Yun-Xing Wang, Jinbu Wang, and Xiaobing Zuo

17 RNA Structure Determination by Structural Probing and Mass Spectrometry: MS3D 361
A.E. Hawkins and D. Fabris

Appendix 391

Index 395

Contributors

Russ B. Altman Department of Bioengineering, Stanford, CA, USA, russ.altman@stanford.edu

Pascal Auffinger Architecture et réactivité de l'ARN, Université de Strasbourg, CNRS, IBMC, Strasbourg, France, p.auffinger@ibmc-cnrs.unistra.fr

Kyle Beauchamp Biophysics Program, Stanford University, Stanford, CA, USA, kyleb@stanford.edu

Michał Boniecki Laboratory of Bioinformatics and Protein Engineering, International Institute of Molecular and Cell Biology, Warsaw, Poland

Christopher Bruns Department of Bioengineering, Stanford, CA, USA

Janusz M. Bujnicki Laboratory of Bioinformatics and Protein Engineering, International Institute of Molecular and Cell Biology, Warsaw, Poland; Laboratory of Structural Bioinformatics, Institute of Molecular Biology and Biotechnology, Faculty of Biology, Adam Mickiewicz University, Poznan, Poland, iamb@genesilico.pl

Song Cao Department of Physics and Department of Biochemistry, University of Missouri, Columbia, MO, USA, caos@missouri.edu

Shi-Jie Chen Department of Physics and Department of Biochemistry, University of Missouri, Columbia, MO, USA, chenshi@missouri.edu

Kelly Damm-Ganamet DNA Software, Inc., Ann Arbor, MI, USA, Kelly@dnasoftware.com

Rhiju Das Biophysics Program, Stanford University, Stanford, CA, USA; Biochemistry Department, Stanford University, Stanford, CA, USA, rhiju@stanford.edu

Feng Ding Department of Biochemistry and Biophysics, School of Medicine, University of North Carolina, Chapel Hill, NC, USA

Nikolay V. Dokholyan Department of Biochemistry and Biophysics, School of Medicine, University of North Carolina, Chapel Hill, NC, USA, dokh@med.unc.edu

Fabian Dreßen Institut für Physikalische Biologie, Heinrich-Heine-Universität Düsseldorf, Düsseldorf, Germany, dressen@biophys.uni-duesseldorf.de

John Eargle Center for Biophysics and Computational Biology, Urbana, IL, USA, eargle@illinois.edu

D. Fabris The RNA Institute, University at Albany, Albany, NY, USA, fabris@albany.edu

Samuel C. Flores Department of Bioengineering, Stanford, CA, USA, samuelfloresc@gmail.com

A.E. Hawkins University of Maryland Baltimore County, Catonsville, MD, USA

Jun Jiang Department of Chemistry, Wayne State University, Detroit, MI, USA, Jonathan@chem.wayne.edu

Magdalena Jonikas Department of Bioengineering, Stanford, CA, USA

Wojciech K. Kasprzak Basic Science Program, SAIC-Frederick, Inc., NCI Frederick, Frederick, MD, USA, kasprzaw@mail.nih.gov

Joy P. Ku Department of Bioengineering, Stanford, CA, USA

Alain Laederach Department of Biology, University of North Carolina at Chapel Hill, Chapel Hill, NC, USA, alain@unc.edu

Neocles B. Leontis Chemistry Department, Bowling Green State University, Bowling Green, OH, USA

Michael Levitt Department of Structural Biology, Stanford School of Medicine, Stanford, CA, USA, michael.levitt@stanford.edu

Kornelia Linnenbrink Institut für Physikalische Biologie, Heinrich-Heine-Universität Düsseldorf, Düsseldorf, Germany, linnenbr@biophys.uni-duesseldorf.de

Paweł Łukasz Laboratory of Bioinformatics and Protein Engineering, International Institute of Molecular and Cell Biology, Warsaw, Poland

Zaida Luthey-Schulten Department of Chemistry, University of Illinois, Urbana, IL, USA, zan@illinois.edu

Joshua S. Martin Department of Biology, University of North Carolina at Chapel Hill, Chapel Hill, NC, USA, jsmartin@bio.unc.edu

Paul Mitiguy Department of Mechanical Engineering, Stanford University, Stanford, CA, USA, mitiguy@stanford.edu

Zheng Ouyang DNA Software, Inc., Ann Arbor, MI, USA, Zheng@dnasoftware.com

Tomasz Puton Laboratory of Bioinformatics and Protein Engineering, International Institute of Molecular and Cell Biology, Warsaw, Poland; Laboratory of Structural Bioinformatics, Institute of Molecular Biology and Biotechnology, Faculty of Biology, Adam Mickiewicz University, Poznan, Poland

Kristian Rother Laboratory of Bioinformatics and Protein Engineering, International Institute of Molecular and Cell Biology, Warsaw, Poland; Laboratory of Structural Bioinformatics, Institute of Molecular Biology and Biotechnology, Faculty of Biology, Adam Mickiewicz University, Poznan, Poland

Magdalena Rother Laboratory of Bioinformatics and Protein Engineering, International Institute of Molecular and Cell Biology, Warsaw, Poland; Laboratory of Structural Bioinformatics, Institute of Molecular Biology and Biotechnology, Faculty of Biology, Adam Mickiewicz University, Poznan, Poland

John SantaLucia Jr. DNA Software, Inc., Ann Arbor, MI, USA; Department of Chemistry, Wayne State University, Detroit, MI, USA, John@dnasoftware.com

Pirro Saro DNA Software, Inc., Ann Arbor, MI, USA, Pirro@dnasoftware.com

Jeanette Schmidt Department of Bioengineering, Stanford, CA, USA

Bruce A. Shapiro Center for Cancer Research Nanobiology Program, National Cancer Institute Frederick, Frederick, MD, USA, shapirbr@mail.nih.gov

Fredrick Sijenyi DNA Software, Inc., Ann Arbor, MI, USA, Fred@dnasoftware.com

Jiří Šponer Institute of Biophysics, Academy of Sciences of the Czech Republic, Brno, Czech Republic; CEITEC – Central European Institute of Technology, Brno, Czech Republic, sponer@ncbr.chemi.muni.cz

Judit E. Šponer Institute of Biophysics, Academy of Sciences of the Czech Republic, Brno, Czech Republic; CEITEC – Central European Institute of Technology, Brno, Czech Republic

Parin Sripakdeevong Biophysics Program, Stanford University, Stanford, CA, USA, sripakpa@stanford.edu

Gerhard Steger Institut für Physikalische Biologie, Heinrich-Heine-Universität Düsseldorf, Düsseldorf, Germany, steger@biophys.uni-duesseldorf.de

Konrad Tomala Laboratory of Bioinformatics and Protein Engineering, International Institute of Molecular and Cell Biology, Warsaw, Poland

Yun-Xing Wang Protein-Nucleic Acid Interaction Section, Structural Biophysics Laboratory, Center for Cancer Research, National Cancer Institute, National Institutes of Health, Frederick, MD, USA, wangyunx@mail.nih.gov

Jinbu Wang Protein-Nucleic Acid Interaction Section, Structural Biophysics Laboratory, Center for Cancer Research, National Cancer Institute, National Institutes of Health, Frederick, MD, USA, wangjinb@mail.nih.gov

Eric Westhof Architecture et Réactivité de l'ARN, Institut de Biologie Moléculaire et Cellulaire du Centre National de la Recherche Scientifique, Université de Strasbourg, Strasbourg, France, e.westhof@ibmc.u-strasbg.fr

Marcus Wood Department of Chemistry, Wayne State University, Detroit, MI, USA, mwoo@chem.wayne.edu

Craig L. Zirbel Department of Mathematics and Statistics, Bowling Green State University, Bowling Green, OH, USA, zirbel@bgsu.edu

Xiaobing Zuo Protein-Nucleic Acid Interaction Section, Structural Biophysics Laboratory, Center for Cancer Research, National Cancer Institute, National Institutes of Health, Frederick, MD, USA, zuox@mail.nih.gov sampling. Some of the problems associated with Monte Carlo moves have been solved in a very recent paper from our group (Sim et al. 2012).

Once one has an ensemble of putative structures, they need to be scored so as to pick out the best ones. Often such scoring is preceded by clustering, aimed at selecting representative structures from each energy basin. Clustering is a surprisingly tricky business, and we are pleased to have been able to develop a new method that seems to aid selection of near-native structures (Sim and Levitt 2011).

In conclusion, I am in complete agreement with the many groups who have contributed to the very impressive book: RNA structure prediction has clearly come of age and promises to make dramatic advances in the next few years. As such the publication of this book on RNA Structure Analysis and Modeling could not have been timed better!

References

Chopra G, Kalisman N, Levitt M (2010) Consistent refinement of submitted models at CASP using a knowledge-based potential. Proteins 78:2668–2678

Holm L, Sander C (1991) Database algorithm for generating protein backbone and side-chain co-ordinates from a C alpha trace application to model building and detection of co-ordinate errors. J Mol Biol 218:183–194

Krivov GG, Shapovalov MV, Dunbrack RL Jr (2009) Improved prediction of protein side-chain conformations with SCWRL4. Proteins 77:778–795

Levitt M (1969) Detailed molecular model for transfer ribonucleic acid. Nature 224:759–763

Levitt M (1972) Folding of nucleic acids. In: Polymerization in Biological Systems, Ciba Foundation Symposium 7, Elsevier, Amsterdam, pp. 146–171

Levitt M (1973) Orientation of double-helical segments in crystals of yeast phenylalanine transfer RNA. J Mol Biol 80:255–263

Levitt M (1992) Accurate modeling of protein conformation by automatic segment matching. J Mol Biol 226:507–533

Levitt M (2001) The birth of computational structural biology. Nat Struct Biol 8:392–393

Levitt M, Warshel A (1975) Computer simulation of protein folding. Nature 253:694–698

Minary P, Levitt M (2010) Conformational optimization with natural degrees of freedom: a novel stochastic chain closure algorithm. J Comp Chem 17:993–1010

Ponder JW, Richards FM (1987) Tertiary templates for proteins. Use of packing criteria in the enumeration of allowed sequences for different structural classes. J Mol Biol 193:775–791

Sim AY, Levitt M (2011) Clustering to identify RNA conformations constrained by secondary structure. Natl Acad Sci USA 108:3590–3595

Sim AY, Levitt, M, Minary, P (2012) Modelling and Design by Hierarchical Natural Moves. Natl Acad Sci USA 109:2890–2895

Simons KT, Kooperberg C, Huang E, David Baker D (1997) Assembly of protein tertiary structures from fragments with similar local sequences using simulated annealing and Bayesian scoring functions. J Mol Biol 268:209–225

Smith M, Brown NL, Air GM, Barrell BG, Coulson AR, Hutchison CA, Sanger F (1977) DNA sequence at the C termini of the overlapping genes A and B in bacteriophage φX174. Nature 265:702–705

Summa CM, Levitt M (2007) Near-native structure refinement using in vacuo energy minimization. Proc Natl Acad Sci USA 104:3177–3182

Tanaka S, Scheraga HA (1976) Medium- and long-range interaction parameters between amino acids for predicting three-dimensional structures of proteins. Macromolecules 9:945–950

Table 2.1 The various prediction programs described in the book with their main elements and outputs

Author	Program	Strategy	Output	CG	KB	MD	MC	2D	3D
Altman	RNABuilder (1)	Method: multiresolution, KB modeling with or without templates and fragment assembly	3D atomic structure model	x	x				x
Altman	NAST	Method: coarse-grained KB modeling with MD sampling	3D coarse-grained structure model	x	x	x			x
Bujnicki	ModeRNA	Input: sequence and 3D structure of template Method: homology and template-based modeling	3D atomic structure model						x
Bujnicki	SimRNA	Input: sequence Method: de novo coarse-grained modeling with Monte Carlo sampling with KB potential	Coarse-grained 3D model from which 3D atomic structure model is built using rebuild RNA	x	x		x		x
Chen	Vfold	Input: sequence and 2D structure Method: lattice-based coarse-grained modeling. MD refinement to obtain atomistic model	Free energies of pseudoknotted structures; coarse-grained 3D structures from which 3D atomic structure model is built	x		x			x
Das	FARNA/FARFAR	Input: sequence and knowledge-based potential Method: de novo modeling using KB potentials and MC sampling	3D atomic structure model		x		x		x
Dokholyan	DMD	Input: sequence and 2D and tertiary constraints Method: coarse-grained, multiscale conformational sampling by discrete MD	Full 3D atomic coordinates deduced from coarse-grained structure model. Folding thermodynamics	x		x			x
Major	MC-FOLD/MC-Sym	KB grammar for fragment assembly	3D atomic structure model		x				x
Santa Lucia	RNA123 (homology modeling module)	Input: sequence and 3D structure template Method: homology modeling with MD energy refinement	3D atomic structure model			x			x

(continued)

Table 2.1 (continued)

Author	Program	Strategy	Output	CG	KB	MD	MC	2D	3D
Shapiro	RNA2D3D	Input: 2D structure. Atomic modeling using fragment assembly and MD energy minimization	3D structures			x			x
Shapiro	MPGAfold	Genetic algorithm to optimize 2D structure	2D structures with pseudoknots					x	
Wang	G2G (global measurements to global structure)	Inputs: sequence and 2D structure; RDC and SAXS measurements method	Global structure of relative helical orientations					x	

Abbreviations: *CG* coarse graining, *KB* knowledge-based, *MD* molecular dynamics, *MC* Monte Carlo, *2D* two-dimensional, *3D* three-dimensional
References: RNABuilder (Flores et al. 2011); NAST (Jonikas et al. 2009); ModeRNA and SimRNA (Rother et al. 2011); Vfold (Cao and Chen 2011); Farna (Das and Baker 2007); FARFAR (Das et al. 2010); DMD (Ding et al. 2008); MC-FOLD/MC-Sym (Parisien and Major 2008); RNA2D3D (Martinez et al. 2008); MPGAfold (Shapiro et al. 2006); G2G (Wang et al. 2009)

of nucleotides), this question is relevant for those portions of RNA sequence that adopt stable architectures, required for their function during at least some period of time. In other words, given a sequence, produce a set of 3D coordinates for the nucleotides, that is biologically relevant and that satisfies the stereochemistry and physical chemistry of RNA molecules.

2.2.1 RNA Modeling Compared to Protein Modeling

In this regard, the parallels and contrasts between RNA and protein structure prediction and folding are apparent. Like proteins, RNA molecules are flexible linear polymers with astronomical conformational possibilities. Unlike proteins, RNA structures generally partition quite cleanly between secondary and tertiary hierarchical levels (Brion and Westhof 1997; Woodson 2010, 2011). Thus, as a rule, the first step in successful 3D modeling of RNA passes through a high-quality prediction of the main secondary structure elements. The state of the art in RNA secondary structure prediction is reviewed by Steger and coauthors in the third chapter of this volume. At the present state of our modeling efforts, the nature of the input data can play a decisive role at this stage of the process. Indeed, despite significant advances in 2D structure prediction, current methods still rely on theoretical approximations and an incomplete set of empirical energy parameters. Thus, working on a single RNA sequence may lead to incorrect evaluation of the importance or the role of one or more structural elements. The idiosyncrasies contained in single sequences can, however, often be ironed out by the use of multiple homologous sequences. Moreover, for RNA molecules, in contrast to proteins, one can obtain many additional experimental data containing much 3D information, using chemical or enzymatic probing and footprinting, small-angle X-ray scattering (SAXS), and cross-linking. The incorporation and computer use of such data changes the tractability of the problem. The chapters by Laederach, Wang and Fabris, and their coauthors (Chapters 15–17) address some of these issues and illustrate the challenges and power of integrating modern experimental data collection with modeling methods.

2.2.2 Defining the Inputs for RNA 3D Modeling

Inputs for the modeling of RNA 3D structure include, in addition to the sequence of the target RNA, the derived secondary structure and the sequences of available homologues, as well as all available experimental data. The database of known RNA 3D structures should also be considered an important resource for 3D modeling. This is especially the case for those approaches relying on a modular view of RNA architecture with the resulting assembly of RNA elements and modules (Jossinet et al. 2010; Westhof et al. 2011).

2.3 3D Modeling Methods and Approaches

A variety of modeling approaches are represented in the contributions to this volume. Some common themes emerge and will be summarized briefly with reference to specific chapters. As will become apparent to readers, promising approaches are rapidly adopted by multiple research groups, although specific implementations vary in ways that are usually not easy to discern. This volume focuses on methods that aim to achieve automaticity in 3D modeling, in the sense that they should require very little human intervention in the modeling process, beyond defining the inputs for the specific problem. The effort, rather, is focused "up front" on designing the algorithms and extracting and compiling relevant knowledge concerning RNA structure from structure databases for automated use by the implemented algorithms.

2.3.1 Homology Modeling

Automated methods generally address one or both of two distinct problems in biological structure prediction, namely, homology modeling and de novo prediction. Homology modeling concerns building atomically accurate 3D models of RNA molecules using at least one homologous 3D structure as template. RNA homology modeling draws on vast experience with protein homology modeling, and so considerable progress has been made already. The contributions of Altman, Bujnicki, and Santa Lucia focus, at least in part, on homology modeling and, between them, exhaustively address the issues involved.

2.3.2 De Novo Modeling

De novo prediction is necessary when no homologous 3D structure is known that can serve as a template for modeling. It is considerably more challenging than homology modeling, as it often requires generating a brand new 3D architecture from any known heretofore. As the goal is to do this without expert human input, the general approach is to generate large numbers of possible architectures and then to evaluate them, using what is already known about RNA structure. Automated, de novo 3D modeling approaches are therefore distinguished operationally by the kind of algorithm employed to generate potential 3D structural models, and also by the nature of the encapsulated knowledge concerning RNA structure that is used to score and rank models to arrive at a small set of predicted 3D structures, or in the favorable case, a single structure. The models generated by conformation-sampling algorithms are called "decoys" by practitioners. For the final output, most programs produce an all-atom predicted structure, which is generally quite "correct" in its local, stereochemical detail, in the sense that bond lengths and angles are within allowed ranges and the model contains no unphysical nonbonded contacts. But this local precision, which most programs achieve routinely, should not mislead users of predicted 3D models into assuming the model is accurate on larger, biologically relevant length scales, ranging from structures of modular motifs to overall folds and architectures.

The contributions of Altman (Chapter 8), Bujnicki (Chapter 5), Chen (Chapter 10), Das (Chapter 4), Dokhalyan (Chapter 9), Santa Lucia (Chapter 6), and Shapiro and their coworkers (Chapter 7) address de novo 3D modeling and among them cover the major methods in use today. All of these methods deploy some kind of algorithm to sample conformation space and some kind of knowledge-based methods to score and rank proposed solutions to the 3D prediction problem. In addition, most approaches rely on some kind of reduced representation of the RNA structure ("coarse graining") to speed up the calculations and allow more thorough exploration of conformational space with available computer resources. Coarse graining is an art that requires striking the right balance between speed of calculation and sufficiently detailed representation of RNA structure to capture the molecular features that stabilize the active conformations. Other ways to speed up conformational sampling involve modification of the algorithms that propagate the dynamics, as represented by the discrete molecular dynamics (DMD) method reported by Dokhalyan and coworkers.

2.3.3 *Defining the Outputs of Different Modeling Approaches*

The outputs of modeling studies depend on the modeling approach and the aim of the study. Indeed, output data can be full atomic coordinates for every single nucleotide or, in the case of coarse-grained methods, coordinates for only a subset of atoms or even a single pseudoatom representing each nucleotide. The different outputs are directly related to the granularity of the modeling approach. Nonetheless, nominally atomic-resolution models, when poorly refined or badly assembled, may be no better or even worse than coarse-grained models, if the characteristic base-pairing and base-stacking interactions of the structures are not represented accurately.

2.3.4 *Precision of Models vs. Accuracy of Models*

There is no necessary correlation between precision and accuracy, and models with comparable precision can differ substantially in the accuracy with which they predict the important interactions between nucleotides that define the RNA 3D structure. Thus, low-precision models can be very accurate (e.g., the original Watson–Crick model for DNA) and highly precise ones can be partly or totally inaccurate and thus misleading. Clearly, less-accurate models may not be at all pertinent for structural biology, while less-precise models can be very rich and enlightening. Still, these considerations should not be taken as license for not using in model building, whenever possible, high-resolution building blocks that are precise with respect to bond lengths and angles within nucleotides, and H-bond distances, van der Waals contacts, and relative orientations within base pairs and other interactions.

2.4 Databases for Extracting Knowledge

All of the precise structural data regarding RNA comes ultimately from atomic-resolution X-ray structures of nucleotides, oligonucleotides, and various biologically relevant structures, ranging in size from individual helical elements to the full ribosome. These data comprise all our basic knowledge of bond lengths, angles, and stereochemistry, as well as interaction preferences, including all types of base pairs and most stacking and base–backbone interactions. This information is used to build force fields and to infer rules for assembly of molecular moieties. These force fields and energetic rules are then used for producing and optimizing structures, sampling the conformational space, or simulating molecular dynamics. The quality and general value of the deduced force fields will strongly depend on the number and variety of structures available. In addition, the quality of the structures is of primary importance; it is directly related to the crystallographic resolution of the X-ray data and on the refinement process since a minor fraction of X-ray structures are obtained at true atomic resolution. One key parameter for compiling reference databases for knowledge extraction is the nonredundancy of the structures that are included in order to avoid bias in the deduced parameters. The chapter by Leontis and Zirbel (Chapter 14) addresses these issues and details a nonredundant database of structures extremely valuable for extracting knowledge about RNA as well as for benchmarking modeling strategies. In this respect, it is worth noting that less than 100 nonredundant RNA structures have been solved at 2-Å resolution or better.

2.5 Evaluating Models or "The Proof of the Pudding Is in the Eating"

As discussed above, 3D models are produced either to monitor our progress in the understanding and use of the physicochemical rules governing RNA architecture or to provide insight and help to experimentalists in the interpretation and meanings of biological data and in the design of new experiments. Although objectives may differ, in every case the models produced should be evaluated to assess their relevance to biological reality. Models that make testable predictions are especially valuable and, as emphasized above, need not be particularly precise. Additional experiments devised on the basis of a given model will provide the relevant tests for evaluating it. Depending on the outcome, the model may be retained and perhaps "tweaked," or it may be rejected and radically revised, leading to new biological insight and further experimental tests. On the other hand, to assess the validity of force fields as well as other empirical assembly rules, precise numerical comparisons have to be performed in a systematic way. This highlights the need for discriminating and meaningful metrics to compare and evaluate predicted vs. experimental structures.

2.5.1 Metrics for Evaluating Models

The most common metric is the root mean square deviations (RMSDs) on corresponding atoms between the predicted and experimental models. RMSDs are easy to compute and yield a simple measure. However, to interpret RMSD values, some critical length scales in RNA structures should be kept in mind for comparison: First, stacking distance between bases is about 3.4 Å; second, successive P–P distance in RNA helices is about 7 Å. While RMSD values below 3.4 Å are of real value, RMSD values beyond 8 Å must be treated with caution. In addition, RMSDs, as generally calculated with rigid-body fitting, spread the errors between two sets of coordinates over the whole ensemble. Consequently, even correctly modeled regions will not superimpose properly and thus will also contribute to the overall RMSD value. Therefore, RMSD values should be supplemented with local structural comparisons, including, for example, the numbers of correct base stackings and of correct Watson–Crick base pairs and, especially for 3D architectures, the number of non-Watson–Crick pairs, correct both with respect to pairing partners and base-pair types (Leontis and Westhof 2001). For a summary of the types of non-Watson–Crick base pairs, see the Appendix of this volume. We stress the importance of predicting the correct non-WC pairings as well as the correct base stackings, both of which are key because there is no three-dimensional architecture without non-Watson–Crick pairs and additional stackings between pairs. While a simple mapping of the 2D structure into a 3D structure does lead to a three-dimensional fold, such a fold will lack the additional stackings or RNA–RNA contacts that are characteristic of the complete 3D architecture. In short, correct predictions imply correct choices of new base stackings between single-stranded nucleotides and helices as well as new long-range base-pair contacts. For these reasons, two new metrics particularly suitable to RNA were introduced: the deformation index and the deformation profile (Parisien et al. 2009). The deformation index monitors the fidelity of the interaction network and encompasses base-stacking and base-pairing interactions within the target structure. The deformation profile highlights dissimilarities between structures at the nucleotide scale for both intradomain and interdomain interactions. These tools demonstrate that there is little correlation between RMSD and interaction network fidelity. To improve force fields or modeling approaches, it is mandatory to assess the origins of the errors. The deformation profile is a very useful tool for identifying the origins of incorrect modeling decisions.

2.5.2 *Necessity for Objective Evaluation of Modeling Efforts: RNA–CASP*

Structure prediction methods for proteins were boosted and consolidated by the CASP project (Critical Assessment of techniques for protein Structure Prediction), a systematic and worldwide evaluation of the predictions of new structures, prior to their publication (Kryshtafovych et al. 2005; Moult et al. 2009). CASP has proven extremely useful, productive, and constructive for benchmarking the progress made in the generation of new ideas and the objective assessment of the newly developed techniques. We believe that setting up a similar process will prove very healthy for the RNA structure-modeling field. To do so, several hurdles need to be overcome. In the case of RNA prediction, two levels would have to be distinguished, namely, the prediction of secondary structure and the modeling of 3D (tertiary) structure. The main issue, however, is how to establish efficient communication between research groups that determine RNA structures, whether at the secondary or tertiary structure levels, and research groups that predict RNA structures, so the latter can register their predictions before the structures are published. Clearly, despite the amazing advances in all aspects of the production of 3D RNA structures by X-ray crystallographic, NMR, or cryo-EM methods, the number of new structures produced per year remains rather low. The proposed process would follow these lines: (1) A structural group working on a new RNA structure (X-ray, NMR, chemical probing, cryoelectron microscopy, or mass spectroscopy) makes known their willingness to "play the game." (2) The group sends the sequence of the RNA under investigation to the coordinator. (3) The coordinator, without disclosing the identity of the experimental laboratory or the function and origin of the RNA, distributes the sequence to the theoreticians ready to tackle the challenge. Each theoretical group must agree not to disclose the sequence or distribute it further or to disclose its own progress or results in any fashion before publication of the structure by the experimental group. (4) The deadline for submitting structure predictions to the coordinator is agreed upon at the outset and generally will coincide with the date the experimental group submits their structures for publication. (5) During a special meeting, the coordinator discloses the theoretical results, and they are compared with the published experimental structures. (6) Special guidelines and rules for the comparisons will be agreed upon before the writing and publication of the analysis. Several laboratories dedicated to RNA bioinformatics around the world have expressed their keen interest to participate in such regularly held contests. The success and real progress generated by CASP in protein structure prediction should encourage us all to pursue this endeavor in the form of an ongoing RNA–CASP process. A first test of RNA–CASP was initiated at the end of 2010 and is now in the process of being published (Cruz et al. 2012).

2.6 Complications Limiting Modeling Approaches

Biological reality is complicated, and the applicability of physicochemical approaches based fundamentally on assumptions of thermodynamic equilibrium should always be properly evaluated as part of the theoretical modeling process. First, RNA molecules begin folding almost immediately as they are transcribed (cotranscriptional folding) so the issue of kinetic vs. equilibrium control in formation of biologically relevant structures is always a real one (Cruz and Westhof 2009). When the first structure to form is not the biologically relevant one, chaperone molecules are observed to play additional important roles. RNA molecules rarely act alone; on the contrary, they almost always act by binding to other RNA molecules or to proteins, and very frequently they bind to both types of macromolecules, if not also to small molecules.

An especially complicated problem is that of "induced fit," which occurs when the conformation adopted by an RNA molecule in isolation is not identical to that found in a complex with a small molecule ligand, antibiotic, or another RNA or a protein (Williamson 2000). Even small ligands, like hydrated magnesium ions, are difficult to treat in an appropriate fashion. Magnesium ions are especially difficult to treat when they bind, not as outer-sphere complexes (with a full share of coordinated water molecules), but instead as an inner-sphere complexes, with the loss of one or more water molecules and direct coordination to the RNA, generally in a state different from the original magnesium-free ion state (see Chapter 11 by P. Auffinger). Treating induced fit, at minimum, requires that the full dynamics of an RNA fragment be known in order to be able to select the proper conformation binding a given ligand. And it is not at all proven that the range of conformations accessible by the usual methods of molecular dynamics simulations, for example, actually covers the states obtained in the presence of the ligand or protein. Thus, one can study the dynamics of the A-site of the ribosome alone or in complex with an antibiotic (because crystal structures exist for all those different states), but the docking of an antibiotic to the A-site starting from an "empty state" (which is not the same as the state of the bound complex minus the ligand) has not been achieved yet (Moitessier et al. 2006).

2.7 Challenges for the Future: Dealing with Massive Data Streams and Connecting to Biology

Several main questions of great potential for biology continue to be actively pursued, and yet we have barely scratched the surface. One is the use of modeling predictions, firstly for searching noncoding RNAs in genomes and secondly for choosing among genomic regions those that are susceptible to fold into architectural domains or fragments (e.g., as riboswitches do). Another major question is the prediction of protein-binding sites along RNA sequences. Some consensus binding sequences are known, but in most cases, only knowledge of the RNA 3D fold allows the full understanding of the binding surface and RNA–protein contacts.

2.8 Conclusion

For modeling to be relevant to twenty-first century biological research, data pipelines need to be developed, maintained, and intelligently monitored to deal with the massive data streams produced by modern high-throughput sequencing methods. This means aiming for full automaticity at all steps of the computations. In this way, one should be able to link computational predictions with the experimental high-throughput technologies being constantly developed and refined. The establishment of such links between experimental and computational high-throughput techniques will bring us closer to the establishment of complete "RNA structuromes" for a given microbial or multicellular organism (Underwood et al. 2010; Weeks et al. 2011).

Acknowledgment NBL expresses his gratitude to Vassiliki Leontis for her support during the preparation of this book. NBL was supported by grants from the National Institutes of Health (grant numbers 1R01GM085328-01A1 to Craig Zirbel and NBL and 2R15GM055898-05 to NBL).

References

Amaral PP, Dinger ME, Mercer TR, Mattick JS (2008) The eukaryotic genome as an RNA machine. Science 319:1787–1789. doi:10.1126/science.1155472

Brion P, Westhof E (1997) Hierarchy and dynamics of RNA folding. Annu Rev Biophys Biomol Struct 26:113–137. doi:10.1146/annurev.biophys.26.1.113

Cruz JA, Westhof E (2009) The dynamic landscapes of RNA architecture. Cell 136:604–609. doi:10.1016/j.cell.2009.02.003

Cruz JA, MF Blanchet, M Boniecki, JM Bujnicki, SJ Chen, S Cao, R Das, F Ding, NV Dokholyan, SC Flores, L Huang, CA Lavender, V Lisi, F Major, K Mikolajczak, DJ Patel, A Philips, T Puton, J Santalucia, F Sijenyi, T Hermann, K Rother, M Rother, A Serganov, M Skorupski, T Soltysinski, P Sripakdeevong, I Tuszynska, KM Weeks, C Waldsich, M Wildauer, NB Leontis and E Westhof (2012). RNA-Puzzles: A CASP-like evaluation of RNA three-dimensional structure prediction. RNA 18:610-625. doi:10.1261/rna.031054.11

Cao S, Chen SJ (2011) Physics-based de novo prediction of RNA 3D structures. J Phys Chem B 115:4216–4226. doi:10.1021/jp112059y

Cruz JA, Blanchet MF, Boniecki M, Bujnicki JM, Chen SJ, Cao S, Das R, Ding F, Dokholyan NV, Flores SC, Huang L, Lavender CA, Lisi V, Major F, Mikolajczak K, Patel DJ, Philips A, Puton T, Santalucia J, Sijenyi F, Hermann T, Rother K, Rother M, Serganov A, Skorupski M, Soltysinski T, Sripakdeevong P, Tuszynska I, Weeks KM, Waldsich C, Wildauer M, Leontis NB, Westhof E (2012) RNA-Puzzles: A CASP-like evaluation of RNA three-dimensional structure prediction. RNA 18:610–625. doi:10.1261/rna.031054.111

Das R, Baker D (2007) Automated de novo prediction of native-like RNA tertiary structures. Proc Natl Acad Sci USA 104:14664–14669. doi:10.1073/pnas.0703836104

Das R, Karanicolas J, Baker D (2010) Atomic accuracy in predicting and designing noncanonical RNA structure. Nat Methods 7:291–294. doi:10.1038/nmeth.1433

Flores SC, Sherman MA, Bruns CM, Eastman P, Altman RB (2011) Fast flexible modeling of RNA structure using internal coordinates. IEEE/ACM Trans Comput Biol Bioinform 8:1247–1257. doi:10.1109/TCBB.2010.104

Hamming RW (1971) Introduction to applied numerical analysis. McGraw-Hill, New York

Hawking SW (2001) The universe in a nutshell. Bantam Books, New York

Jossinet F, Ludwig TE, Westhof E (2010) Assemble: an interactive graphical tool to analyze and build RNA architectures at the 2D and 3D levels. Bioinformatics 26:2057–2059. doi:10.1093/bioinformatics/btq321

Jonikas MA, Radmer RJ, Laederach A, Das R, Pearlman S, Herschlag D, Altman RB (2009) Coarse-grained modeling of large RNA molecules with knowledge-based potentials and structural filters. RNA 15:189–199. doi:10.1261/rna.1270809

Kryshtafovych A, Venclovas C, Fidelis K, Moult J (2005) Progress over the first decade of CASP experiments. Proteins 61(Suppl 7):225–236. doi:10.1002/prot.20740

Leontis NB, Westhof E (2001) Geometric nomenclature and classification of RNA base pairs. RNA 7:499–512

Moitessier N, Westhof E, Hanessian S (2006) Docking of aminoglycosides to hydrated and flexible RNA. J Med Chem 49:1023–1033. doi:10.1021/jm0508437

Moult J, Fidelis K, Kryshtafovych A, Rost B, Tramontano A (2009) Critical assessment of methods of protein structure prediction – Round VIII. Proteins 77(Suppl 9):1–4. doi:10.1002/prot.22589

Martinez HM, Maizel JV Jr, Shapiro BA (2008) RNA2D3D: a program for generating, viewing, and comparing 3-dimensional models of RNA. J Biomol Struct Dyn 25:669–683

Parisien M, Cruz JA, Westhof E, Major F (2009) New metrics for comparing and assessing discrepancies between RNA 3D structures and models. RNA 15:1875–1885. doi:10.1261/rna.1700409

Parisien M, Major F (2008) The MC-Fold and MC-Sym pipeline infers RNA structure from sequence data. Nature 452:51–55. doi:10.1038/nature06684

Rother M, Milanowska K, Puton T, Jeleniewicz J, Rother K, Bujnicki JM (2011) ModeRNA server: an online tool for modeling RNA 3D structures. Bioinformatics 27:2441–2442. doi:10.1093/bioinformatics/btr400

Shapiro BA, Kasprzak W, Grunewald C, Aman J (2006) Graphical exploratory data analysis of RNA secondary structure dynamics predicted by the massively parallel genetic algorithm. J Mol Graph Model 25:514–531. doi:10.1016/j.jmgm.2006.04.004

Taft RJ, Pheasant M, Mattick JS (2007) The relationship between non-protein-coding DNA and eukaryotic complexity. BioEssays 29:288–299. doi:10.1002/bies.20544

Underwood JG, Uzilov AV, Katzman S, Onodera CS, Mainzer JE, Mathews DH, Lowe TM, Salama SR, Haussler D (2010) FragSeq: transcriptome-wide RNA structure probing using high-throughput sequencing. Nat Methods 7:995–1001. doi:10.1038/nmeth.1529

Wang J, Zuo X, Yu P, Xu H, Starich MR, Tiede DM, Shapiro BA, Schwieters CD, Wang YX (2009) A method for helical RNA global structure determination in solution using small-angle x-ray scattering and NMR measurements. J Mol Biol 393:717–734. doi:10.1016/j.jmb.2009.08.001

Weeks KM, Mauger DM (2011) Exploring RNA structural codes with SHAPE chemistry. Acc Chem Res 44:1280–1291. doi:10.1021/ar200051h

Westhof E, Romby P (2010) The RNA structurome: high-throughput probing. Nat Methods 7:965–967. doi:10.1038/nmeth1210-965

Westhof E, Masquida B, Jossinet F (2011) Predicting and modeling RNA architecture. Cold Spring Harbor Perspect Biol 3:doi:10.1101/cshperspect.a003632

Williamson JR (2000) Induced fit in RNA-protein recognition. Nat Struct Biol 7:834–837. doi:10.1038/79575

Woodson SA (2010) Compact intermediates in RNA folding. Annu Rev Biophys 39:61–77. doi:10.1146/annurev.biophys.093008.131334

Woodson SA (2011) RNA folding pathways and the self-assembly of ribosomes. Acc Chem Res. doi:10.1021/ar2000474

Chapter 3
Methods for Predicting RNA Secondary Structure

Kornelia Aigner, Fabian Dreßen, and Gerhard Steger

Abstract The formation of RNA structure is a hierarchical process: the secondary structure builds up by thermodynamically favorable stacks of base pairs (helix formation) and unfavorable loops (non-Watson–Crick base pairs; hairpin, internal, and bulge loops; junctions). The tertiary structure folds on top of the thermodynamically optimal or close-to-optimal secondary structure by formation of pseudoknots, base triples, and/or stacking of helices. In this chapter, we will concentrate on available algorithms and tools for calculating RNA secondary structures as the basis for further prediction or experimental determination of higher order structures. We give an introduction to the thermodynamic RNA folding model and an overview of methods to predict thermodynamically optimal and suboptimal secondary structures (with and without pseudoknots) for a single RNA. Furthermore, we summarize methods that predict a common or consensus structure for a set of homologous RNAs; such methods take advantage of the fact that the structures of noncoding RNAs are more conserved and more critical for their biological function than their sequences.

3.1 Introduction

In this review, we will concentrate on software tools intended for prediction of secondary structure(s) of a given RNA sequence. The first such computational tool available was mfold (Zuker and Stiegler 1981); in the past 30 years, however, it was improved and refined several times (Zuker 2003). It is still commonly used, but it is now replaced by the UNAfold package (Markham and Zuker 2008), which includes several features not available in mfold. The two major alternative packages of

K. Aigner • F. Dreßen • G. Steger (✉)
Institut für Physikalische Biologie, Heinrich-Heine-Universität Düsseldorf, Düsseldorf, Germany
e-mail: aigner@biophys.uni-duesseldorf.de; dressen@biophys.uni-duesseldorf.de; steger@biophys.uni-duesseldorf.de

N. Leontis and E. Westhof (eds.), *RNA 3D Structure Analysis and Prediction*, Nucleic Acids and Molecular Biology 27, DOI 10.1007/978-3-642-25740-7_3,

comparable or even greater scope are the Vienna RNA (Hofacker 2003) and the RNAstructure (Reuter and Mathews 2010) packages. All rely on a simplifying thermodynamic model of nearest-neighbor interaction; we will briefly summarize this model in Sect. 3.2.1. In Sect. 3.2.2, we present some of the available tools.

Because all tools use the same basic thermodynamic model and associated thermodynamic parameters, they "know" about special features of certain loops: for example, parameters of thermodynamically extra-stable hairpin loops (for a review, see Varani 1995) or small internal loops with non-Watson–Crick base pairs are taken into account (e.g., see Xia et al. 1997), but no tool mentions such details in its output. More complex arrangements, for example, stacking of helices in multi-branched loops, are not taken into account, by and large, because of the increased computational complexity and the lack of relevant parameters. Furthermore, all of the abovementioned tools disregard pseudoknots, which are important structural features in many noncoding as well as messenger RNAs. Thus, we will turn to the prediction of pseudoknotted RNA structures in Sect. 3.3.

In those cases where a set of two or more homologous RNA sequences is available, comparative sequence analysis methods can be applied to predict a consensus structure common to all sequences in the set. Such approaches, which we review in Sect. 3.4, are based on the observation that in many cases, RNA secondary and tertiary structures are more conserved than primary sequence and are of greater importance for the biological function.

We apologize to all authors whose methods and tools we have not mentioned in this review for lack of space.

3.2 RNA Secondary Structure Prediction Based on Thermodynamics

3.2.1 Overview of RNA Secondary Structure Formation

A secondary structure of an RNA sequence R consists of base stacks and loops. It is defined—at least in the context of this chapter—as

$$R = r_1, r_2, \ldots, r_N,$$

with the indices $1 \leq i \leq N$ numbering the nucleotides $r_i \in \{A, U, G, C\}$ in the $5' \rightarrow 3'$ direction. Base pairs are denoted by $r_i{:}r_j$ or, for short, $i{:}j$ with $1 \leq i < j \leq N$. Allowed base pairs are *cis*-Watson–Crick (WC; A:U, U:A, G:C, C:G) and wobble pairs (G:U, U:G). Formation of base pairs belonging to a given secondary structure is restricted by

$$j \geq 4 + i, \tag{3.1}$$

which gives the minimum size of a hairpin loop, and the order of two base pairs $i{:}j$ and $k{:}l$ has to satisfy

$$i = k \quad \text{and} \quad j = l, \tag{3.2}$$

or

$$i < j < k < l, \tag{3.3}$$

or

$$i < k < l < j. \tag{3.4}$$

Condition (3.2) allows for neighboring base pairs but disallows any triple strand formation; a base triple $j{:}k{:}l$ would force $i = k$ and $j \neq l$. Condition (3.3) allows for formation of several hairpin loops in a structure. Condition (3.4) explicitly disallows "tertiary" interactions; such interactions do, in fact, occur in many RNAs, for example, in pseudoknots (see Sect. 3.3).

Structure formation—from an unfolded, random coil structure, C, into the folded structure, S—is a standard equilibrium reaction with a temperature-dependent equilibrium constant, K:

$$\begin{aligned} &\mathrm{C} \rightleftharpoons \mathrm{S}, \\ &K = \frac{[\mathrm{S}]}{[\mathrm{C}]}, \\ &\Delta G_T^0 = -RT \ln K = \Delta \mathrm{H}^0 - T \cdot \Delta S^0. \end{aligned}$$

At the denaturation temperature $T_{\mathrm{m}} = \Delta H^0 / \Delta S^0$ (melting temperature or midpoint of transition), the folded structure S has the same concentration as the unfolded structure ($K = 1; \Delta G_{T_{\mathrm{m}}}^0 = 0$). This is only true if the structure S denatures in an all-or-none transition. In most cases, however, structural rearrangements and/or partial denaturation take place prior to complete denaturation, as temperature is increased.

The number of possible secondary structures of a single sequence grows exponentially ($\approx 1.8^N$) with the sequence length N (Waterman 1995). Accordingly, all possible structures S_i of a single sequence coexist in solution with concentrations dependent on their free energies $\Delta G^0(\mathrm{S}_i)$; that is, each structure is present as a fraction given by (3.5):

$$f_{\mathrm{S}_i} = \exp\left(\frac{-\Delta G_T^0(\mathrm{S}_i)}{RT}\right) / Q. \tag{3.5}$$

The partition function, Q, for the ensemble of all possible structures, is given by (3.6):

$$Q = \sum_{\text{all structures } \mathrm{S}_i} \exp\left(\frac{-\Delta G_T^0(\mathrm{S}_i)}{RT}\right). \tag{3.6}$$

The structure of lowest free energy is called the optimal structure or structure of minimum free energy (mfe). It is possible for a single sequence to fold into quite different structures with nearly identical energies. This is of special biological relevance for RNA switches (Garst and Batey 2009; Nagel and Pleij 2002). Thus, one should not assume that an RNA folds into a single, static structure.

The free energy ΔG_T^0 of a single structure S is calculated as a sum over the free energy contributions (enthalpy ΔH^0 and entropy ΔS^0) of all structural elements i and j of S at temperature T:

$$\Delta G_T^0 = \sum_i \left(\Delta H_{\text{stack}}^0 - T\Delta S_{\text{stack}}^0\right) + \sum_j \left(\Delta H_{\text{loop}}^0 - T\Delta S_{\text{loop}}^0\right). \tag{3.7}$$

In this calculation, a nearest-neighbor model with base-pair stacking and loop formation is assumed to be sufficient. Energetic contributions of adjacent base pairs are favorable ($\Delta G^0 < 0$) due to their stacking on top of each other to form regular helices. Formation of loops is often but not necessarily unfavorable ($\Delta G^0 > 0$); exact values depend on loop type, nucleotides neighboring the loop-closing base pair(s), as well as on the exact sequence of the loop and whether the loop nucleotides form a stable, structured motif. Loop types are classified according to the number of loop-closing base pair(s): a single base pair closes hairpin loops, two base pairs close bulge loops (with no nucleotides in one strand) and interior loops (with symmetric or asymmetric numbers of nominally unpaired nucleotides in both strands), and three or more base pairs close multiloops (also called bifurcations or junctions). Note that for a given interior loop of n nucleotides, there are up to $6 \times 6 \times n^4$ different sequence combinations with possibly different energetic contributions, when taking into account the six possibilities for each of the closing base pairs. The parameter set measured by the group of D. Turner is used almost universally (Mathews et al. 1999, 2004; Xia et al. 1998). Parameters are known only within certain error limits; because these errors are smallest near $T = 37°\text{C}$, mostly $\Delta G_{37°\text{C}}^0$ values are reported.

A loop should not be thought of as a floppy structural element: in many cases, loop nucleotides form distinct structures due to stacking and/or non-Watson–Crick (non-WC; Leontis et al. 2002; Stombaugh et al. 2009) interactions with other loop nucleotides. Famous examples are loop E of eukaryotic 5S rRNA and the multiloop of tRNA. Eukaryal loop E, which is the same as the sarcin/ricin loop, is an asymmetric internal loop of four and five bases in its parts; all nucleotides are involved in non-WC interactions including one triple-base interaction (Wimberly et al. 1993). In tRNA, stacking of multiloop-closing base pairs across the multiloop is a major energetic contribution to the stability of the cloverleaf and is critical for formation of the tRNA tertiary structure.

Compensation of the negatively charged phosphate backbone of nucleic acids by positively charged counter ions M^+ leads to stabilization of structural elements according to

$$\begin{gathered} \mathrm{C} + n \cdot \mathrm{M}^+ \rightleftharpoons \mathrm{S} \\ K = \frac{[\mathrm{S}]}{[\mathrm{C}] \cdot [\mathrm{M}]^n}. \end{gathered} \tag{3.8}$$

From this expression, a logarithmic dependence between denaturation temperature T_{m} and salt concentration (ionic strength) follows

$$\frac{\mathrm{d}T_{\mathrm{m}}}{\mathrm{d}\ln[\mathrm{M}]} = -n\frac{RT_{\mathrm{m}}^2}{\Delta H^0}. \tag{3.9}$$

All thermodynamic parameters for RNA structure formation were determined in 1 M NaCl. This is not far from the ionic conditions in cells, except when specific interactions with divalent cations play a role (Draper 2008; Ramesh and Winkler 2010). If necessary, however, values for the ionic strength dependence of a structure or a structural element may be found in the literature, including functions for G:C-content of the RNA, or for dependence upon various types of buffers (e.g., TRIS/borate) and cosolvents like formamide or urea (Klump 1977; Michov 1986; McConaughy et al. 1969; Record and Lohman 1978; Riesner and Steger 1990; Sadhu and Gedamu 1987; Shelton et al. 1999; Steger et al. 1980).

3.2.2 *Tools for RNA Secondary Structure Prediction Based on Thermodynamics*

Most users seek to predict the mfe structure for a given (single) sequence. For the answer, the most widely used tools (see Table 3.1) rely on (3.7); that is, their basic algorithm solves the optimization problem of finding the mfe structure of a given single sequence in the haystack of thermodynamically possible secondary structures via dynamic programming (Bellman and Kalaba 1960; Nussinov et al. 1978; Zuker and Stiegler 1981). The computational effort grows with the cube of the sequence length N, that is, $O(N^3)$.

All tools except UNAfold rely on the same recent set of thermodynamic parameters, which only allows for calculation of $\Delta G^0_{37^\circ\mathrm{C}}$; UNAfold uses a parameter set of enthalpy and entropy values that makes possible calculations at any relevant temperature.

Knowledge of the mfe structure might not be sufficient due to several reasons:

- Minor errors in the thermodynamic parameter set can lead to incorrect prediction of the mfe structure; nonetheless, the "true" mfe structure may be one of the suboptimal structures close in energy to the calculated mfe structure. While a further improvement of the accuracies of experimentally determined parameters is unlikely, improvements in secondary structure prediction by statistical evaluations of known structures look promising (Andronescu et al. 2007; Wu et al. 2009).
- Quite often, the mfe structure only accounts for a very tiny fraction of all possible structures; that is, a cluster of suboptimal structures, which are nearly identical to each other but different from the mfe structure, might account for a much higher fraction of all possible structures and thus be of higher (biological) relevance.
- Usually, it is assumed that structure formation is a hierarchical process: the tertiary structure builds on top of the most favorable secondary structure(s) (Brion and

Table 3.1 Tools for prediction of RNA secondary structure

Package name	Address[a]	Reference
UNAfold	C: dinamelt.bioinfo.rpi.edu/download.php[b,c,d]	Markham and Zuker (2008)
	W: dinamelt.bioinfo.rpi.edu/	Markham and Zuker (2005)
Mfold[e]	C: mfold.bioinfo.rpi.edu/download/[b,c]	Zuker (1989)
	W: mfold.bioinfo.rpi.edu/cgi-bin/rna-form1.cgi	Zuker (2003)
Vienna RNA	C: www.tbi.univie.ac.at/~ivo/RNA/[b,d]	Hofacker et al. (1994)
	W: rna.tbi.univie.ac.at/	Hofacker (2003)
RNAstructure	C: rna.urmc.rochester.edu/RNAstructure.html[b,c,d]	Reuter and Mathews (2010)
RNAshapes	C: bibiserv.techfak.uni-bielefeld.de/rnashapes/[b,c,d]	Steffen et al. (2006)
	W: bibiserv.techfak.uni-bielefeld.de/rnashapes/	Giegerich et al. (2004)
CentroidFold	C: http://www.ncrna.org/centroidfold/	Hamada et al. (2009)
	W: http://www.ncrna.org/centroidfold/	
Sfold	W: http://sfold.wadsworth.org/srna.pl	Chan et al. (2005)

The Vienna RNA, RNAstructure, and UNAfold packages include, for example, programs for prediction of the mfe structure and for partition function folding of a single sequence and for bimolecular structure prediction

[a]C, source code and binary are available at given address; W, address of Web service

[b]Unix, Linux

[c]MacOSX

[d]Windows

[e]Mfold is replaced by UNAfold

Westhof 1997; Cho et al. 2009; Tinoco and Bustamante 1999). But as long as the free energy of the RNA decreases upon tertiary structure formation, this structure may also able to fold starting from suboptimal secondary structures.

- Site-specific binding of multivalent ions, small molecules, or macromolecules, including proteins or RNAs, might influence the process of structure formation. Note, however, that the thermodynamic stability of even short RNA helices is larger than that of most proteins.

Consequently, the programs of Table 3.1, also predict individual, suboptimal secondary structures (like mfold, which is able to generate the thermodynamically best structure for each admissible base pair) or, alternatively, predict the probability of any base pair possible for a certain sequence by using (3.5) and (3.6) (like RNAfold); the prediction is usually represented as a dot plot (see Fig. 3.1b). This base-pairing probability matrix is easily converted to a plot showing the probability of each nucleotide to be paired or unpaired; this allows, for example, for comparison to chemical or enzymatic mapping data. In case certain bases are experimentally known to pair or to remain unpaired, mfold as well as RNAfold allow imposition of the corresponding constraints, so that the predicted structures and the dot plot satisfy the known constraints (see help and man pages of mfold and RNAfold, respectively, and Steger 2004).

Enumeration of all secondary structures is possible (Waterman and Byers 1985; Wuchty et al. 1999), but one must be aware of the huge number of structures that result, many of which are very similar to each other. In many cases, the algorithm implemented in RNAshapes (see Table 3.1) is more useful: it classifies all

a

```
                     CCCAAGCGCAAAGAUACCUGGGGAGACCCGAUAAGGGGUCCGCCCCCCUAUCUCAAGCGCGGGACACA
-30.70  (0.5217548)  (((..((((..(((((...((((.(((((......)))))..))))..)))))...)))))))..... 0.7931457  []
-29.20  (0.0457596)  (((..((((..(((((..((((....))))....((((....))))..)))))...)))))))..... 0.2064049  [[][]]
```

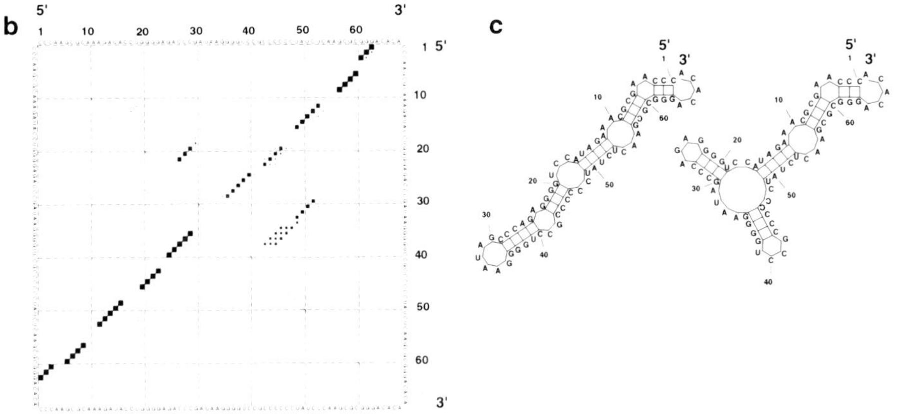

Fig. 3.1 Predicted secondary structures of an artificial RNA. (**a**) Output of RNAshapes: structures that are optimal in their shape class, from *left* to *right*: free energy (in kcal/mol at 37°C), probability of structure, structure in *bracket-dot* representation, probability of all structures in that shape class, and shape representation. (**b**) Dot plot produced by RNAfold: the mfe structure is represented in the *lower left triangle* and the probability of all possible base pairs in the *top right triangle*; the area of each dot is proportional to the pairing probability of this base pair. (**c**) Planar drawings of the mfe (*left*) and the optimal Y-shaped structure by ConStruct. The second line in (**a**), the *dots* in the *lower left triangle* of (**b**), and the *left drawing* in (**c**) represent the identical structure (with minimum of free energy); the third line in (**a**) and the *right drawing* in (**c**) represent also the identical, suboptimal structure, which consists of base pairs shown in the *upper right triangle* of (**b**)

structures into "abstract shapes" and predicts a "shape representative" (shrep) of each "shape class"; shreps differ significantly from each other (for a deeper insight into these terms see Giegerich et al. 2004). For example, the sequence shown in Fig. 3.1a is able to fold into two different shape classes; one is a stem loop and the other is a Y-shaped structure.

An alternative approach is to extract from the partition function the structure of "maximum expected accuracy" by maximizing the sum of the probabilities of base-paired (BP) and single-stranded (SS) nucleotides:

$$\sum_{(i,j)\in \mathrm{BP}} \gamma \cdot 2\mathrm{p}_{\mathrm{bp}}(i,j) + \sum_{k\in \mathrm{SS}} p_{\mathrm{SS}}(k). \tag{3.10}$$

Equation 3.10 indicates that the pairing probabilities can be weighted by a factor γ. This approach, including prediction of suboptimal structures, is implemented in MaxExpect (Lu et al. 2009), which is part of RNAstructure (see Table 3.1). MaxExpect was shown to improve the percentage of predicted pairs that are in known structures to the same level of sensitivity as free energy minimization (Lu et al. 2009). Similar approaches are implemented in CentroidFold and Sfold (see Table 3.1).

3.3 Pseudoknots

A pseudoknot is an RNA structure characterized by WC base pairing between nucleotides in a loop with complementary residues outside the loop. In contrast to proteins (Taylor 2007), no knots are known in RNA. Pseudoknots are a tertiary structural motif that occurs widely in RNA. They were first detected nearly 30 years ago as part of tRNA-like structures in plant viral RNAs (Rietveld et al. 1982). Some pseudoknots play a role in ribosomal frameshifting, while others are essential for the three-dimensional topology (and function) of many structured RNAs. In the following, we will give a description of pseudoknots and sequence constraints on their biophysical stability.

Databases on structural, functional, and sequence data related to RNA pseudoknots are maintained by PseudoBase (http://www.ekevanbatenburg.nl/PKBASE/PKBABOUT.HTML; van Batenburg et al. 2001) and PseudoBase++ (http://pseudobaseplus.utep.edu/; Taufer et al. 2008).

3.3.1 Conformation

A classical or H-type (hairpin-type) pseudoknot consists of two helical regions named S1 and S2 (or H1 and H2) and three loop regions L1, L2, and L3 (see Fig. 3.2). In sequence, the serial arrangement of these elements is S1, L1, S2, L2, S1′ (complement of S1), L3, and S2′ (complement of S2). The crossing order $S1<S2<S1'<S2'$ fulfills the definition of base pairs in a tertiary structure [see Sect. 3.2.1, (3.4)]. In many cases, the loop region L2 is absent and the two helices coaxially stack as shown in Fig. 3.2.

In the classical pseudoknot, the loops L1–L3 contain only unpaired nucleotides. There are, however, more complicated pseudoknots, in which these regions contain structured parts, including non-WC pairs and base triples; an example of a double pseudoknot with a stem-loop structure in L3 is shown in Fig. 3.3.

In the absence of L2, the helices S1 and S2 generally stack coaxially forming a structure closely resembling an uninterrupted A-form helix. Consequently, the loops L1 and L3 are not equivalent: L1 crosses the deep (major) groove and L3 the shallow (minor) groove of the double helix (see Fig. 3.2 right and Fig. 3.4 left). In a WC base pair, the distance between the phosphates (P'_0 and P_0 in Fig. 3.4) is about 1.7 nm; this distance can be bridged, for example, by a minimum of three nucleotides in a hairpin loop. In an RNA double helix, the minimal distance between a given phosphate on one strand and a phosphate on the opposite strand is about 1 nm when crossing the deep groove (e.g., P'_0 to P_{-7}; see Fig. 3.4). The smallest distance bridging the shallow groove is about 1.1 nm, the distance between P_0' and P_2 or P_3. These distances fit well to the sizes of small pseudoknots with coaxial helix stacking: 3–7 base pairs in stem regions bridged by loops of at least 2 nucleotides. A loop L2 and smaller L1 and/or L3 tend to introduce a bend in between the two stems. The shallow and wide minor groove

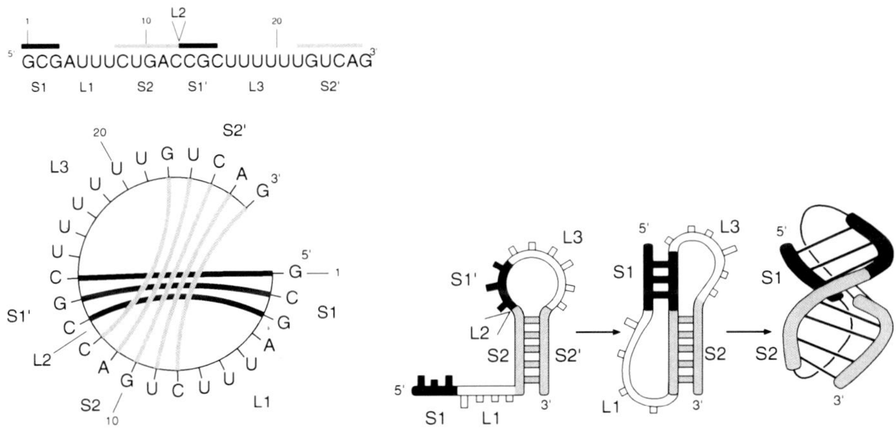

Fig. 3.2 Principle of RNA pseudoknotting. *Top left*: The sequence consists of complementary regions S1 and S1′ (*black boxes*) and S2 and S2′ (*gray boxes*) with intervening loop regions L1, L2, and L3; in this example, L2 is absent. *Bottom left*: In this circular graph, the two helical regions S1 and S2 lead to crossing lines connecting the base pairs. *Bottom right*. Formation of a pseudoknot is sketched as a series of steps consisting of formation of a hairpin with helix S2 and hairpin loop S1′ and L2, formation of the second helix S1, and finally a rotation of one of the helices by 180° which leads to coaxial stacking of the two helices. Modified from Pleij et al. (1985)

```
           10        20        30        40        50        60        70
  123456789|123456789|123456789|123456789|123456789|123456789|123456789|123
1 ACAGAGCGCGUACUGUCUGACGACGUAUCCGCGCGGACUAGAAGGCUGGUGCCUCGUCCAACAAAUAGAUACA
2 ((((.[[[[[..)))).............]]]]].......................................
3 <S1> <-S2>  <s1>             <s2->
4 .....[[[[[..............(((((]]]]].................................))))).
5      <-S2>              <-S3><s2->                                 <s3->
6 ((((.[[[[[..))))........(((((]]]]](((((.....((((....)))).))))).......))))).
```

Fig. 3.3 Example of a double pseudoknot with base pairs in loop regions. As shown in lines 2 and 3, the first pseudoknot consists of helices S1 (nucleotides 1–4 paired to 16–13) and S2 (6–10 with 34–30) with loops L1 (nucleotide 5), L2 (11–12), and L3 (17–29). As shown in lines 4 and 5, the second pseudoknot consists of S2 (6–10 with 34–30) and S3 (25–29 with 72–68) with L1 (11–24) and L3 (35–67). Line 6 summarizes both pseudoknots and shows the additional stem-loop in loop region 35–67. Example modified from PseudoBase PKB173 (van Batenburg et al. 2001)

of S1 allows for tertiary contacts, triple pairs, and hydrogen bonds between nucleotides of S1 with those of L3 (Batey et al. 1999; Nissen et al. 2001).

3.3.2 Thermodynamic Parameters for Pseudoknots

Our knowledge of thermodynamic parameters for pseudoknot formation is low. According to the end-to-end distances of a stem (see P–P distances in Fig. 3.4), energies are neither linearly dependent on loop length nor on stem length. Experimental determination of the parameters is quite complex due to the huge number of feasible pseudoknots with variable combinations of stem and loop lengths and sequence, and

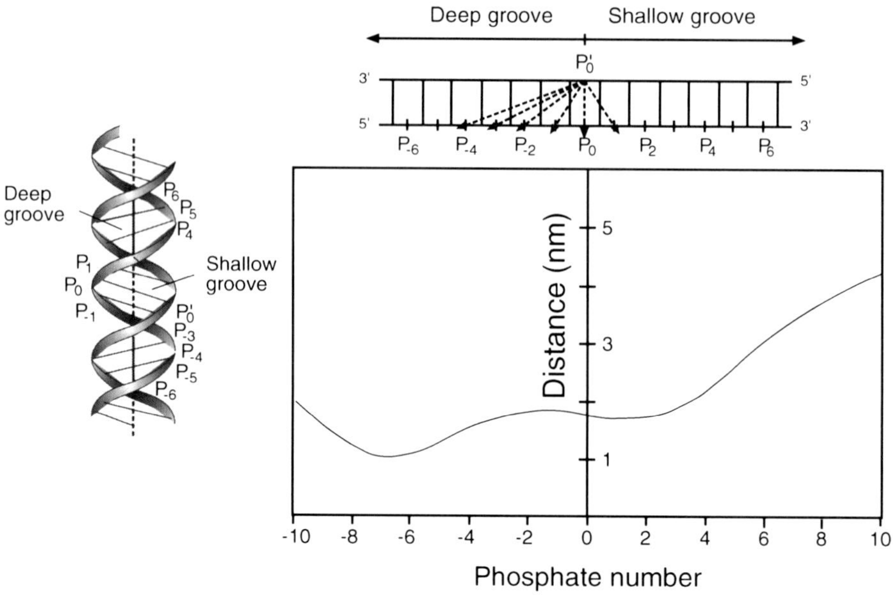

Fig. 3.4 Distance constraints on pseudoknots. Shown are the minimum distances between a certain phosphate P_0' to phosphates P located opposite on the other strand. Indices are negative for phosphates located in 5′ direction of the opposing strand and positive in 3′ direction. *Left*: Three-dimensional model of a helix. *Top*: Two-dimensional model of a helix; the *arrows* symbolize the distance from P_0' to the corresponding phosphate. *Bottom*: A graph with distances from P_0' to phosphates in the opposing strand. According to Pleij et al. (1985)

the difficulties inherent in evaluating parameters from overlapping and/or coupled unfolding transitions by optical melting or calorimetry (Gluick and Draper 1994; Gultyaev et al. 1999; Nixon and Giedroc 1998, 2000; Qiu et al. 1996; Soto et al. 2007; Theimer and Giedroc 1999, 2000; Theimer et al. 1998; Wyatt et al. 1990).

The total free energy of a pseudoknot is assumed to be the sum over free energies of stems, coaxial stacking, loop lengths and sequences, tertiary interactions, and assembling (Liu et al. 2010):

$$\Delta G^0_{\text{pseudoknot}} = \sum \Delta G^0_{\text{stems}} + \Delta G^0_{\text{coaxial stacking}} - T \sum \Delta S^0_{\text{loops}} + \Delta G^0_{\text{loop sequences}} + \Delta G^0_{\text{tertiary interactions}} + \Delta G^0_{\text{assemble}}.$$

$\Delta G^0_{\text{stems}}$ and $\Delta G^0_{\text{coaxial stacking}}$ can be calculated with experimentally determined nearest-neighbor parameters (Xia et al. 1998) and coaxial stacking parameters (Walter et al. 1994), respectively. Several computational models have been established for the remaining parameters (Cao and Chen 2006, 2009; Dirks and Pierce 2003, 2004; Gultyaev et al. 1999; Rivas and Eddy 1999). The models particularly determine the loop length dependence of $\Delta S^0_{\text{loops}}$. Taking into account volume exclusion effects of the loop strands and considering the loop length with respect to the length of the associated stem resulted in improvement of the models (statistical polymer model; Cao and Chen 2006, 2009). Details of $\Delta G^0_{\text{loop sequences}}$ are currently neither determined experimentally nor does a computational model exist. This energy does, however, contribute to the total energy, even if the loop sequence is not involved in tertiary interactions, as demonstrated experimentally by Liu et al. (2010). $\Delta G^0_{\text{tertiary interactions}}$ accounts for possible interactions between loops and stems; it is currently neither determined experimentally nor does a computational model exist. In some cases, tertiary interactions are more favorable than the maximum number of canonical base pairs (Liu et al. 2010). $\Delta G^0_{\text{assemble}}$ is assessed to account for the entropy change as the two subunits (the two stems with their associated loops) are assembled into the pseudoknot (Cao and Chen 2006).

3.3.3 Ionic Strength Dependence of Pseudoknots

For their formation, most pseudoknots need a relatively high ionic strength including the presence of divalent cations like Mg^{2+} (Gluick et al. 1997). Considering the structure of a simple pseudoknot, as depicted in Fig. 3.2 right, the reason for this is quite obvious: the stabilizing interactions realized upon formation of stems 1 and 2 and stacking of stem 1 on stem 2 are partly counteracted by the necessary loop formation and the close approach of four negatively charged phosphate backbones. To compensate for this increased charge density, "diffuse" (fully hydrated) Mg^{2+} ions seem generally to be sufficient; binding of dehydrated ions (inner-sphere complexes) to specific positions is not necessary (Soto et al. 2007).

3.3.4 Prediction Methods for Pseudoknots

None of the tools mentioned in Sect. 3.2.2 are capable of predicting pseudoknots or any form of tertiary interactions due to the restrictions in their dynamic programming algorithms. Expanding these algorithms to general pseudoknot prediction is difficult; actually, Lyngsø and Pedersen (2000) have proven that the general problem of predicting RNA secondary structures containing pseudoknots is NP complete for a large class of reasonable models of pseudoknots. Thus, several heuristic approaches were developed. In the following, we will mention several of the recent tools able to detect pseudoknots and other tertiary interactions in structure predictions; an incomplete list of available tools is summarized in Table 3.2.

PKNOTS, developed by Rivas and Eddy (1999), is the first dynamic programming algorithm that finds optimal, pseudoknotted RNA structures. Due to its computational effort of $O(N^6)$, its use is restricted to short sequences.

pknotsRG (Reeder and Giegerich 2004; Reeder et al. 2007) predicts the structure of an RNA sequence, possibly containing pseudoknots. Energies of possible

Table 3.2 Tools for prediction of tertiary interactions

Name	Address[a]	Effort[b]	Reference
ConStruct	C: http://www.biophys.uni-duesseldorf.de/construct3/		Wilm et al. (2008b)
DotKnot	W: http://dotknot.csse.uwa.edu.au		Sperschneider and Datta (2010)
HotKnots	C: http://www.cs.ubc.ca/labs/beta/Software/HotKnots		Ren et al. (2005)
	W: http://www.rnasoft.ca/cgi-bin/RNAsoft/HotKnots/hotknots.pl		
HPknotter	W: http://bioalgorithm.life.nctu.edu.tw/HPKNOTTER/		Huang et al. (2005)
ILM	C: http://www.cse.wustl.edu/~zhang/projects/rna/ilm/	$O(N^3) - O(N^4)$	Ruan et al. (2004)
	W: http://cic.cs.wustl.edu/RNA/		
KNetFold	W: http://knetfold.abcc.ncifcrf.gov/	$O(n^2)^c$	Bindewald and Shapiro (2006)
KnotSeeker	C: http://knotseeker.csse.uwa.edu.au/download.html		Sperschneider and Datta (2008)
	W: http://knotseeker.csse.uwa.edu.au/		
NUPACK	W: http://nupack.org/	$O(n^5)$	Dirks and Pierce (2003)
	C: http://nupack.org/downloads		
PKNOTS	C: ftp://selab.janelia.org/pub/software/pknots/	$\geq O(N^6)$	Rivas and Eddy (1999)
pknotsRG	C: http://bibiserv.techfak.uni-bielefeld.de/download/tools/pknotsrg.html		Reeder et al. (2007)
	W: http://bibiserv.techfak.uni-bielefeld.de/pknotsrg	$O(N^4)$	Reeder and Giegerich (2004)
PSTAG	C: http://phmmts.dna.bio.keio.ac.jp/pstag/download.html	$O(on^4 + mn^5)^d$	Matsui et al. (2005)
	W: http://phmmts.dna.bio.keio.ac.jp/pstag/		
vsfold5	W: http://www.rna.it-chiba.ac.jp/~vsfold/vsfold5	$O(N^{4.7})$	Dawson et al. (2007)

[a]W, address of Web service; C, code is available at given address
[b]Computing effort; N, length of sequence
[c]n, length of alignment
[d]n, length of unfolded pair of sequences; m, o, nodes on structure tree

structures are calculated as sum of the energies of the two pseudoknot helices and some not described loop folding energies. Suboptimal foldings up to a user-defined energy threshold can be enumerated, and for large scale analysis, a fast sliding window mode is available.

ILM (Ruan et al. 2004) combines dynamic programming (mainly maximizing number of base pairs) and comparative information to find iteratively high-scoring helices, adds them to the structure, and removes the corresponding sequence segments from the sequence. Due to the removal step, there is no restriction on the type of pseudoknot. The thermodynamic approach uses energy parameters for helix stacking from the Vienna package.

HotKnots (Ren et al. 2005) expands the idea of ILM by considering several alternative secondary structures and returning a fixed number of suboptimal folding scenarios. The program uses Turner parameters (Mathews et al. 1999; Serra and Turner 1995) together with those of Dirks and Pierce (2004) and Cao and Chen (2006) for pseudoknotted loops to determine the energy of a structure. According to its authors, HotKnots outperforms STAR (Gultyaev 1991), PKNOTS, pknotsRG, and ILM.

KnotSeeker was described by Sperschneider and Datta (2008) as capable of detecting pseudoknots in long RNA sequences. The algorithm combines the output of several known programs for prediction in a serial fashion. According to the authors, KnotSeeker has higher sensitivity and specificity in detection of pseudoknots than pknotsRG, ILM, and HPknotter (Huang et al. 2005).

DotKnot (Sperschneider and Datta 2010) predicts a wide class of pseudoknots including bulged stems (not accessible for pknotsRG) by consulting probability dot plots, from which probable stems are inferred. These are assembled to compose pseudoknot candidates by employing bulge-loop and multiloop dictionaries. After free energy calculations with one of three different energy models, chosen according to the length of the interhelix loop, "reliable" pseudoknots are retained. This approach also manages long sequences with complex pseudoknotted structures.

The authors of each of the abovementioned programs tested their programs with restricted datasets, and the programs are not benchmarked by an independent group. However, new experimental results on free energies for specific pseudoknots from Liu et al. (2010) show that (1) it is not sufficient to calculate pseudoknot energies just by summing nearest-neighbor interactions within the component helices; (2) conformational entropy parameters for loops give the best approximation to loop entropies; (3) the lack of parameters for tertiary interactions is best compensated for by building as many *cis* WC base pairs as possible, although crystal structures show that these are sometimes replaced by favorable tertiary interactions.

3.4 Prediction of Consensus Structures

The accuracy of (mfe) secondary structure prediction for a single RNA sequence is relatively low. This is due to several factors including simplifications in the underlying model, uncertainties of the energy parameters (especially with stacking in larger loops and junctions), ignorance of kinetic factors (which are of increasing importance with increasing sequence length), and disregard of energy contributions of tertiary interactions. Values of accuracy for predicting correct base pairs range from as low as $(45 \pm 16)\%$ up to $(83 \pm 22)\%$ mostly depending on the tested sequence families (see Doshi et al. 2004; Mathews et al. 1999; Wilm et al. 2006, 2008b). A formidable improvement in prediction accuracy can be achieved, however, by using the additional information from sufficiently diverged homologous sequences. This approach is based on the fact that the secondary and tertiary structure of a noncoding RNA changes more slowly than the sequence during evolution. Mutations in base-paired regions are mainly compensated by further mutations that retain the pairing scheme. Due to the isostericity of all WC pairs (and other groups of non-WC pairs; see Leontis et al. 2002; Stombaugh et al. 2009), the structure common to homologous RNAs can easily be conserved while their sequences might differ from each other to a large extent.

The common structure for a set of homologous sequences is called the consensus structure. To find it, one would like to perform simultaneously a sequence and structure alignment, which has a prohibitive computational cost of $O(N^{3m})$ for m sequences of length N (Sankoff 1985). Hence, several simplifying and more pragmatic approaches for consensus structure prediction have been developed (see Table 3.3) that can be classified as follows (Gardner and Giegerich 2004):

1. Align the sequences first and then predict the structure common to the aligned sequences (Bernhart et al. 2008; Bindewald and Shapiro 2006; Wilm et al. 2008b). For the primary alignment step, pure sequence alignment programs or one of the sequence + structure alignment programs (see below) can be used. Dynamic programming (Bernhart et al. 2008; Wilm et al. 2008b) (for secondary structure prediction) or "maximum weighted matching" (for secondary structure prediction including pseudoknots or base triples; Tabaska et al. 1998; Wilm et al. 2008b) might be used in the structure prediction step given the fixed alignment. Several RNA sequence + structure editors are available (e.g., Griffiths-Jones 2004; Jossinet and Westhof 2005; Seibel et al. 2006; Wilm et al. 2008b) that allow a user to refine the initial alignment.
2. Predict structures for all single sequences and then align these structures (Dalli et al. 2006; Höchsmann et al. 2004; Moretti et al. 2008; Xu et al. 2007).
3. Align and predict structures at the same time, using heuristics and/or restrict the alignment to two sequences to lower the computing cost of Sankoff's algorithm (Bauer et al. 2007; Harmanci et al. 2007, 2008; Hofacker et al. 2004; Holmes 2005; Katoh and Toh 2008; Kiryu et al. 2007; Lindgreen et al. 2007; Perriquet et al. 2003; Torarinsson et al. 2007; Will et al. 2007; Yao et al. 2005).

This separation of approaches should not to be taken too strictly; for example, several of the Sankoff-like approaches first restrict the sequence + structure search space by taking into account a sequence alignment and partition functions for the individual sequences. Other approaches do also exist: for example, RNAcast predicts an abstract shape common to all sequences (Reeder and Giegerich 2005), where each shape of an RNA molecule comprises a class of similar structures and has a representative structure of minimal free energy within the class. That is, RNAcast predicts a consensus structure but does not align the sequences.

In general, all methods for consensus structure prediction outperform the single-sequence methods, but several prerequisites have to be met:

Table 3.3 Tools for prediction of RNA consensus secondary structure

Name	Address[a]	Reference
CARNAC	C: http://bioinfo.lifl.fr/RNA/carnac/index.php	Perriquet et al. (2003)
	W: http://bioinfo.lifl.fr/RNA/carnac/carnac.php	Touzet and Perriquet (2004)
CMfinder	C: http://bio.cs.washington.edu/yzizhen/CMfinder/	Yao et al. (2005)
	W: http://wingless.cs.washington.edu/htbin-post/unrestricted/CMfinderWeb/CMfinderInput.pl	
Consan	C: http://selab.janelia.org/software.html	Eddy and Dowell (2006)
ConStruct	C: http://www.biophys.uni-duesseldorf.de/construct3/	Wilm et al. (2008b)
Dynalign	C: http://rna.urmc.rochester.edu/dynalign.html	Harmanci et al. (2007)
foldalignM	C: http://foldalign.ku.dk/software/index.html	Torarinsson et al. (2007)
KNetFold	C: http://www-lmmb.ncifcrf.gov/~bshapiro/downloader_v1/register.php	Bindewald and Shapiro (2006)
	W: http://knetfold.abcc.ncifcrf.gov/	
LARA	C: https://www.mi.fu-berlin.de/w/LiSA/	Bauer et al. (2007)
LocARNA	C: http://www.bioinf.uni-freiburg.de/Software/LocARNA/	Will et al. (2007)
MAFFT	W: http://align.bmr.kyushu-u.ac.jp/mafft/online/server/	Katoh and Toh (2008)
MASTR	C: http://mastr.binf.ku.dk/	Lindgreen et al. (2007)
Murlet	W: http://murlet.ncrna.org/murlet/murlet.html	Kiryu et al. (2007)
MXSCARNA	C: http://www.ncrna.org/software/mxscarna/download/	Tabei et al. (2008)
	W: http://mxscarna.ncrna.org/mxscarna/mxscarna.html	
PARTS	C: http://rna.urmc.rochester.edu/	Harmanci et al. (2008)
Pfold	W: http://www.daimi.au.dk/~compbio/rnafold/	Knudsen and Hein (2003)
PMcomp/PMmulti	C: http://www.tbi.univie.ac.at/~ivo/RNA/PMcomp/	Hofacker et al. (2004)
	W: http://rna.tbi.univie.ac.at/cgi-bin/pmcgi.pl	
R-Coffee	C: http://www.tcoffee.org/Projects_home_page/r_coffee_home_page.html	Wilm et al. (2008a)
	W: http://www.tcoffee.org/	Moretti et al. (2008)
RNAalifold	C: http://www.tbi.univie.ac.at/~ivo/RNA/	Hofacker et al. (2002)
	W: http://rna.tbi.univie.ac.at/cgi-bin/RNAalifold.cgi	Bernhart et al. (2008)
RNAcast	C: http://bibiserv.techfak.uni-bielefeld.de/rnacast/	Reeder and Giegerich (2005)
	W: http://bibiserv.techfak.uni-bielefeld.de/rnashapes/submission.html	
RNAforester	C: http://bibiserv.techfak.uni-bielefeld.de/rnaforester/	Höchsmann et al. (2004)
	W: http://bibiserv.techfak.uni-bielefeld.de/rnaforester/submission.html	
RNAmine	W: http://rnamine.ncrna.org/rnamine/	Hamada et al. (2006)
RNASampler	C: http://ural.wustl.edu/~xingxu/RNASampler/index.html	Xu et al. (2007)
SCARNA	W: http://www.scarna.org/scarna/	Tabei et al. (2006)
SimulFold	C: http://www.cs.ubc.ca/~irmtraud/simulfold/	Meyer and Miklós (2007)
StemLoc	C: http://biowiki.org/StemLoc	Holmes (2005)

(continued)

Table 3.3 (continued)

Name	Address[a]	Reference
StrAl	C: http://www.biophys.uni-duesseldorf.de/stral/about.php W: http://www.biophys.uni-duesseldorf.de/stral/advancedForm.php	Dalli et al. (2006)
WAR	W: http://genome.ku.dk/resources/war/	Torarinsson and Lindgreen (2008)

[a]C, source code is available at given address; W, address of Web service

- The performance of most (iterative) programs improves with an increasing number of input sequences and decreasing identities of sequences. Optimal values might be five sequences with an average pairwise sequence identity (APSI) of 55–70%.
- Only the structure alignment programs (approach 3 or RNAcast) might give reasonable results for a sequence set with an APSI below 55%, but most of these programs are very demanding in computer resources.
- While even a single compensating base-pair change might hint to a certain structure, a pure statistical analysis [e.g., via information theory (Chiu and Kolodziejczak 1991; Wilm et al. 2008b); for other methods see Gruber et al. (2008)] needs more than ten sequences and still does not reach the accuracy of thermodynamic-based approaches.

3.5 Conclusions

In concluding this review, we propose the following approach for constructing an RNA alignment for consensus structure prediction:

1. We assume at least one and probably no more than a few closely related sequences are known.
2. First, use pure sequence search methods (like BLAST) to find more homologues of the sequence(s) from *step 1*. Due to the use of pure sequence search, the found homologues will be closely related to the already known sequences. For an overview and benchmark of selected RNA search tools, see Freyhult et al. (2007).
3. Next, create an alignment of the sequences and a consensus structure using an alignment program appropriate for the lengths and number of sequences; for example, MAFFT (in mode Q-INS-i; Katoh and Toh 2008) or StrAl (Dalli et al. 2006) accepts more and longer sequences than StemLoc (Holmes 2005) or LocARNA (Will et al. 2007). This preliminary consensus structure should be checked for consistency (and refined accordingly) by means of ConStruct (Wilm et al. 2008b) or RNAalifold (Bernhart et al. 2008).
4. Use the preliminary consensus structure to create either a pattern (see, e.g., Dsouza et al. 1997; Gautheret and Lambert 2001; Gräf et al. 2006; Macke et al. 2001; Mosig et al. 2009) or a covariance model (see Klein and Eddy 2003;

Nawrocki and Eddy 2007). Use either model to search more specifically for further members of the RNA group under inspection. Alternatively, reiterate from step 2.

5. Check the refined model for consistency with ConStruct or RNAalifold using thermodynamics and covariation analysis. If this gives new information—especially in terms of tertiary interactions and/or base triples—reiterate from step 4, otherwise, this final model could be refined further by verification from wet lab experiments.

If additional experimental data is available, for example, from chemical or enzymatic mapping (Ehresmann et al. 1987; Tullius and Greenbaum 2005), the initial structure prediction by RNAfold or mfold can accordingly be constrained and thus incorporated into the model (Deigan et al. 2009). If in addition information on the three-dimensional structure of one of the sequences from the set is available from X-ray or NMR analysis, the use of an editor like S2S (Jossinet and Westhof 2005) is advantageous.

Acknowledgments We thank Jana Sperschneider for useful discussions.

References

Andronescu M, Condon A, Hoos H, Mathews D, Murphy K (2007) Efficient parameter estimation for RNA secondary structure prediction. Bioinformatics 23:i19–i28, http://dx.doi.org/10.1093/bioinformatics/btm223

Batey R, Rambo R, Doudna J (1999) Tertiary motifs in RNA structure and folding. Angew Chem Int Ed Engl 38:2326–2343, http://dx.doi.org/10.1002/(SICI)1521-3773(19990816)38:16<2326::AID-ANIE2326>3.0.CO;2-3

Bauer M, Klau G, Reinert K (2007) Accurate multiple sequence-structure alignment of RNA sequences using combinatorial optimization. BMC Bioinformatics 8:271, http://dx.doi.org/10.1186/1471-2105-8-271

Bellman R, Kalaba R (1960) On *k*th best policies. SIAM J Appl Math 8:582–588, http://dx.doi.org/10.1137/0108044

Bernhart S, Hofacker I, Will S, Gruber A, Stadler P (2008) RNAalifold: improved consensus structure prediction for RNA alignments. BMC Bioinformatics 9:474, http://dx.doi.org/10.1186/1471-2105-9-474

Bindewald E, Shapiro B (2006) RNA secondary structure prediction from sequence alignments using a network of *k*-nearest neighbor classifiers. RNA 12:342–352, http://dx.doi.org/10.1261/rna.2164906

Brion P, Westhof E (1997) Hierarchy and dynamics of RNA folding. Annu Rev Biophys Biomol Struct 26:113–137, http://dx.doi.org/10.1146/annurev.biophys.26.1.113

Cao S, Chen S (2006) Predicting RNA pseudoknot folding thermodynamics. Nucleic Acids Res 34:2634–2652, http://dx.doi.org/10.1093/nar/gkl346

Cao S, Chen S (2009) Predicting structures and stabilities for H-type pseudoknots with interhelix loops. RNA 15:696–706, http://dx.doi.org/10.1261/rna.1429009

Chan C, Lawrence C, Ding Y (2005) Structure clustering features on the Sfold Web server. Bioinformatics 21:3926–3928, http://dx.doi.org/10.1093/bioinformatics/bti632

Chiu D, Kolodziejczak T (1991) Inferring consensus structure from nucleic acid sequences. Comput Appl Biosci 7:347–352, http://dx.doi.org/doi:10.1093/bioinformatics/7.3.347

Cho S, Pincus D, Thirumalai D (2009) Assembly mechanisms of RNA pseudoknots are determined by the stabilities of constituent secondary structures. Proc Natl Acad Sci USA 106:17349–17354, http://dx.doi.org/10.1073/pnas.0906625106
Dalli D, Wilm A, Mainz I, Steger G (2006) STRAL: Progressive alignment of non-coding RNA using base pairing probability vectors in quadratic time. Bioinformatics 22:1593–1599, http://bioinformatics.oxfordjournals.org/cgi/reprint/22/13/1593, http://dx.doi.org/10.1093/bioinformatics/btl142
Dawson W, Fujiwara K, Kawai G (2007) Prediction of RNA pseudoknots using heuristic modeling with mapping and sequential folding. PLoS ONE 2:e905, http://dx.doi.org/10.1371/journal.pone.0000905
Deigan K, Li T, Mathews D, Weeks K (2009) Accurate SHAPE-directed RNA structure determination. Proc Natl Acad Sci USA 106:97–102, http://dx.doi.org/10.1073/pnas.0806929106
Dirks R, Pierce N (2003) A partition function algorithm for nucleic acid secondary structure including pseudoknots. J Comput Chem 24:1664–1677, http://dx.doi.org/10.1002/jcc.10296
Dirks R, Pierce N (2004) An algorithm for computing nucleic acid base-pairing probabilities including pseudoknots. J Comput Chem 25:1295–1304, http://dx.doi.org/10.1002/jcc.20057
Doshi K, Cannone J, Cobaugh C, Gutell R (2004) Evaluation of the suitability of free-energy minimization using nearest-neighbor energy parameters for RNA secondary structure prediction. BMC Bioinformatics 5:105, http://dx.doi.org/10.1186/1471-2105-5-105
Draper D (2008) RNA folding: thermodynamic and molecular descriptions of the roles of ions. Biophys J 95:5489–5495, http://dx.doi.org/10.1529/biophysj.108.131813
Dsouza M, Larsen N, Overbeek R (1997) Searching for patterns in genomic data. Trends Genet 13:497–498, http://dx.doi.org/10.1016/S0168-9525(97)01347-4
Eddy S, Dowell R (2006) Efficient pairwise RNA structure prediction and alignment using sequence alignment constraints. BMC Bioinformatics 7:400, http://www.biomedcentral.com/1471-2105/7/400
Ehresmann C, Baudin F, Mougel M, Romby P, Ebel JP, Ehresmann B (1987) Probing the structure of RNAs in solution. Nucleic Acids Res 15:9109–9128, http://dx.doi.org/10.1093/nar/15.22.9109
Freyhult E, Bollback J, Gardner P (2007) Exploring genomic dark matter: a critical assessment of the performance of homology search methods on noncoding RNA. Genome Res 17:117–125, http://dx.doi.org/10.1101/gr.5890907
Gardner P, Giegerich R (2004) A comprehensive comparison of comparative RNA structure prediction approaches. BMC Bioinformatics 5:140, http://dx.doi.org/10.1186/1471-2105-5-140
Garst A, Batey R (2009) A switch in time: detailing the life of a riboswitch. Biochim Biophys Acta 1789:584–591, http://dx.doi.org/10.1016/j.bbagrm.2009.06.004
Gautheret D, Lambert A (2001) Direct RNA motif definition and identification from multiple sequence alignments using secondary structure profiles. J Mol Biol 313:1003–1011, http://dx.doi.org/10.1006/jmbi.2001.5102
Giegerich R, Voss B, Rehmsmeier M (2004) Abstract shapes of RNA. Nucleic Acids Res 32:4843–4851, http://dx.doi.org/10.1093/nar/gkh779
Gluick T, Draper D (1994) Thermodynamics of folding a pseudoknotted mRNA fragment. J Mol Biol 241:246–262, http://dx.doi.org/10.1006/jmbi.1994.1493
Gluick T, Gerstner R, Draper D (1997) Effects of Mg^{2+}, K^{+}, and H^{+} on an equilibrium between alternative conformations of an RNA pseudoknot. J Mol Biol 270:451–463, http://dx.doi.org/10.1006/jmbi.1997.1119
Gräf S, Teune JH, Strothmann D, Kurtz S, Steger G (2006) A computational approach to search for non-coding RNAs in large genomic data. In: Nellen W, Hammann C (eds) Small RNAs: analysis and regulatory functions, vol 17, Nucleic acids and molecular biology series. Springer, Berlin, pp 57–74, http://www.springer.com/life+sci/biochemistry+and+biophysics/book/978-3-540-28129-0
Griffiths-Jones S (2004) RALEE-RNA ALignment editor in Emacs. Bioinformatics 21:257–259, http://dx.doi.org/10.1093/bioinformatics/bth489

Gruber A, Bernhart S, Hofacker I, Washietl S (2008) Strategies for measuring evolutionary conservation of RNA secondary structures. BMC Bioinformatics 9:122, http://dx.doi.org/10.1186/1471-2105-9-122

Gultyaev A (1991) The computer simulation of RNA folding involving pseudoknot formation. Nucleic Acids Res 19:2489–2494, http://dx.doi.org/10.1093/nar/19.9.2489

Gultyaev A, van Batenburg F, Pleij C (1999) An approximation of loop free energy values of RNA H-pseudoknots. RNA 5:609–617, http://dx.doi.org/10.1017/S135583829998189X

Hamada M, Tsuda K, Kudo T, Kin T, Asai K (2006) Mining frequent stem patterns from unaligned RNA sequences. Bioinformatics 22:2480–2487, http://dx.doi.org/10.1093/bioinformatics/btl431

Hamada M, Kiryu H, Sato K, Mituyama T, Asai K (2009) Prediction of RNA secondary structure using generalized centroid estimators. Bioinformatics 25:465–473, http://dx.doi.org/10.1093/bioinformatics/btn601

Harmanci A, Sharma G, Mathews D (2007) Efficient pairwise RNA structure prediction using probabilistic alignment constraints in Dynalign. BMC Bioinformatics 8:130, http://www.biomedcentral.com/1471-2105/8/130

Harmanci A, Sharma G, Mathews D (2008) PARTS: probabilistic alignment for RNA joinT secondary structure prediction. Nucleic Acids Res 36:2406–2417, http://dx.doi.org/10.1093/nar/gkn043

Höchsmann M, Voss B, Giegerich R (2004) Pure multiple RNA secondary structure alignments: a progressive profile approach. IEEE/ACM Trans Comput Biol Bioinform 1:53–62, http://dx.doi.org/10.1109/TCBB.2004.11

Hofacker I (2003) Vienna RNA secondary structure server. Nucleic Acids Res 31:3429–3431, http://dx.doi.org/10.1093/nar/gkg599

Hofacker I, Fontana W, Stadler P, Bonhoeffer S, Tacker M, Schuster P (1994) Fast folding and comparison of RNA structures. Monatsh Chem 125:167–188, http://www.springerlink.com/content/p88384567740kn15/fulltext.pdf

Hofacker I, Fekete M, Stadler P (2002) Secondary structure prediction for aligned RNA sequences. J Mol Biol 319:1059–1066, http://dx.doi.org/10.1016/S0022-2836(02)00308-X

Hofacker I, Bernhart S, Stadler P (2004) Alignment of RNA base pairing probability matrices. Bioinformatics 20:2222–2227, http://bioinformatics.oxfordjournals.org/cgi/reprint/20/14/2222

Holmes I (2005) Accelerated probabilistic inference of RNA structure evolution. BMC Bioinformatics 6:73, http://www.biomedcentral.com/1471-2105/6/73

Huang C, Lu C, Chiu H (2005) A heuristic approach for detecting RNA H-type pseudoknots. Bioinformatics 21:3501–3508, http://dx.doi.org/10.1093/bioinformatics/bti568

Jossinet F, Westhof E (2005) Sequence to structure (S2S): display, manipulate and interconnect RNA data from sequence to structure. Bioinformatics 21:3320–3321, http://dx.doi.org/10.1093/bioinformatics/bti504

Katoh K, Toh H (2008) Improved accuracy of multiple ncRNA alignment by incorporating structural information into a MAFFT-based framework. BMC Bioinformatics 9:212, http://dx.doi.org/10.1186/1471-2105-9-212

Kiryu H, Tabei Y, Kin T, Asai K (2007) Murlet: a practical multiple alignment tool for structural RNA sequences. Bioinformatics 23:1588–1598, http://bioinformatics.oxfordjournals.org/cgi/reprint/23/13/1588.pdf

Klein R, Eddy S (2003) RSEARCH: finding homologs of single structured RNA sequences. BMC Bioinformatics 4:44, http://www.biomedcentral.com/1471-2105/4/44

Klump H (1977) Thermodynamic values of the helix-coil transition of DNA in the presence of quaternary ammonium salt. Biochim Biophys Acta: Nucleic Acids and Protein Synthesis 475:605–610, http://dx.doi.org/10.1016/0005-2787(77)90321-5

Knudsen B, Hein J (2003) Pfold: RNA secondary structure prediction using stochastic context-free grammars. Nucleic Acids Res 31:3423–3428, http://nar.oxfordjournals.org/cgi/reprint/31/13/3423.pdf

Leontis N, Stombaugh J, Westhof E (2002) The non-Watson-Crick base pairs and their associated isostericity matrices. Nucleic Acids Res 30:3497–3531, http://dx.doi.org/10.1093/nar/gkf481

Lindgreen S, Gardner P, Krogh A (2007) MASTR: multiple alignment and structure prediction of non-coding RNAs using simulated annealing. Bioinformatics 23:3304–3311, http://dx.doi.org/10.1093/bioinformatics/btm525

Liu B, Shankar N, Turner D (2010) Fluorescence competition assay measurements of free energy changes for RNA pseudoknots. Biochemistry 49:623–634, http://dx.doi.org/10.1021/bi901541j

Lu Z, Gloor J, Mathews D (2009) Improved RNA secondary structure prediction by maximizing expected pair accuracy. RNA 15:1805–1813, http://dx.doi.org/10.1261/rna.1643609

Lyngsø R, Pedersen C (2000) RNA pseudoknot prediction in energy-based models. J Comput Biol 7:409–427, http://dx.doi.org/10.1089/106652700750050862

Macke T, Ecker D, Gutell R, Gautheret D, Case D, Sampath R (2001) RNAMotif, an RNA secondary structure definition and search algorithm. Nucleic Acids Res 29:4724–4735, http://dx.doi.org/10.1093/nar/29.22.4724

Markham N, Zuker M (2005) DINAMelt web server for nucleic acid melting prediction. Nucleic Acids Res 33:W577–W581, http://dx.doi.org/10.1093/nar/gki591

Markham N, Zuker M (2008) UNAFold: software for nucleic acid folding and hybridization. Methods Mol Biol 453:3–31, http://dx.doi.org/10.1007/978-1-60327-429-6_1

Mathews D, Sabina J, Zuker M, Turner D (1999) Expanded sequence dependence of thermodynamic parameters improves prediction of RNA secondary structure. J Mol Biol 288:911–940, http://dx.doi.org/10.1006/jmbi.1999.2700

Mathews D, Disney M, Childs J, Schroeder S, Zuker M, Turner D (2004) Incorporating chemical modification constraints into a dynamic programming algorithm for prediction of RNA secondary structure. Proc Natl Acad Sci USA 101:7287–7292, http://dx.doi.org/10.1073/pnas.0401799101

Matsui H, Sato K, Sakakibara Y (2005) Pair stochastic tree adjoining grammars for aligning and predicting pseudoknot RNA structures. Bioinformatics 21:2611–2617, http://dx.doi.org/10.1093/bioinformatics/bti385

McConaughy B, Laird C, McCarthy B (1969) Nucleic acid reassociation in formamide. Biochemistry 8:3289–3295, http://dx.doi.org/10.1021/bi00836a024

Meyer I, Miklós I (2007) SimulFold: simultaneously inferring RNA structures including pseudoknots, alignments, and trees using a Bayesian MCMC framework. PLoS Comput Biol 3:e149, http://dx.doi.org/10.1371/journal.pcbi.0030149

Michov B (1986) Specifying the equilibrium constants in Tris-borate buffers. Electrophoresis 7:150–151, http://dx.doi.org/10.1002/elps.1150070310

Moretti S, Wilm A, Higgins D, Xenarios I, Notredame C (2008) R-Coffee: a web server for accurately aligning noncoding RNA sequences. Nucleic Acids Res 36:W10–W13, http://dx.doi.org/10.1093/nar/gkn278

Mosig A, Zhu L, Stadler P (2009) Customized strategies for discovering distant ncRNA homologs. Brief Funct Genomic Proteomic 8:451–460, http://dx.doi.org/10.1093/bfgp/elp035

Nagel J, Pleij C (2002) Self-induced structural switches in RNA. Biochimie 84:913–923, http://dx.doi.org/10.1016/S0300-9084(02)01448-7

Nawrocki E, Eddy S (2007) Query-dependent banding (QDB) for faster RNA similarity searches. PLoS Comput Biol 3:e56, http://dx.doi.org/10.1371/journal.pcbi.0030056

Nissen P, Ippolito J, Ban N, Moore P, Steitz T (2001) RNA tertiary interactions in the large ribosomal subunit: the A-minor motif. Proc Natl Acad Sci USA 98:4899–4903, http://dx.doi.org/10.1073/pnas.081082398

Nixon P, Giedroc D (1998) Equilibrium unfolding (folding) pathway of a model H-type pseudoknotted RNA: the role of magnesium ions in stability. Biochemistry 37:16116–16129, http://dx.doi.org/10.1021/bi981726z

Nixon P, Giedroc D (2000) Energetics of a strongly pH dependent RNA tertiary structure in a frameshifting pseudoknot. J Mol Biol 296:659–671, http://dx.doi.org/10.1006/jmbi.1999.3464

Nussinov R, Pieczenik G, Griggs J, Kleitman D (1978) Algorithms for loop matchings. SIAM J Appl Math 35:68–82, http://dx.doi.org/10.1137/0135006

Perriquet O, Touzet H, Dauchet M (2003) Finding the common structure shared by two homologous RNAs. Bioinformatics 19:108–116, http://bioinformatics.oxfordjournals.org/cgi/reprint/19/1/108.pdf

Pleij C, Rietveld K, Bosch L (1985) A new principle of RNA folding based on pseudoknotting. Nucleic Acids Res 13:1717–1731, http://dx.doi.org/10.1093/nar/13.5.1717

Qiu H, Kaluarachchi K, Du Z, Hoffman D, Giedroc D (1996) Thermodynamics of folding of the RNA pseudoknot of the T4 gene 32 autoregulatory messenger RNA. Biochemistry 35:4176–4186, http://dx.doi.org/10.1021/bi9527348

Ramesh A, Winkler W (2010) Magnesium-sensing riboswitches in bacteria. RNA Biol 7:77–83, http://dx.doi.org/10.4161/rna.7.1.10490

Record M, Lohman T (1978) A semiempirical extension of polyelectrolyte theory to the treatment of oligoelectrolytes: Application to oligonucleotide helix-coil transitions. Biopolymers 17:159–166, http://dx.doi.org/10.1002/bip.1978.360170112

Reeder J, Giegerich R (2004) Design, implementation and evaluation of a practical pseudoknot folding algorithm based on thermodynamics. BMC Bioinformatics 5:104, http://dx.doi.org/10.1186/1471-2105-5-104

Reeder J, Giegerich R (2005) Consensus shapes: an alternative to the Sankoff algorithm for RNA consensus structure prediction. Bioinformatics 21:3516–3523, http://dx.doi.org/10.1093/bioinformatics/bti577

Reeder J, Steffen P, Giegerich R (2007) pknotsRG: RNA pseudoknot folding including near-optimal structures and sliding windows. Nucleic Acids Res 35:W320–W324, http://dx.doi.org/10.1093/nar/gkm258

Ren J, Rastegari B, Condon A, Hoos H (2005) HotKnots: heuristic prediction of RNA secondary structures including pseudoknots. RNA 11:1494–1504, http://dx.doi.org/10.1261/rna.7284905

Reuter J, Mathews D (2010) RNAstructure: software for RNA secondary structure prediction and analysis. Bioinformatics 11:129, http://dx.doi.org/10.1186/1471-2105-11-129

Riesner D, Steger G (1990) Viroids and viroid-like RNA. In: Saenger W (ed) Nucleic acids, subvolume d, physical data II, theoretical investigations, Landolt-Börnstein, group vii biophysics, vol 1. Springer, Berlin, pp 194–243

Rietveld K, Van Poelgeest R, Pleij C, Van Boom J, Bosch L (1982) The tRNA-like structure at the 3′ terminus of turnip yellow mosaic virus RNA. Differences and similarities with canonical tRNA. Nucleic Acids Res 10:1929–1946, http://dx.doi.org/10.1093/nar/10.6.1929

Rivas E, Eddy S (1999) A dynamic programming algorithm for RNA structure prediction including pseudoknots. J Mol Biol 285:2053–2068, http://dx.doi.org/10.1006/jmbi.1998.2436

Ruan J, Stormo G, Zhang W (2004) An iterated loop matching approach to the prediction of RNA secondary structures with pseudoknots. Bioinformatics 20:58–66, http://bioinformatics.oxfordjournals.org/cgi/reprint/20/1/58.pdf

Sadhu C, Gedamu L (1987) In vitro synthesis of double stranded RNA and measurement of thermal stability: effect of base composition, formamide and ionic strength. Biochem Int 14:1015–1022

Sankoff D (1985) Simultaneous solution of the RNA folding, alignment and protosequence problems. SIAM J Appl Math 45:810–825, http://dx.doi.org/10.1137/0145048

Seibel P, Müller T, Dandekar T, Schultz J, Wolf M (2006) 4SALE-a tool for synchronous RNA sequence and secondary structure alignment and editing. BMC Bioinformatics 7:498, http://dx.doi.org/10.1186/1471-2105-7-498

Serra M, Turner D (1995) Predicting thermodynamic properties of RNA. Methods Enzymol 259:242–261

Shelton V, Sosnick T, Pan T (1999) Applicability of urea in the thermodynamic analysis of secondary and tertiary RNA folding. Biochemistry 38:16831–16839, http://dx.doi.org/10.1021/bi991699s

Soto A, Misra V, Draper D (2007) Tertiary structure of an RNA pseudoknot is stabilized by "diffuse" Mg^{2+} ions. Biochemistry 46:2973–2983, http://dx.doi.org/10.1021/bi0616753

Sperschneider J, Datta A (2008) KnotSeeker: heuristic pseudoknot detection in long RNA sequences. RNA 14:630–640, http://dx.doi.org/10.1261/rna.968808

Sperschneider J, Datta A (2010) DotKnot: pseudoknot prediction using the probability dot plot under a refined energy model. Nucleic Acids Res 38:e103, http://dx.doi.org/10.1093/nar/gkq021
Steffen P, Voss B, Rehmsmeier M, Reeder J, Giegerich R (2006) RNAshapes: an integrated RNA analysis package based on abstract shapes. Bioinformatics 22:500–503, http://bioinformatics.oxfordjournals.org/cgi/reprint/22/4/500.pdf
Steger G (2004) Secondary structure prediction. In: Bindereif A, Hartmann R, Schön A, Westhof E (eds) Handbook of RNA biochemistry. Wiley-VCH, Weinheim, pp 513–535, http://www.wiley-vch.de/publish/en/books/bySubjectCH00/bySubSubjectCHB1/3-527-30826-1/?sID=18eede5181376097ccb16fc47772d157
Steger G, Müller H, Riesner D (1980) Helix-coil transitions in double-stranded viral RNA: fine resolution melting and ionic strength dependence. Biochim Biophys Acta 606:274–284
Stombaugh J, Zirbel C, Westhof E, Leontis N (2009) Frequency and isostericity of RNA base pairs. Nucleic Acids Res 37:2294–2312, http://dx.doi.org/10.1093/nar/gkp011
Tabaska J, Cary R, Gabow H, Stormo G (1998) An RNA folding method capable of identifying pseudoknots and base triples. Bioinformatics 14:691–699, http://bioinformatics.oxfordjournals.org/cgi/reprint/14/8/691
Tabei Y, Tsuda K, Kin T, Asai K (2006) SCARNA: fast and accurate structural alignment of RNA sequences by matching fixed-length stem fragments. Bioinformatics 22:1723–1729, http://bioinformatics.oxfordjournals.org/cgi/reprint/22/14/1723
Tabei Y, Kiryu H, Kin T, Asai K (2008) A fast structural multiple alignment method for long RNA sequences. BMC Bioinformatics 9:33, http://dx.doi.org/10.1186/1471-2105-9-33
Taufer M, Licon A, Araiza R, Mireles D, van Batenburg F, Gultyaev A, Leung M (2008) PseudoBase++: an extension of PseudoBase for easy searching, formatting and visualization of pseudoknots. Nucleic Acids Res 37:D127–D135, http://dx.doi.org/10.1093/nar/gkn806
Taylor W (2007) Protein knots and fold complexity: some new twists. Comput Biol Chem 31:151–162, http://dx.doi.org/10.1016/j.compbiolchem.2007.03.002
Theimer C, Giedroc D (1999) Equilibrium unfolding pathway of an H-type RNA pseudoknot which promotes programmed -1 ribosomal frameshifting. J Mol Biol 289:1283–1299, http://dx.doi.org/10.1006/jmbi.1999.2850
Theimer C, Giedroc D (2000) Contribution of the intercalated adenosine at the helical junction to the stability of the gag-pro frameshifting pseudoknot from mouse mammary tumor virus. RNA 6:409–421, http://dx.doi.org/10.1017/S1355838200992057
Theimer C, Wang Y, Hoffman D, Krisch H, Giedroc D (1998) Non-nearest neighbor effects on the thermodynamics of unfolding of a model mRNA pseudoknot. J Mol Biol 279:545–564, http://dx.doi.org/10.1006/jmbi.1998.1812
Tinoco I, Bustamante C (1999) How RNA folds. J Mol Biol 293:271–281, http://dx.doi.org/10.1006/jmbi.1999.3001
Torarinsson E, Lindgreen S (2008) WAR: webserver for aligning structural RNAs. Nucleic Acids Res 36:W79–W84, http://dx.doi.org/10.1093/nar/gkn275
Torarinsson E, Havgaard J, Gorodkin J (2007) Multiple structural alignment and clustering of RNA sequences. Bioinformatics 23:926–932, http://bioinformatics.oxfordjournals.org/cgi/reprint/23/8/926.pdf
Touzet H, Perriquet O (2004) CARNAC: folding families of related RNAs. Nucleic Acids Res 32:W142–W145, http://dx.doi.org/10.1093/nar/gkh415
Tullius T, Greenbaum J (2005) Mapping nucleic acid structure by hydroxyl radical cleavage. Curr Opin Chem Biol 9:127–134, http://dx.doi.org/10.1016/j.cbpa.2005.02.009
van Batenburg F, Gultyaev A, Pleij C (2001) PseudoBase: structural information on RNA pseudoknots. Nucleic Acids Res 29:194–195, http://dx.doi.org/10.1093/nar/29.1.194
Varani G (1995) Exceptionally stable nucleic acid hairpins. Annu Rev Biophys Biomol Struct 24:379–404, http://dx.doi.org/10.1146/annurev.bb.24.060195.002115
Walter A, Turner D, Kim J, Lyttle M, Muller P, Mathews D, Zuker M (1994) Coaxial stacking of helixes enhances binding of oligoribonucleotides and improves predictions of RNA folding. Proc Natl Acad Sci USA 91:9218–9222

Waterman M (1995) Introduction to computational biology. Maps, sequences and genomes. Chapman & Hall, London

Waterman M, Byers T (1985) A dynamic programming algorithm to find all solutions in a neighborhood of the optimum. Math Biosci 77:179–188, http://dx.doi.org/10.1016/0025-5564(85)90096-3

Will S, Reiche K, Hofacker I, Stadler P, Backofen R (2007) Inferring noncoding RNA families and classes by means of genome-scale structure-based clustering. PLoS Comput Biol 3:e65, http://dx.doi.org/10.1371%2Fjournal.pcbi.0030065

Wilm A, Mainz I, Steger G (2006) An enhanced RNA alignment benchmark for sequence alignment programs. Algorithms Mol Biol 1:19, http://dx.doi.org/10.1186/1748-7188-1-19, http://www.biomedcentral.com/content/pdf/1748-7188-1-19.pdf

Wilm A, Higgins D, Notredame C (2008a) R-Coffee: a method for multiple alignment of non-coding RNA. Nucleic Acids Res 36:e52, http://dx.doi.org/10.1093/nar/gkn174

Wilm A, Linnenbrink K, Steger G (2008b) ConStruct: improved construction of RNA consensus structures. BMC Bioinformatics 9:219, http://dx.doi.org/10.1186/1471-2105-9-219

Wimberly B, Varani G, Tinoco I (1993) The conformation of loop E of eukaryotic 5S ribosomal RNA. Biochemistry 32:1078–1087, http://dx.doi.org/10.1021/bi00055a013

Wu J, Gardner D, Ozer S, Gutell R, Ren P (2009) Correlation of RNA secondary structure statistics with thermodynamic stability and applications to folding. J Mol Biol 391:769–783, http://dx.doi.org/10.1016/j.jmb.2009.06.036

Wuchty S, Fontana W, Hofacker I, Schuster P (1999) Complete suboptimal folding of RNA and the stability of secondary structures. Biopolymers 49:145–165, http://www3.interscience.wiley.com/journal/40003742/abstract?CRETRY=1&SRETRY=0

Wyatt J, Puglisi J, Tinoco I (1990) RNA pseudoknots. Stability and loop size requirements. J Mol Biol 214:455–470, http://dx.doi.org/10.1016/0022-2836(90)90193-P

Xia T, McDowell J, Turner D (1997) Thermodynamics of nonsymmetric tandem mismatches adjacent to G.C base pairs in RNA. Biochemistry 36:12486–12497, http://dx.doi.org/10.1021/bi971069v

Xia T, SantaLucia J, Burkard M, Kierzek R, Schroeder S, Jiao X, Cox C, Turner D (1998) Thermodynamic parameters for an expanded nearest-neighbor model for formation of RNA duplexes with Watson-Crick base pairs. Biochemistry 37:14719–14735, http://dx.doi.org/10.1021/bi9809425

Xu X, Ji Y, Stormo G (2007) RNA sampler: a new sampling based algorithm for common RNA secondary structure prediction and structural alignment. Bioinformatics 23:1883–1891, http://dx.doi.org/10.1093/bioinformatics/btm272

Yao Z, Weinberg Z, Ruzzo W (2005) CMfinder-a covariance model based RNA motif finding algorithm. Bioinformatics 22:445–452, http://dx.doi.org/10.1093/bioinformatics/btk008

Zuker M (1989) On finding all suboptimal foldings of an RNA molecule. Science 244:48–52, http://dx.doi.org/10.1126/science.2468181

Zuker M (2003) Mfold web server for nucleic acid folding and hybridization prediction. Nucleic Acids Res 31:3406–3415, http://dx.doi.org/10.1093/nar/gkg595

Zuker M, Stiegler P (1981) Optimal computer folding of large RNA sequences using thermodynamics and auxiliary information. Nucleic Acids Res 9:133–148, http://nar.oxfordjournals.org/cgi/reprint/9/1/133

Chapter 4
Why Can't We Predict RNA Structure At Atomic Resolution?

Parin Sripakdeevong, Kyle Beauchamp, and Rhiju Das

Abstract No existing algorithm can start with arbitrary RNA sequences and return the precise three-dimensional structures that ensure their biological function. This chapter outlines current algorithms for automated RNA structure prediction (including our own FARNA–FARFAR), highlights their successes, and dissects their limitations, using a tetraloop and the sarcin/ricin motif as examples. The barriers to future advances are considered in light of three particular challenges: improving computational sampling, reducing reliance on experimentally solved structures, and avoiding coarse-grained representations of atomic-level interactions. To help meet these challenges and better understand the current state of the field, we propose an ongoing community-wide CASP-style experiment for evaluating the performance of current structure prediction algorithms.

4.1 RNA as a Model System

Predicting the three-dimensional structures of biopolymers from their primary sequence remains an unsolved but foundational problem in theoretical biophysics. This problem lies at the frontier of modern biological inquiry, encompassing questions from folding of individual protein and RNA domains to the fiber assembly of histone-compacted DNA genomes. However, a predictive, atomic-resolution

Sripakdeevong and Beauchamp are equally contributing authors.

P. Sripakdeevong • K. Beauchamp
Biophysics Program, Stanford University, Stanford, CA, USA
e-mail: sripakpa@stanford.edu; kyleb@stanford.edu

R. Das (✉)
Biophysics Program, Stanford University, Stanford, CA, USA

Biochemistry Department, Stanford University, Stanford, CA, USA
e-mail: rhiju@stanford.edu

N. Leontis and E. Westhof (eds.), *RNA 3D Structure Analysis and Prediction*,
Nucleic Acids and Molecular Biology 27, DOI 10.1007/978-3-642-25740-7_4,

understanding of these three-dimensional processes is presently out of reach. Attaining such an understanding will likely require simple starting points, and we view the folding of small RNA systems as the most tractable of these unsolved puzzles.

Beyond validating and refining our physical understanding of biomolecule behavior, a general algorithm to model RNA structure would have immediate practical implications. Riboswitches, ribozymes, and new classes of functional noncoding RNAs are being discovered rapidly, through RNA secondary structure prediction algorithms, bioinformatic tools, and a large suite of experimental approaches. Accurate and fast tools for predicting three-dimensional structure would not only accelerate these discoveries but also lead to richer experimentally testable hypotheses for how these molecules sense the cellular state and bind recognition partners. Furthermore, accurate three-dimensional RNA models would expand the use of RNA as a designer molecule, with potential applications ranging from the control of organisms [see, e.g., (Win et al. 2009)], the engineering of nano-scaffolds [see, e.g., (Jaeger and Chworos 2006)], the development of aptamer-based therapeutics [see, e.g., (Nimjee et al. 2004)], and the emerging fields of nucleic acid computation and logic [see, e.g., (Stojanovic and Stefanovic 2003)].

This chapter discusses the present state of computational modeling of three-dimensional RNA structure, highlighting successes and describing the barriers to future progress. Our hope is that dissecting the limitations of the field will hasten the development of atomic accuracy methods for modeling RNA structures without extensive experimental input.

4.2 Is the RNA Structure Prediction Problem Well Defined?

RNA, despite its small four-letter alphabet, is now recognized to perform a multitude of roles in the cell, including information transfer, catalysis (Nissen et al. 2000), gene regulation, and ligand sensing (Mandal and Breaker 2004). The attainment of a small set of unique three-dimensional states has been a hallmark of previously characterized functional biomolecules, from catalytic proteins to information-storing DNA double helices. Do RNA molecules of the same type have structures agreeing at atomic resolution, up to the fluctuations expected of a biomolecule in solution? Is the information necessary to specify these structures contained in the RNA sequence alone?

We now know that the answer to both questions is "yes" for a broad range of natural and in vitro selected RNA sequences, although there are also examples of both unstructured RNAs and RNAs guided into functional conformations by partners [induced fit; see, e.g., (Ferre-D'amare and Rupert 2002; Hainzl et al. 2005)]. In the 1960s, studies of transfer RNA sequences defined a conserved secondary structure [see, e.g., (Holley et al. 1965; Shulman et al. 1973 ; Rich and RajBhandary 1976)]—the pattern of classic Watson–Crick base pairs—and then defined interhelical tertiary interactions mediated by noncanonical base–base contacts [see, e.g., (Levitt 1969; Kim et al. 1974)]. These pioneering studies established a paradigm of theoretical investigation and experimental decipherment that has been followed for each novel class of RNAs that has been discovered in subsequent decades. In many respects, it now appears that RNA is easier to fold than other biopolymers. For example, unlike proteins, which typically require at least a dozen residues to form well-defined structures, the simplest RNAs with well-defined, recurrent structures are as small as eight residues (Jucker et al. 1996).

These simple molecules include hairpin loops, single strands that fold back on themselves to form short Watson–Crick helices. In some cases, the loops contain only four nonhelical bases—the so-called tetraloops (Varani 1995), with two classes, UUCG and GCAA (with their respective homologues), being the most extensively studied (see Fig. 4.1a) (Antao and Tinoco 1992; Jucker and Pardi 1995; Molinaro and Tinoco 1995; Jucker et al. 1996; Correll et al. 2003). These motifs have been observed in isolation, as single strands of RNA (Jucker et al. 1996), and as segments within larger RNA structures. Spectroscopic, crystallographic, and thermodynamic experiments indicate that these tetraloops form stable structures that are largely conserved among homologous sequences.

Larger RNA systems exhibit well-defined three-dimensional folds as well, and work over the last decade has yielded a rich trove of crystallographic structures of ligand binding aptamers, riboswitches, and ribozymes. Most famously, ribosomal subunits of several organisms have been crystallized by several groups (Ban et al. 2000; Wimberly et al. 2000; Harms et al. 2001; Yusupov et al. 2001), and the resulting structures are remarkably similar. For example, the conserved structural core shared by the respective 16 S and 23 S rRNAs of *Escherichia coli* and *Thermus thermophilus*, two bacteria that diverged early in evolution, comprises 90% or more of these molecules, despite extensive sequence differences (Zirbel et al. 2009). Similar stories of intricate structures shared across homologues are now plentiful [see, e.g., (Lehnert et al. 1996; Golden et al. 1998, 2004; Adams et al. 2004; Batey et al. 2004; Serganov et al. 2004)]. Such structural conservation implies that the structure prediction problem is a meaningful one for functional RNA sequences—the three-dimensional structures of these molecules are well defined and indeed critical for understanding their biological function and evolution.

This chapter focuses on recent ideas for predicting the structure of an RNA sequence without experimental input. RNA secondary structures have been routinely ascertained prior to atomic-resolution experiments, often making use of phylogenetic covariation studies or easily obtained chemical footprinting profiles (Staehelin et al. 1968; Nussinov and Jacobson 1980; Zuker and Stiegler 1981). We therefore focus on the more difficult problem of modeling three-dimensional structures, especially regions involving noncanonical base–base and base–backbone interactions. Further questions, such as the existence of alternative structures, the thermodynamics of these different states, the kinetics of self-assembly, and binding to proteins and other macromolecular partners, are also important for understanding the biological behavior of RNA. While some current modeling approaches provide partial answers to these questions (Bowman et al. 2008; Ding et al. 2008), few rigorous experimental comparisons of simulations and nonstructural experimental

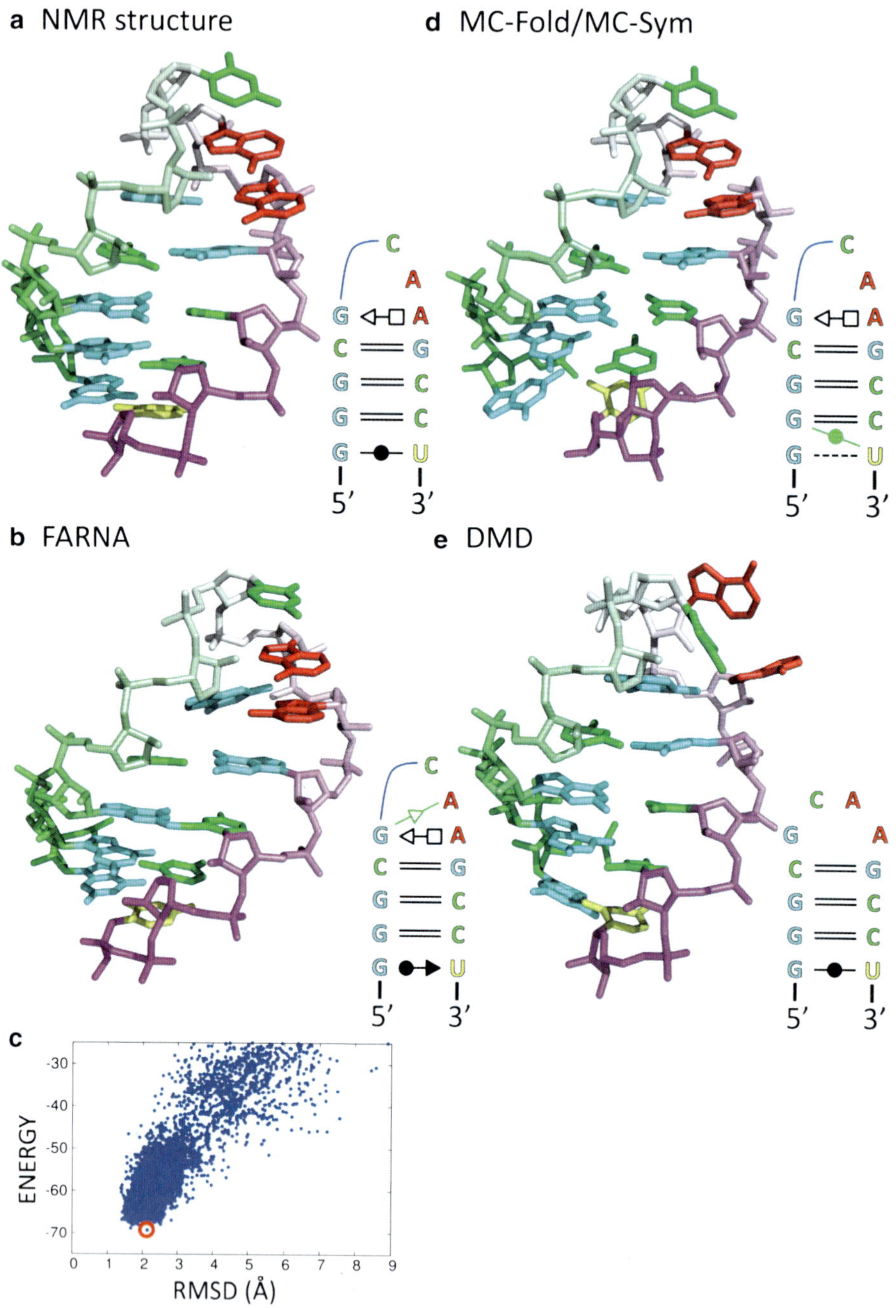

Fig. 4.1 Models of the GCAA tetraloop structure. Secondary structure annotations follow the convention of Leontis and Westhof (2001) and were prepared with the aid of RNAmlView (Yang et al. 2003) and FR3D (Sarver et al. 2008). (**a**) NMR structure (PDB: 1ZIH). (**b**) Model with the lowest energy score among 5,000 FARNA models (2.1 Å RMSD). (**c**) Plot of FARNA energy

data (e.g., folding rates) have been reported. As with the much longer-studied but still unsolved problem of protein folding, we feel that the RNA structure prediction problem—involving comparison of hundreds of predicted atomic-level RNA coordinates to high resolution experimental models—currently provides the most appropriate test of computational approaches.

4.3 3D RNA Modeling Inspired by Protein Structure Prediction

Following efforts by several labs to produce manual 3D modeling packages [see, e.g., (Mueller and Brimacombe 1997; Massire and Westhof 1998; Martinez et al. 2008)], several automated modeling algorithms have become available. The methods differ greatly in their search methods and also in the assumptions made to approximate the physics of RNA self-assembly. Each algorithm offers a partial solution to the RNA tertiary folding problem; within the proper domain of application, each method reproduces existing experimental structures for at least some small systems. In this section, we first focus on the fragment assembly approaches studied in our group.

Our approaches draw inspiration from the most successful strategies taken in "knowledge-based" protein structure modeling: they make full use of approximate sequence homology, known structural motifs, and PDB-derived base-pair contact distributions. Fragment Assembly of RNA (FARNA) directly applies the Rosetta approach for de novo protein modeling (Das and Baker 2007) to RNA, a Monte Carlo conformational search making use of trinucleotide fragments drawn from a ~3,000-nucleotide crystal structure of the large ribosomal subunit (Ban et al. 2000).

The assembly is guided by a coarse-grained scoring function, with parameters ascertained from the same ribosome crystal structure. The choice of using a knowledge-based potential was motivated by two considerations, both based on past experience with 3D protein modeling. First, we expected that deriving such a term from the database would ensure inclusion of physical terms that might be incorrectly modeled in a bottom-up, "physics-based" derivation of the potential. For example, high-level effects of base aromaticity on hydrogen bond strength and the influence of the hydrophobic effect remain difficult to compute and to calibrate, as they are for proteins (Simons et al. 1997).

The base–base interaction potential dominates the FARNA scoring function. We constructed this potential for each base interacting with the others, inspired by

Fig. 4.1 (continued) score vs. RMSD to the NMR structure for all 5,000 FARNA models. The lowest energy score model is highlighted with a *red circle*. (**d**) Model with the lowest RMSD among 3 MC-Sym models (1.7 Å RMSD). The lowest energy secondary structure as determined by MC-Fold was used as MC-Sym's input. (**e**) Model with the lowest RMSD among 20 DMD models (1.9 Å RMSD). Reported RMSDs were calculated over all heavy atoms with respect to the first member of the NMR ensemble

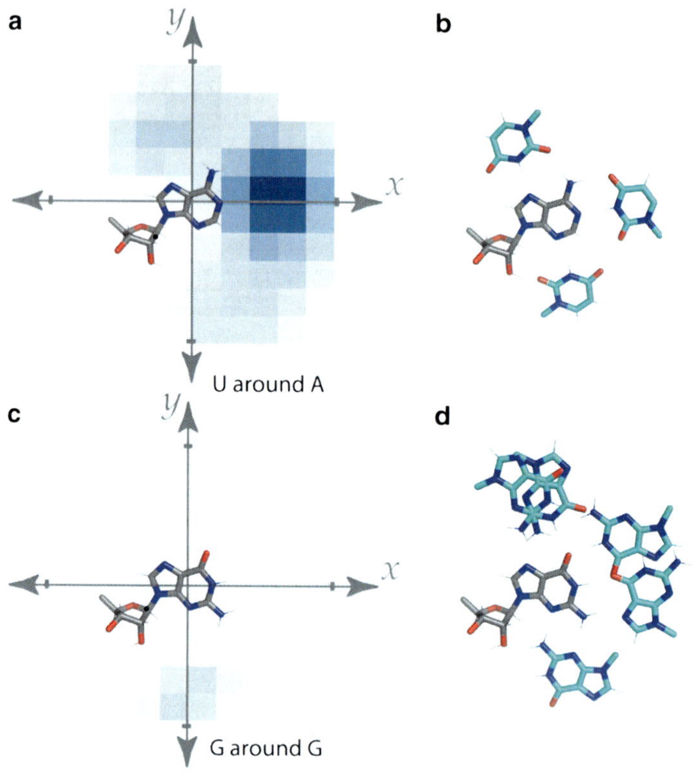

Fig. 4.2 Comparisons of the knowledge-based potential used in Fragment Assembly of RNA (FARNA) to base–base orientations generated by enumerative sampling. The distribution of uracil bases around adenosine (filtered for configurations in which the base normals are antiparallel), based on (**a**) the crystal structure of the large ribosomal subunit (PDB: 1JJ2), as used in the FARNA scoring function (Das and Baker 2007); and (**b**) a calculation enumerating all physically reasonable base–base orientations, scored with the high-resolution Rosetta force field. The three common antiparallel A-U configurations are seen with both approaches. In contrast, the distribution of parallel guanosine–guanosine base pairs as inferred from the ribosome (**c**) does not recapitulate all physically allowed configurations (**d**). Additionally, counts in (c) have been scaled by 8-fold but are still barely visible

previous studies on classifying these interactions (Leontis and Westhof 2001; Sykes and Levitt 2005). After fixing one nucleobase at the origin, a total of six rigid body degrees of freedom describe the other base's orientation, three translational and three rotational. However, if this six-dimensional space is binned, the available statistics for base-pairing orientations in experimental structures is sparse; some bins have only one or two instances, and derived potentials can be noisy. A desire for a smooth landscape during the coarse Monte Carlo search led us to choose a two-dimensional reduction. (A similar choice was made in the Rosetta low-resolution potential for protein beta strand-pairings.) Thus, base pairing frequencies were tallied as a function of x and y, i.e., the displacement of the centroid of the second base along directions parallel to the first base's plane (cf. Fig. 4.2a, b).

Following a common (but not formally rigorous) recipe (Simons et al. 1997), a scoring function was derived by taking the log-ratio of the observed frequencies of these base–base orientations generated in de novo decoys compared to the frequencies seen in the ribosome structure. Separate terms favoring the appropriate base stagger (z) and colinearity of base normals were also implemented. Adding these terms one-by-one to further favor "RNA-like" base–base arrangements led empirically to more accurate conformations of a test hairpin loop (Das and Baker 2007), at the expense of added computation to sample the more complex energy landscape. In fact, we also tested a higher dimensional representation including base–base rotation information that we expected to give better accuracy (parameterized on x and y, as before, but also the base–base "twist" in the x–y plane), but fragment assembly with thousands of Monte Carlo cycles was unable to efficiently sample even simple hairpin loop conformations in this more complex energy landscape.

Besides base pairing, a second critical term was a potential increasing with decreasing distance separating two atoms, preventing them from overlapping. The functional form matches that successfully used in protein low-resolution modeling. Explicitly, the form is proportional to $(d^2 - d_0^2)^2$ for distances below a cutoff d_0 (3–5 Å, parameterized from the distance of closest approach seen in the ribosome crystal structure). Two other terms had less effect: i) a compaction term, proportional to radius-of-gyration, favors the well-packed conformations characteristic of experimentally observed RNA structures, but such conformations are already well-favored by the base–base interaction potential. ii) a base-stacking term favors base stacks that have colinear base normal vectors; here, the stacking geometries already ensured by constructing models from ribosome fragments made the additional potential largely superfluous.

As should be apparent from the description above, derivation of a knowledge-based scoring function is a heuristic procedure, and the best test of such potentials is whether they result in more accurate de novo models. In favorable cases (under 20 residues), FARNA can sample and select out moderate resolution (2–4 Å all-atom root-mean-square-deviation, RMSD) models, as is illustrated for the GCAA hairpin loop in Fig. 4.1b, c (PDB: 1ZIH) (Jucker et al. 1996). Nevertheless, many contain steric clashes and poorly optimized hydrogen bonds. Furthermore, in larger systems, the scoring function fails to discriminate these <4 Å accuracy conformations from nonnative decoys, although the accuracy can be improved by using experimental data (Das et al. 2008).

As with Rosetta approaches for protein structure prediction, the FARNA approach to RNA modeling is computationally expensive. The computational time to create a single model for a 12-nucleotide motif like the GCAA hairpin loop is approximately 10 s on an Intel Xeon 2.33 GHz processor; typical runs, however, involve the generation of at least 5,000 models, requiring 14 CPU-hours. The computational expense for generating single models of larger sequences scales approximately as the number of nucleotides.

The most rigorous test of FARNA has been the blind modeling of a 74 nucleotide RNA transcript containing three stems from a bacterial ribosome, for

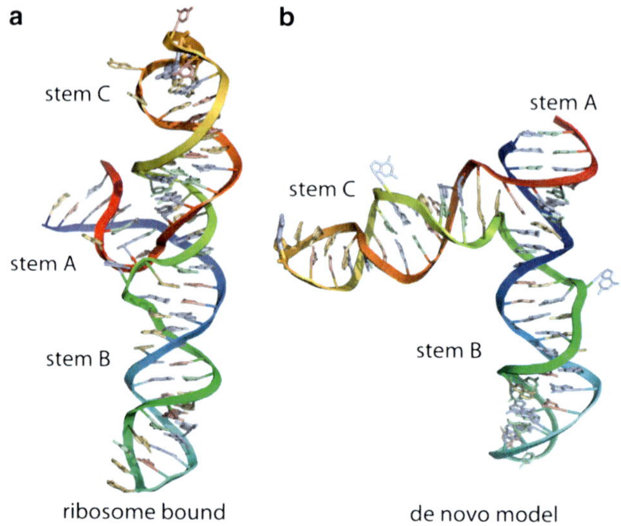

Fig. 4.3 De novo modeling of target T33 in the Critical Assessment of PRotein Interactions (CAPRI) trials, the complex of an rRNA segment and a methyltransferase (Fleishman et al. 2010) (**a**) The previously available structure of the three-helix junction in the context of the *E. coli* ribosome. (**b**) Representative de novo model generated by Fragment Assembly of RNA (FARNA) suggested a large conformational change, with additional support from full-atom refinement as well as low-resolution docking simulations with the protein target (not shown). The subsequently released crystallographic model of the RlmAII-bound RNA confirmed the conformational rearrangement but cannot be presented here because the coordinates are not yet publicly available

Critical Assessment of Prediction of Interactions (CAPRI) target T33, a complex of this RNA and a methyltransferase. Biochemical data suggested a large conformational difference between the structure of this RNA when bound to RlmAII compared to its known structure within the ribosome (Fig. 4.3a). We therefore applied automated de novo modeling to the RNA, with the hopes of selecting an accurate conformation through post facto docking to the protein component.

Although the modeling did not converge at high resolution (<2 Å RMSD), low energy configurations shared an overall global fold that was distinct from the ribosome-bound fold, especially in the helix–helix geometry at the molecule's three-way junction (Fig. 4.3b). Subsequent release of the protein-bound RNA crystallographic model revealed that a conformational rearrangement indeed occurs. The blind prediction was accurate at modest resolution, 5.4 Å RMSD over C4′ atoms (residues 694–702, 730–737, and 759–767), compared to 12.4 Å in the previously available ribosome-bound conformation (Fleishman et al. 2010). (The unavailability of the crystallographic coordinates to the public at the time of writing preclude presentation of the protein-bound model in this chapter.)

4.4 A Wealth of 3D RNA Modeling Approaches

There are now several algorithms for de novo modeling of RNA structure in addition to the fragment assembly approach described in the previous section, spanning a spectrum from more knowledge-based methods to more physics-based methods. Before discussing limitations of our fragment assembly approach, we survey these alternative methods, comparing results on one widely modeled sequence, the GCAA tetraloop, and, in the next section, the sarcin–ricin loop.

Like FARNA, the accuracy of the MC-Fold/MC-Sym pipeline (Parisien and Major 2008) depends on the available set of experimentally solved RNA structures. MC-Fold uses small RNA building blocks (nucleotide cyclic motifs, NCM) that are pieced into a two-dimensional representation of the RNA. The result is essentially an extended secondary structure (2D–3D) that includes both canonical and noncanonical non-Watson–Crick base pairs; it is the optimum of a Bayesian scoring function derived from the previously tallied frequencies of NCMs in experimental structures. This two-dimensional model is then submitted to MC-Sym (Major et al. 1991), a pioneering modeling method that generates three-dimensional structures consistent with the inputted secondary structure, often with outstanding accuracy (better than 2 Å all-atom RMSD; Fig. 4.1d). Like FARNA, the MC-Fold/MC-Sym method does not require prior determination of secondary structure or experimental constraints but can accept and benefit from these additional data (McGraw et al. 2009).

While both FARNA and MC-Fold/MC-Sym reward previously seen base-pairing geometries, several algorithms are less reliant on the existing databases. For instance, discrete molecular dynamics (DMD) (Ding et al. 2008) uses an efficient molecular dynamics engine to sample coarse-grained RNA structures. In numerous cases, DMD achieves native-like structures (~4 Å C4′ RMSD, with some models as low as 1.9 Å; Fig. 4.1e) without explicit calibration on any RNA conformations aside from canonical helices; energetic parameters are calibrated to classic thermodynamic experiments on RNA helix formation (Xia et al. 1998). Similarly, NAST (Jonikas et al. 2009a, b) uses coarse-grained molecular dynamics with a force field parameterized to reproduce canonical helices. Both of these approaches use molecular dynamics strategies that, by construction, do not explicitly model noncanonical regions. The accuracy of these methods can be improved through the use of experimental constraints (Gherghe et al. 2009; Jonikas et al. 2009a, b).

At the other end of the spectrum, all-atom molecular dynamics approaches do not make use of information from structural databases, aside from corrections to the underlying energy function to stabilize experimental conformations (Foloppe and MacKerell 2000). For example, all-atom molecular dynamics simulations have been used (Sorin et al. 2002; Bowman et al. 2008; Garcia and Paschek 2008) to investigate small RNA hairpin loops. Experimental hairpin loop structures appear to be stable in several solvation models (Sorin et al. 2002; Bowman et al. 2008), and the native-like secondary structure appears to be reachable from randomized conformations. Nevertheless, the precise details of noncanonical loop geometry may not be recapitulated by presently available force fields [see, e.g., (Ditzler et al. 2010)].

4.5 Case Study: Sarcin–Ricin Loop Suggests Limitations of Current Methods

The preceding survey of methods suggests that residue-level, and occasionally atomic-level, accuracy can be achieved in three-dimensional RNA modeling by a multitude of approaches. Yet, the RNA structure prediction problem is far from solved. The computational methods described so far cannot reliably produce high-quality models of an arbitrary RNA, a point we demonstrate with a long-studied model system. The structure of the sarcin–ricin loop, revealed in exquisite detail by X-ray crystallography (Figs. 4.4a and 4.5, PDB: 1Q9A), contains a tightly intermeshed array of hydrogen bonds. Within the seven nucleotides that form the core of this motif, there are 11 hydrogen bonds present (6 base–base, 4 base–phosphate, and 1 base–sugar), resulting in an average of 1.57 hydrogen-bonds per nucleotide, greater even than the 1.50 hydrogen-bonds per nucleotide in a repeating GC helix. The structural stability of this small motif has made it a paradigmatic system for experimental studies (Endo et al. 1991; Seggerson and Moore 1998) and a test case for evaluating current computational algorithms.

Applying FARNA to the sarcin–ricin loop yields mixed results. While FARNA produces native-like models (under 2 Å RMSD), the knowledge-based scoring function fails to distinguish these models from incorrect models (Fig. 4.4b, c). The same motif proves a challenge for other algorithms as well. Each of the top 20 models produced by MC-Fold has an incorrect base pair, suggesting limitations in the knowledge-based scoring function (Fig. 4.4d). As was the case with FARNA, several of the poorer scoring MC-Fold models (Fig. 4.4e) contain the correct base pairs and topology. Interestingly, using slightly different homologous sequences leads to better performance with both FARNA (unpublished data, PS, RD) and MC-Fold/MC-Sym (Parisien and Major 2008).

DMD, followed by all-atom reconstruction (Sharma et al. 2008), likewise cannot reproduce this structure at high resolution (Fig. 4.4f). These three algorithms use very different modeling strategies for RNA conformational sampling—a smoothed energy landscape (FARNA), a two-dimensional NCM description (MC-Fold), and a coarse-grained representation (DMD). Yet, all three algorithms fail on the same model system. A final class of algorithms, all-atom molecular dynamics simulations, satisfy a basic consistency check: the sarcin–ricin loop structure is stable in several tested force fields over the nanosecond time scale (Spackova and Sponer 2006). However, such simulations have not yet been carried out on the long timescales necessary for folding and discriminating complex RNA structures de

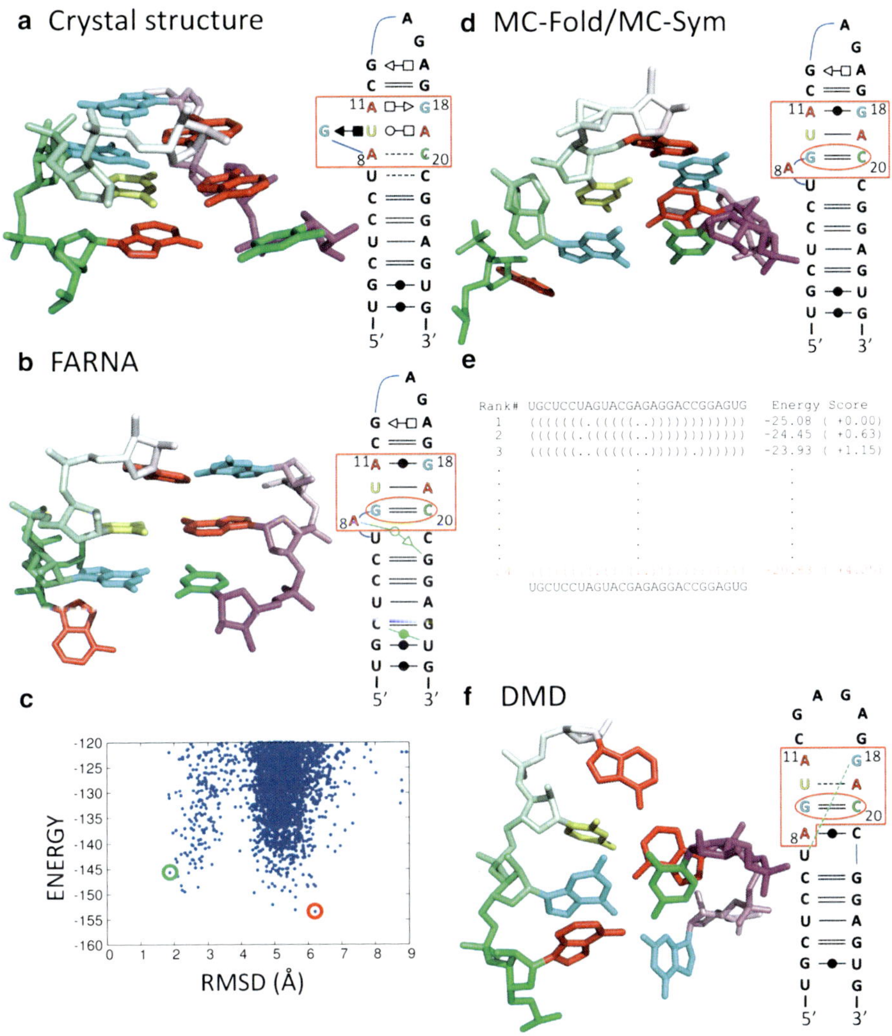

Fig. 4.4 Models of the sarcin–ricin loop structure. The depicted three-dimensional structures focus on the bulged-G motif region (*red box* in secondary structure) where all three algorithms fail. Interestingly, all three algorithms predict that G9 and C20 form a Watson–Crick base pair (*red circle*) which is absent in the crystal structure. (**a**) Crystal structure (PDB: 1Q9A). (**b**) Model with the lowest energy score among 50,000 FARNA models (6.2 Å RMSD). (**c**) The knowledge-based FARNA scoring function incorrectly ranks the non-native model (*red*) as having better energy score than the near-native model (*green*). (**d**) Model with the lowest RMSD among 100 MC-Sym models (3.8 Å RMSD). (**e**) The lowest energy secondary structure as determined by MC-Fold was used as MC-Sym's input. This first-ranked secondary structure contains incorrect base-pairings, and the native (correct) secondary structure is ranked #24 (*red*) by MC-Fold. (**f**) Model with the lowest RMSD among 20 DMD models (4.7 Å RMSD). Reported RMSDs were calculated over all heavy atoms with respect to the crystal structure

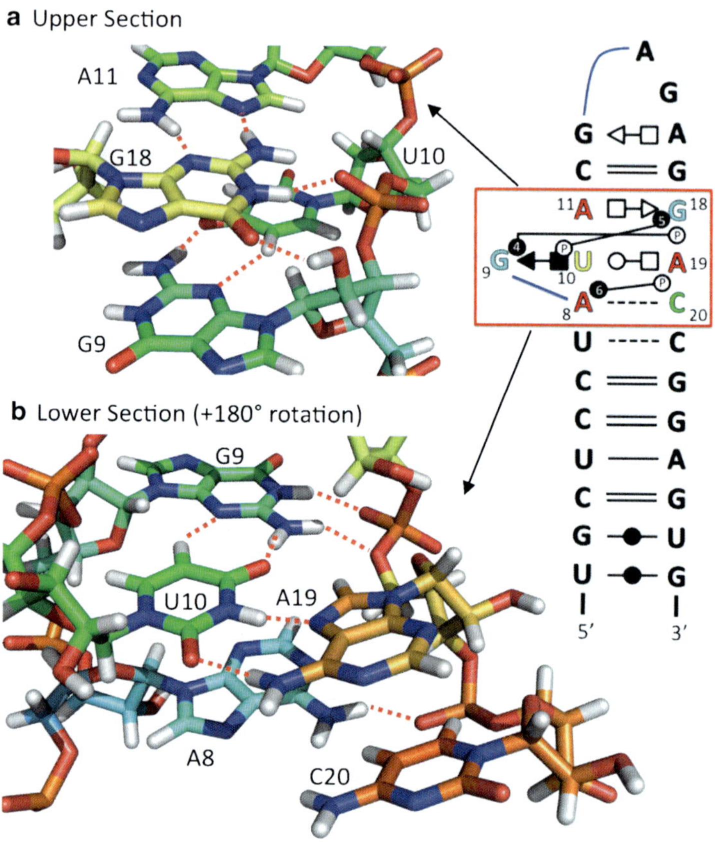

Fig. 4.5 Hydrogen bonding network at the bulged-G motif in the crystal structure of the sarcin–ricin loop (PDB: 1Q9A). Base-phosphate hydrogen bonds in the bulged-G motif region are annotated following a recently proposed convention (Zirbel et al. 2009). (**a**) Upper section of the motif. (**b**) Lower section of the motif. To display all the hydrogen bonds, the views in (**a**) and (**b**) are rotated by 180° with respect to each other. The experimentally observed hydrogen bonds are shown as *red dashed lines*. Within the seven nucleotides which form the core of this motif (A8-G9-U10-A11/G18-A19-C20), there are 11 unique hydrogen bonds (6 base–base, 4 base–phosphate, and 1 base–sugar), averaging to 1.57 hydrogen bonds per nucleotide. This is slightly greater than the number of hydrogen bonds in a repeating GC helix (1.50 hydrogen bonds per nucleotide). In contrast, none of the models generated by DMD, FARNA, or MC-FOLD shown in Fig. 4.4 have greater than 1.0 hydrogen bonds per nucleotide in this same region. Furthermore, in the models generated by these three algorithms, very few of the hydrogen bonds are of the base–phosphate or base–sugar type

novo (Pérez et al. 2007). Thus, it is presently unclear whether (and at what resolution) molecular dynamics can recapitulate larger experimental structures in simulations started from random conformations; indeed, a few cautionary tales have suggested that noncanonical motifs are unstable in existing MD force fields (Fadrná et al. 2009).

4.6 What Are the Bottlenecks?

The situation in RNA structure prediction bears some similarities to the state of protein modeling. Several algorithms are able to reproduce a handful of known small structures at reasonable resolution. Nevertheless, foundational bottlenecks prevent the prediction of known complex structures, as described above, despite the diversity of approaches being applied. The failure on known structures lowers our confidence that the existing approaches can be used to accurately predict new structures. Here, we describe three hypotheses for bottlenecks that need to be overcome.

4.6.1 Computational Sampling

Despite their differences, all RNA modeling algorithms proposed to date share a major difficulty in computational sampling, especially if they seek atomic resolution. For example, the trial runs on the sarcin/ricin loop above were made possible by the small size (<30 residues) of the tested motif; modeling of larger segments of the ribosome, much less the entire large ribosomal subunit, is presently difficult. The root of this problem was first discussed more than 40 years ago, when Levinthal noted that the conformational space available to a biomolecule is astronomical (10^{100}) and grows exponentially with the number of residues (Levinthal 1968). Forty years later, algorithms—for protein as well as RNA structure modeling—continue to face Levinthal's Paradox, as they typically involve a near-random search through conformation space.

As noted above, the difficulty of conformational sampling has prevented all-atom molecular dynamics approaches from demonstrating de novo recapitulation of RNA structure at high resolution. Other approaches are less expensive, but still require high performance computing. For example, FARNA calculations, even though constrained by experimental data, required approximately 10,000 CPU-hours on the Rosetta@Home distributed computing project to model the P4–P6 domain of the *Tetrahymena* ribozyme (160 residues) to ~13 Å accuracy (Das and Baker 2007). The barrier of ~100 nucleotides is particularly worrisome because several of the most biologically and medically important RNAs—including viral genomes (Watts et al. 2009) and untranslated regions of mRNA transcripts [see, e.g., (Penny et al. 1996; Birney et al. 2007)]—can exceed thousands of residues in length. Several methods (MC-Sym, DMD, NAST) appear significantly less expensive than FARNA; nevertheless, each algorithm is expected to encounter a conformational sampling bottleneck for some length of RNA. Detailed presentations of these length limits are not yet available but would certainly be valuable for users of the algorithms.

4.6.2 Overeliance on Existing Structures

One common approach to ameliorate the computational sampling bottleneck is to restrict the search to torsion angles of base pairing combinations drawn from the

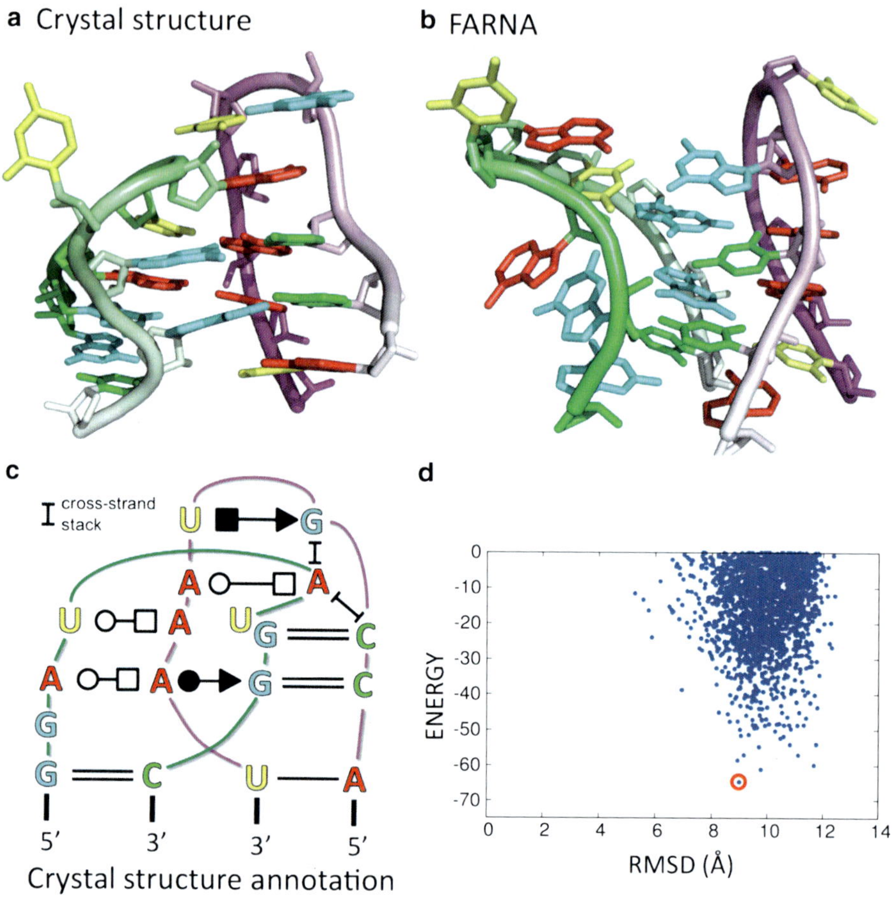

Fig. 4.6 The L2–L3 tertiary interaction from the purine riboswitch is poorly sampled using FARNA. (**a**) The crystal structure (PDB: 2EEW). (**b**) Model with the lowest energy score among 5,000 FARNA models. (**c**) Secondary structure annotation of the crystal structure. (**d**) Plot of FARNA energy score vs. RMSD to the crystal structure shows that FARNA fails to generate models that are closer than 5 Å RMSD. Examination of the large ribosomal subunit fragment library reveals the lack of near-native fragments at many nucleotide positions

experimental database of known RNA structures. On one hand, the success with the tetraloop motif can perhaps be attributed to not just its simplicity but its overall frequency in the database of experimental RNA structures. On the other hand, we might expect poor performance on novel sequence motifs if they exhibit torsional geometries that are at low frequency or are absent in current databases.

As an extreme illustration, the FARNA method fails to recover high resolution models for motifs such as the kissing loop from the purine-binding riboswitch, unless this structure or its homologues are included as sources of fragments (Fig. 4.6). The intricate base-pairing and base-stacking network formed by the two loops requires the individual nucleotides to adopt highly specific conformations. Consistent with this observation, the backbone conformations of three nucleotides in this motif are not on the list of commonly observed backbone rotamers compiled by the RNA Ontology Consortium (Richardson et al. 2008). The lack of native-like fragments in the fragment library prevents FARNA from generating models that are within 5 Å RMSD of the crystal structure.

Beyond adversely limiting the conformational space, reliance on existing structures can also cause problems in ranking models by available scoring functions. A widely studied RNA motif involves a quadruplex of G nucleotides forming parallel base pairs [see, e.g., (Mashima et al. 2009)], yet available ribosome crystal structures contain no such guanosine arrangements. This leads to a known deficiency in FARNA (cf. Fig. 4.2c, d); namely, some known G–G base interactions are not rewarded by the FARNA scoring function, and quadruplexes cannot be modeled. The scoring function can be reparameterized with the entire nonredundant crystallographic RNA data set (rather than just a single ribosomal structure), but will continue to miss important interactions, such as those involving protonated C's or A's, which are important for stabilizing RNA motifs but that are rare in the entire set of RNA structures. Consequently, while the ultimate goal of structure prediction is to model motifs that have not yet been observed experimentally, it is these novel structures that knowledge-based algorithms have the most difficulty predicting.

4.6.3 Simplified Representation

Perhaps the central shared bottleneck of the various de novo approaches discussed so far is the use of a simplified representation. Searching RNA conformations in all-atom detail requires attention to hydrogen bonds and packing interactions at the Angstrom level; algorithms to directly and efficiently sample conformations at this level of detail are not available. Instead, as is the case in protein structure modeling, de novo RNA modeling algorithms typically resort to a coarse-grained representation to carry out large-scale conformational search. In FARNA, the energy function is highly smoothed; MC-Fold uses a two-dimensional secondary-structure-like representation; and NAST and DMD both use reduced-atom models to accelerate molecular dynamics sampling.

These methods inevitably neglect certain details of RNA structure—but in the case of the sarcin–ricin bulged-G motif, these details cannot safely be ignored. Because the FARNA scoring function represents base-pairing and base-stacking interactions at the base-centroid level, the atomic details of individual hydrogen bonds are not represented, leading to an inability to select the native conformation of this motif (Fig. 4.4c).

The discrimination of realistic RNA conformations should be possible using all-atom physical potentials, and indeed such potentials have been the key feature in recent blind de novo Rosetta predictions of protein structure at near-atomic accuracy (Rohl et al. 2004). For RNA, the backbone torsional combinations seen in real structures are those physically allowed by sterics and torsional constraints; further, base pairing patterns follow the "laws" of hydrogen bonding as well (Leontis and

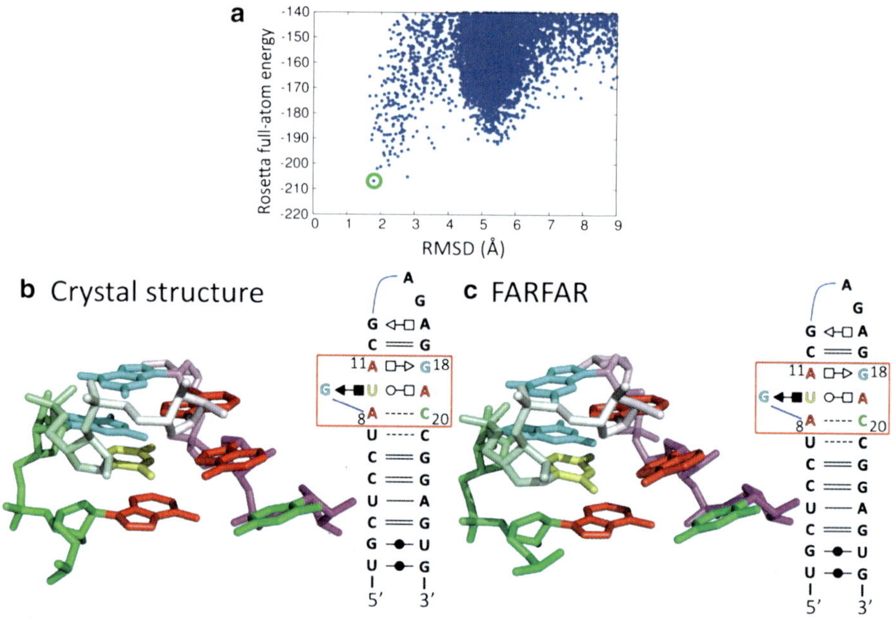

Fig. 4.7 Successful modeling of the sarcin–ricin loop with Fragment Assembly of RNA with Full-Atom Refinement (FARFAR). Previous attempts to model the sarcin–ricin loop with FARNA failed due to inaccuracies in the knowledge-based FARNA scoring function (see Fig. 4.4). (**a**) In contrast, the Rosetta full-atom force field used in FARFAR more accurately models the energetics of the hydrogen bonding network in the bulged-G motif. After the 50,000 FARNA models were refined (minimized) and scored in the full-atom force field, the near-native model (*green*) was correctly ranked as the lowest energy state. (**b**, **c**) This near-native model has a 1.798 Å RMSD with respect to the whole sarcin–ricin crystal structure and even a lower local RMSD of 1.038 Å when aligned just over the bulged-G motif nucleotides

Westhof 2001). (See also Fig. 4.2b, d, which were generated by exhaustively sampling base–base arrangements, scored with the Rosetta full-atom potential.) Recognizing the power of all-atom potentials, several groups have explored the refinement of automatically generated pools of low-resolution structures with all-atom potentials (Sharma et al. 2008; Jonikas et al. 2009a, b).

In our own recent work (Das et al. 2010) we have found that the full-atom Rosetta RNA force field can correctly refine and discriminate near-native structures for more than a dozen noncanonical motifs, including the bulged-G region of the sarcin–ricin loop structure (Fig. 4.7). The Rosetta RNA force field is essentially the same as used in protein structure prediction, with physics-based van der Waals (Rohl et al. 2004) and hydrogen bonding terms (Kortemme et al. 2003) supplemented with a torsional potential inferred from the ribosome α, β, γ, δ, ε, ζ, and χ torsion angles; a desolvation penalty for polar groups that models how neighbor atoms occlude water; and a weak carbon–hydrogen bonding term. (The latter two terms appear to improve protein structure prediction as well.) The overall

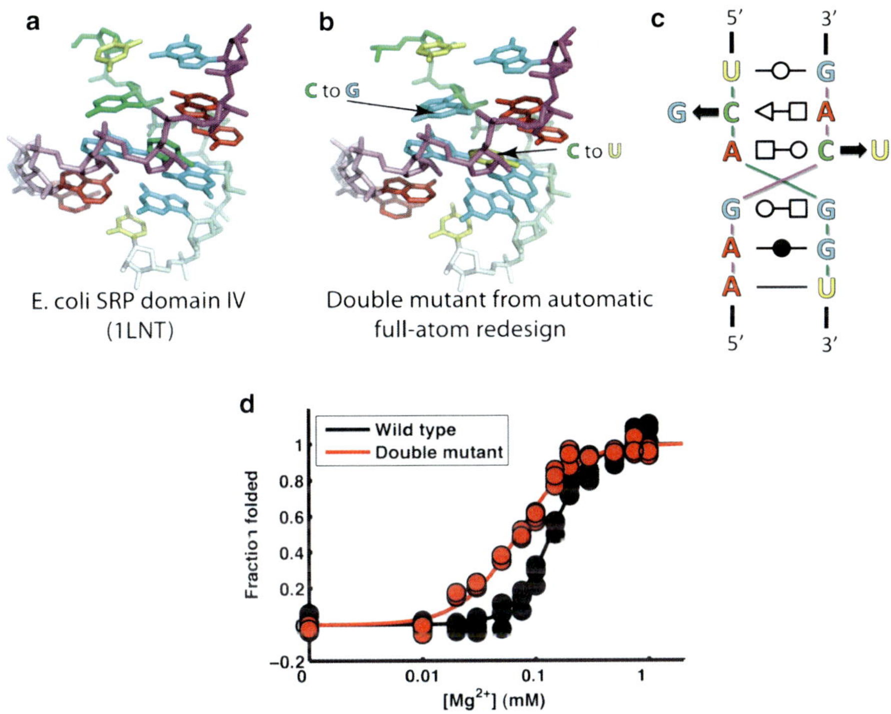

Fig. 4.8 Automated "redesign" of an RNA noncanonical motif. (**a**) The crystal structure of the most conserved domain of the signal recognition particle RNA (PDB: 1LNT). Sidechains from this structure were completely stripped away and then rebuilt, sampling all possible sequences, guided by the Rosetta full-atom force field. (**b**) Two mutations (*highlighted with arrows*) were discovered in the library of designs that occurred more frequently than in an alignment of natural sequences from all three kingdoms of life. (**c**) The corresponding secondary structure of the wild type and the mutant. (**d**) Experimental structure mapping measurements verify the stabilization of the motif by the two mutations (less Mg^{2+} required for folding) (Das et al. 2010)

structure modeling procedure (Fragment Assembly of RNA with Full-Atom Refinement, FARFAR) doubles the time of the previous FARNA method, to 21 s on an Intel Xeon 2.33 GHz processor for the 12-residue GCAA hairpin loop. Further independent tests of the approach, involving the "re-design" of RNA sequences that stabilize known backbone conformations, gave higher native sequence recoveries than low resolution potentials (Das et al. 2010). Most importantly, the calculations gave blind predictions for thermostabilizing noncanonical mutations that were validated in subsequent experiments (Fig. 4.8). We hope that the free availability of these algorithms to academic users (as part of the Rosetta software suite) will encourage their testing and development in the broader community.

The demonstrations of atomic accuracy structural modeling and design are exciting steps, but also confirm that conceptual advances in conformational sampling are much needed. In the published benchmark (Das et al. 2010), the structures of half of the 32 noncanonical motifs could be recovered at atomic accuracy. For most cases in the other half, sampled models produced scores worse than the experimental structure, indicating that conformational sampling was not efficient. In particular, motifs beyond approximately 12 residues in length are still difficult to sample at the Angstrom-level resolution required for high accuracy discrimination; a similar conformational sampling issue remains the major bottleneck in de novo prediction of protein structure and docking (Das et al. 2009; Kim et al. 2009; Raman et al. 2009). We also expect there to be missing physics in the all-atom Rosetta energy function, due to the neglect of explicit metal ions and water, of terms to modulate the strength of base stacking, of long-range electrostatic effects, and of conformational entropy. However, more effective conformational search procedures will be needed to establish whether these effects are critical for discriminating high-accuracy models from nonnative models. Based on the three major issues discussed above, we are currently focusing on approaches that enumeratively search realistic conformations of biomolecules, are independent of previously solved structures, and bypass coarse-grained search stages.

4.7 Future Directions/Community Wide RNA Experiments

Given the promising algorithms currently under development, it is reasonable to expect improved de novo methods for (small) RNA structure prediction in the next few years. However, once such novel algorithms are developed, they must be rigorously tested before they will be accepted and used by the wider RNA community.

In particular, the present cycle of algorithm development, testing, and publication inevitably pressures scientists to present their results in an optimistic and sometimes uncritical fashion. A blind, CASP-style competition to systematically assess the performance of RNA 3D structure prediction algorithms will therefore be crucial for future progress. We pledge – and request the cooperation of other experimentalists – to make available, prior to publication of an experimental atomic-resolution RNA structure, the nucleotide sequence of the solved molecule and to provide a deadline for modelers to submit solutions. We expect that objective evaluation of such trials will encourage thoughtful and open discussion of the strengths and limitations of current approaches and engender new collaborations between modelers and experimentalists.

For a CASP-style experiment to be interesting and useful, truly novel targets must be included. We note that at least three classes of such targets are already available to the modeling community. First, the growing interest in functional RNAs has led to crystallographic analyses of several new, large riboswitches. Although the sizes of these RNAs (>100 residues) puts them out of the reach of current algorithms, submotifs (such as internal loops and junctions) may fold in a manner largely independent of ligand binding or other interactions. The L2–L3 motif from the adenine riboswitch is such an example and is stable independent of

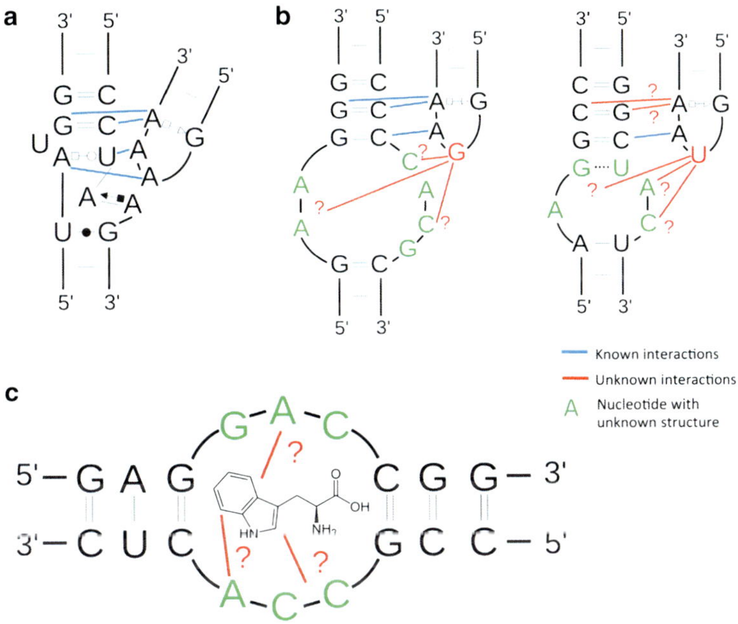

Fig. 4.9 Examples of novel RNA motifs with unknown structure. These motifs' small size, lack of homology to known structures, and potential ease of experimental validation make them ideal tests for computational algorithms. (**a**) The naturally occurring GAAA/11-nucleotide tetraloop receptor with known experimental structure (Pley et al. 1994; Cate et al. 1996; Ye et al. 2008). (**b**) In vitro selected tetraloop receptor motifs (Costa and Michel 1997). The binding affinities and specificities of these tetraloop receptors are markedly different from those of known naturally occurring tetraloop receptors. It is postulated that these novel binding specificity patterns are due to some (as yet undetermined) interactions between the second base of the tetraloop (*red*) and the nucleotides in the asymmetric loop of the receptor (*green*). (**c**) Binding site of an in vitro selected L-tryptophan binding aptamer (Majerfeld and Yarus 2005)

adenine binding (Mandal and Breaker 2004; Serganov et al. 2004). Biochemical identification of these subpuzzles, combined with the rapid rate at which these functional molecules are being crystallized, suggest that they are excellent targets for blind prediction.

Using in vitro evolution to redesign existing motifs provides another class of novel targets (Fig. 4.9a). A compelling example comes from the determination of optimal receptor motifs that specifically bind GNRA tetraloops (Costa and Michel 1997; Geary et al. 2007). These new motifs are less than a dozen residues in size, and most have presently unknown structure, making them ideal targets for current modeling approaches. Furthermore, these targets offer the prospect of rapid experimental validation. Given the growing number of ribozyme and riboswitch structures with the classic 11-nucleotide receptor motif for GAAA, the experimental structures of the alternate tetraloop-receptors may be attained by their substitution into RNA scaffolds that are known to be crystallizable (Pley et al. 1994; Cate et al. 1996; Ye et al. 2008).

Finally, there is a large body of work focused on sequences that bind small and large molecules, again isolated through in vitro evolution. Many of these functional sequences are small—only 13 nucleotides in the case of an L-tryptophan aptamer (Majerfeld and Yarus 2005) (Fig. 4.9b)—again bringing them close to the reach of all-atom computational modeling. Furthermore, their small size should permit their rapid experimental characterization by modern NMR approaches, ensuring a nearly unending supply of targets for blind trials.

4.8 Conclusions

Predicting the structure of an arbitrary RNA sequence remains an unsolved problem. A number of algorithms can rightly claim success in specific cases, including some blind tests; but a general solution has yet to appear, even for small sequences. Present methods are limited by computational sampling, over-reliance on previously solved experimental structures, and the use of coarse-grained or reduced representations. Recent progress, especially in all-atom refinement and design, makes us particularly excited about the future; a solution to RNA structure prediction appears more and more feasible. We propose that the time is ripe for the creation of a community-wide CASP-style experiment, where groups compete to produce blind models of RNAs about to be solved by crystallography or NMR. The prospect of such blind trials bodes well for the maturing and eventual practical impact of the RNA structure prediction field.

Note added in proof Since the time of writing (2010), we have described a method called stepwise assembly that appears to resolve the conformational sampling bottleneck for small RNA loops (Sripakdeevong et al. 2011). Further, we and others have initiated *RNA-Puzzles*, a series of community-wide blind trials for RNA structure prediction (Cruz et al. 2012).

References

Adams P et al (2004) Crystal structure of a self-splicing group I intron with both exons. Nature 430 (6995):45–50

Antao VP, Tinoco I Jr (1992) Thermodynamic parameters for loop formation in RNA and DNA hairpin tetraloops. Nucleic Acids Res 20(4):819

Ban N et al (2000) The complete atomic structure of the large ribosomal subunit at 2.4 A resolution. Science 289(5481):905–920

Batey R et al (2004) Structure of a natural guanine-responsive riboswitch complexed with the metabolite hypoxanthine. Nature 432(7015):411–415

Birney E et al (2007) Identification and analysis of functional elements in 1% of the human genome by the ENCODE pilot project. Nature 447(7146):799–816

Bowman GR et al (2008) Structural insight into RNA hairpin folding intermediates. J Am Chem Soc 130(30):9676–9678

Cate JH et al (1996) Crystal structure of a group I ribozyme domain: principles of RNA packing. Science 273(5282):1678–1685

Richardson JS et al (2008) RNA backbone: consensus all-angle conformers and modular string nomenclature (an RNA Ontology Consortium contribution). RNA 14(3):465
Rohl CA et al (2004) Protein structure prediction using Rosetta. Methods Enzymol 383:66–93
Sarver M et al (2008) FR3D: finding local and composite recurrent structural motifs in RNA 3D structures. J Math Biol 56(1–2):215–252
Seggerson K, Moore PB (1998) Structure and stability of variants of the sarcin-ricin loop of 28 S rRNA: NMR studies of the prokaryotic SRL and a functional mutant. RNA 4(10):1203–1215
Serganov A et al (2004) Structural basis for discriminative regulation of gene expression by adenine-and guanine-sensing mRNAs. Chem Biol 11(12):1729–1741
Sharma S et al (2008) iFoldRNA: three-dimensional RNA structure prediction and folding. Bioinformatics 24(17):1951
Shulman RG et al (1973) Determination of secondary and tertiary structural features of transfer RNA molecules in solution by nuclear magnetic resonance. Proc Natl Acad Sci USA 70 (7):2042
Simons KT et al (1997) Assembly of protein tertiary structures from fragments with similar local sequences using simulated annealing and Bayesian scoring functions. J Mol Biol 268 (1):209–225
Sorin EJ et al (2002) RNA simulations: probing hairpin unfolding and the dynamics of a GNRA tetraloop. J Mol Biol 317(4):493–506
Spackova N, Sponer J (2006) Molecular dynamics simulations of sarcin-ricin rRNA motif. Nucleic Acids Res 34(2):697
Sripakdeevong P et al (2011) An enumerative stepwise ansatz enables atomic-accuracy RNA loop modeling. Proc Natl Acad Sci USA 108(51):20573–20578
Staehelin M et al (1968) Structure of a mammalian serine tRNA. Nature 219(5161):1363–1365
Stojanovic MN, Stefanovic D (2003) A deoxyribozyme-based molecular automaton. Nat Biotechnol 21(9):1069–1074
Sykes MT, Levitt M (2005) Describing RNA structure by libraries of clustered nucleotide doublets. J Mol Biol 351(1):26–38
Varani G (1995) Exceptionally stable nucleic acid hairpins. Annu Rev Biophys Biomol Struct 24 (1):379–404
Watts JM et al (2009) Architecture and secondary structure of an entire HIV-1 RNA genome. Nature 460(7256):711–716
Wimberly BT et al (2000) Structure of the 30 S ribosomal subunit. Nature 407(6802):327–339
Win MN et al (2009) Frameworks for programming biological function through RNA parts and devices. Chem Biol 16(3):298–310
Xia T et al (1998) Thermodynamic parameters for an expanded nearest-neighbor model for formation of RNA duplexes with Watson-Crick base pairs. Biochemistry 37(42):14719–14735
Yang H et al (2003) Tools for the automatic identification and classification of RNA base pairs. Nucleic Acids Res 31(13):3450
Ye JD et al (2008) Synthetic antibodies for specific recognition and crystallization of structured RNA. Proc Natl Acad Sci USA 105(1):82
Yusupov MM et al (2001) Crystal structure of the ribosome at 5.5 A resolution. Science 292 (5518):883–896
Zirbel CL et al (2009) Classification and energetics of the base-phosphate interactions in RNA. Nucleic Acids Res 37(15):4898–4918
Zuker M, Stiegler P (1981) Optimal computer folding of large RNA sequences using thermodynamics and auxiliary information. Nucleic Acids Res 9(1):133

Chapter 5
Template-Based and Template-Free Modeling of RNA 3D Structure: Inspirations from Protein Structure Modeling

Kristian Rother, Magdalena Rother, Michał Boniecki, Tomasz Puton, Konrad Tomala, Paweł Łukasz, and Janusz M. Bujnicki

Abstract In analogy to proteins, the function of RNA depends on its structure and dynamics, which are encoded in the linear sequence. While there are numerous methods for computational prediction of protein 3D structure from sequence, there have been however very few such methods for RNA. This chapter discusses template-based and template-free approaches for macromolecular structure prediction, with special emphasis on comparison between the already tried-and-tested methods for protein structure modeling and the very recently developed "protein-like" modeling methods for RNA. As examples, we briefly review our recently developed tools for RNA 3D structure prediction, including ModeRNA (template-based or comparative/homology modeling) and SimRNA (template-free or de novo modeling).

ModeRNA requires, as an input, atomic 3D coordinates of a template RNA molecule and a user-specified sequence alignment between the target to be modeled and the template. It can model posttranscriptional modifications, a functionally important feature analogous to posttranslational modifications in proteins. It can model the structures of RNAs of essentially any length, provided that a starting template is known.

SimRNA can fold RNA 3D structure starting from sequence alone. It is based on a coarse-grained representation of the polynucleotide chains (only three atoms per

---

Kristian Rother, Magdalena Rother, and Michał Boniecki contributed equally to the manuscript.

K. Rother • M. Rother • T. Puton • J.M. Bujnicki (✉)
Laboratory of Bioinformatics and Protein Engineering, International Institute of Molecular and Cell Biology, ul. Ks. Trojdena 4, 02-109 Warsaw, Poland

Laboratory of Structural Bioinformatics, Institute of Molecular Biology and Biotechnology, Faculty of Biology, Adam Mickiewicz University, ul. Umultowska 89, 61-614 Poznan, Poland
e-mail: iamb@genesilico.pl

M. Boniecki • K. Tomala • P. Łukasz
Laboratory of Bioinformatics and Protein Engineering, International Institute of Molecular and Cell Biology, ul. Ks. Trojdena 4, 02-109 Warsaw, Poland

N. Leontis and E. Westhof (eds.), *RNA 3D Structure Analysis and Prediction*, Nucleic Acids and Molecular Biology 27, DOI 10.1007/978-3-642-25740-7_5,

nucleotide) and uses a Monte Carlo sampling scheme to generate moves in the 3D space, with a statistical potential to estimate the free energy. The current implementation based on simulated annealing is able to find native-like conformations for RNAs <100 nt in length, with multiple runs required to fold long sequences.

5.1 Introduction

RNAs and proteins are linear polymers composed of a limited set of building blocks (ribonucleotide and amino acid residues, respectively). Despite the fundamental chemical differences of these building blocks, the higher-order structure of RNA and protein molecules can be described with similar terms. Each residue comprises two parts: one is common to the given type of a macromolecule and is used to form a continuous "backbone" and the other is variable and forms a "side chain". The order of building blocks held together by covalent bonds is called the primary structure, the local conformation of the chain stabilized mostly by hydrogen bonds (by the backbone in proteins and by side chains in RNA) is the secondary structure, while the path of the chain in three dimensions resulting from various long-range interactions is the tertiary structure. Most RNA and protein molecules fold spontaneously into complex three-dimensional shapes that provide a framework for their biological functions. These functions typically involve interactions with various molecules in the cell, including other proteins and RNAs. Thus, the function of both proteins and RNAs depends on the three-dimensional structure and dynamics, which in turn is encoded in the linear sequence of individual molecules.

The knowledge of structure is very important for the understanding of RNA and protein function. However, experimental sequence determination of genes and entire genomes, from which the sequences of RNAs and proteins can be reliably inferred, is much cheaper and simpler than experimental determination of structures. As a consequence, the rate of macromolecular structure determination lags behind the rate of determination of new sequences and the gap between the number of known structures and known sequences continues to widen. It is unlikely and unnecessary that structures will be solved experimentally for all protein and RNA molecules. Understanding of the "1D-3D code" provides an opportunity for theoretical prediction of protein and RNA structures from their sequences. This has proven to be a very difficult task; however, a few successful strategies have been identified, which now allow for reasonably accurate (practically useful) predictions of 3D structures. Most methods have been developed initially for proteins only; the reader is referred to numerous review articles and books devoted to this topic, e.g., the volume edited by one of the authors of this article (Bujnicki 2008) or the series of articles in the special issue of proteins devoted to the CASP experiment (Moult et al. 2009). However, essentially the same principles have been recently demonstrated to be applicable for modeling of RNAs.

5.2 Classification of Methods for Macromolecular 3D Structure Prediction

Methods for 3D structure prediction can be divided into those based on "first principles," i.e., the fundamental laws of physics that govern the process of folding, and those based on information about other structures available in databases.

5.2.1 Template-Free, Ab Initio Structure Prediction

One approach to 3D structure prediction is based on the thermodynamic hypothesis formulated by Anfinsen, according to which the native structure of a protein corresponds to the global minimum of the free energy of the system comprising the macromolecule (Anfinsen 1973). Accordingly, physics-based methods model the process of folding by simulating the conformational changes of a macromolecule while it searches for the state of minimal free energy [review: (Hardin et al. 2002)]. The "score" of each conformation is calculated as the true physical energy based on the interactions within the macromolecule and between the macromolecule and the solvent (Scheraga 1996). Since the same basic laws of physics apply to all types of molecules, one can postulate that analogous methods could work for RNA as well.

The ab initio approach is, however, plagued by serious problems. In particular, the atomic model of the molecular structure has a large number of degrees of freedom ($N_{\mathrm{atoms}} \times 3$–6), which makes the search space enormous, and the function with which to calculate the energy of the system is very complex. As a result, both the sampling and energy calculations are very costly in terms of computational power required. Typically, the free energy landscape is extremely rugged, i.e., it possesses multiple local minima, and it is essentially impossible to perform an exhaustive evaluation of all these minima to identify the one with the globally lowest value. Further, some of the components of the free energy function (e.g., the entropy) are very difficult to calculate and may be not inferred accurately for large molecules. For these reasons, the use of ab initio methods is limited to very small molecules, and even then the user cannot be sure whether a native-like conformation has been generated during the folding simulation and whether it was scored better than the less native-like ones. To increase the efficiency of computations, full-atom models may be replaced by coarse-grained models, which treat groups of atoms as single interaction centers, so that a smaller number of interactions need to be evaluated [review: (Tozzini 2009)]. An example of a coarse-grained method for ab initio RNA 3D structure modeling is DMD (Ding et al. 2008). However, it must be emphasized that simplification of the model and the energy function usually leads to reduced accuracy. As of today, it is not practical to expect that a folding simulation for a protein or RNA molecule comprising more than 100 residues would confidently predict a native-like structure with a correctly estimated energy.

5.2.2 *Template-Based Structure Prediction*

At the other end of the methodological spectrum are approaches based on the principles of evolution. After experimental determination of the first handful of protein structures, it became clear that evolutionarily related (homologous) proteins usually retain the same three-dimensional fold (i.e., the 3D arrangement and connectivity of elements of secondary structure) despite the accumulation of divergent mutations (Chothia and Lesk 1986). It was also found that structural divergence is much slower than sequence divergence, although these two features are strongly correlated. Thus, methods have been developed to align the sequence of one protein (a target) to the structure of another protein (a template), model the overall fold of the target based on that of the template, and infer how the target structure will change due to substitutions, insertions, and deletions (indels), as compared with the template [reviews: (Cohen-Gonsaud et al. 2004; Krieger et al. 2003)]. The process of identification of a structurally related template has been termed "fold recognition", while the transformation of atomic coordinates of the template structure into the target has been typically referred to as "homology modeling" or "comparative modeling" (the latter takes into account a possibility that the template does not have to be homologous, as long as it is structurally similar to the target). This entire approach has been termed "template-based modeling."

Comparative analyses of evolutionarily related RNAs [see e.g., (Dror et al. 2005)] revealed patterns of conservation that are analogous to those observed in proteins: the secondary and tertiary structure is usually more conserved than sequence, and core regions important to stability and function tend to be more conserved at all levels. In general, it can be stated that in families of homologous RNAs, the 3D fold is often conserved, and alignment of sequences and secondary structure patterns can be used to recognize such structural conservation, enabling template-based modeling.

Template-based modeling has two main limitations. First, the modeling of the "target" structure starts with another known structure of a structurally similar molecule to be used as a "template"; hence, if such a structure does not exist or cannot be identified reliably, then the model cannot be built or almost certainly will be completely wrong. Further, each element of the target sequence must be aligned to the structurally equivalent element in the template sequence/structure. In particular, homologous residues should be aligned with each other. High sequence similarity is not a prerequisite for template-based modeling. In fact, it is possible to create good homology models even if the sequence identity between the target and the template is zero (Chothia and Gerstein 1997). However, on the average, molecules with higher sequence similarity tend to exhibit more similar structures (Chothia and Lesk 1986). Besides, for highly similar sequences, it is generally easier to generate a correct alignment (to find homologous residues between the target and the template). Therefore, using templates with higher sequence similarity is recommended. Apart from sequence divergence, structures may also change because of environmental factors, e.g., the binding of other molecules or the composition of the solution (salt, pH)

(Kumar et al. 2000). This is particularly true for RNA, where the binding of metal ions is often a key factor enabling a stable tertiary structure (Pyle 2002). It is generally the responsibility of the user of the homology modeling software to choose a template, whose biological state corresponds best to the desired biological state of the target to be modeled. With an incorrectly chosen template and/or wrong alignment, the model will be always very far from the native structure. These limitations concern all homology modeling tools, as templates and alignments are always necessary in this approach (Fiser et al. 2002).

Finally, it must be noted that like proteins, homologous RNAs need not retain the same structure in all details. Topological variability (e.g., preserving the overall 3D structure while changing the pattern of secondary structure elements) has been observed in many protein families (Grishin 2001), as well as in RNA families, with one prominent example being the RNA subunit of RNase P from *Escherichia coli* (type A) and *Bacillus subtilis* (type B) (Krasilnikov et al. 2004). However, methods for automated template-based modeling of macromolecules assume that the overall fold is conserved between the template and the target, and special intervention of the user is usually required to model topological variations. There exist methods for interactive (user-guided) modeling of macromolecular structures based on assembly of large fragments derived from various structures that are homologous to different parts of the template and may or may not be homologous to each other. The approach that allows the user to rearrange and recombine multiple template structures has been particularly widely used in the RNA modeling field, with methods such as S2S/Assemble (Jossinet et al. 2010; Jossinet and Westhof 2005) or ERNA-3D (Zwieb and Muller 1997). However, similar methods have been also applied to model protein structures (Kosinski et al. 2003) [review: (Bujnicki 2006)]. Thus, it can be concluded that comparative modeling of proteins and RNAs presents analogous opportunities and challenges.

5.2.3 *Template-Free, De Novo Structure Prediction*

In the protein structure prediction field, the most successful approach combines the features of physics-based folding and the use of previously solved structures as templates. According to the recent editions of the CASP benchmark, the best methods do simulate the folding but use a simplified (coarse-grained) model and a scoring function that replaces (at least partially) the physical energy with terms describing the frequency of occurrence of certain features in the database. These methods, exemplified by ROSETTA (Simons et al. 1997), TASSER (Zhang and Skolnick 2004a), and CABS (Kolinski and Bujnicki 2005), improve the efficiency of the conformational search by using small fragments derived from other (not necessarily homologous) known structures and/or by discretizing the search space from continuous to lattice-based [review: (Bujnicki 2006)]. This variant of template-free structure prediction is often termed "de novo modeling." A number of methods based on similar principles have been recently proposed also for RNA

3D modeling, including FARNA/FARFAR (Das and Baker 2007; Das et al. 2010) and MC-Fold|MC-Sym (Parisien and Major 2008).

De novo methods for structure prediction share many problems with the ab initio approach, including a high computational cost of the conformational sampling and uncertainty as to which of the large number of alternative conformations generated is the most native-like structure.

5.3 ModeRNA, a New Method for Template-Based RNA Structure Modeling

Inspired by the SWISS-MODEL method for protein structure modeling (Schwede et al. 2003), we have developed ModeRNA, a scriptable tool for automatic prediction of RNA 3D structures by template-based modeling (Rother et al. 2011b).

As a minimal input, ModeRNA requires the 3D coordinates of a template structure and a pairwise sequence alignment between the sequences of the template and the target RNA to be modeled. The problem of obtaining the sequence alignment is discussed in the following section. For each position in the target–template sequence alignment, ModeRNA infers a set of operations necessary to generate the model of the target from the structure of the template. These include copying coordinates of residues that are invariant between the target and the template, introducing substitutions for aligned residues that differ, adding or removing posttranscriptional modifications, processing, insertions/deletions, and adding structural fragments for short regions without a template. ModeRNA generates coordinates of the modeled target RNA and a report with detailed information about all steps of the modeling process. Figure 5.1 shows an example of a model built with ModeRNA, compared to the native structure.

The main advantages of ModeRNA are that it can be run in a fully automated mode, it is very fast and can be used in a batch mode (e.g., to model hundreds or thousands of structures from one RNA family), and it can model RNAs with modified nucleotides that are specified in the sequences to be modeled. There is no restriction on size of the molecules to be modeled. The ModeRNA software and detailed descriptions of commands, examples, as well as a tutorial can be found on the Bujnicki laboratory website, URL: http://iimcb.genesilico.pl/moderna.

ModeRNA is implemented in Python, free for all users, and released under the GPL open source license, which means that it can be customized and integrated with other software. Implementations for UNIX (including Mac OS X) and Windows systems are available. The program requires Python and the Biopython library (Cock et al. 2009). Several functions for numerical calculations that are part of the PyCogent library (Knight et al. 2007) have been included in the ModeRNA code. For Windows, an executable version is available that does not require installation of any additional software. ModeRNA does not require large computational resources, for instance, one tRNA model can be built in 2–20 s on a standard PC with one 2.4 GHz processor, with

template, and the resulting ends must be sealed to restore the continuity of the backbone. Such operations may involve the replacement of a longer segment of sequence (e.g., a loop) by a shorter one that has a more extended conformation.

Indel modeling in ModeRNA follows the fragment insertion approach, similar to the one widely used in comparative modeling of proteins (Michalsky et al. 2003), and implemented, e.g., in SWISS-MODEL (Schwede et al. 2003). A similar approach has been validated as a reliable method for modeling of 3–32-nt-long loops by another group (Schudoma et al. 2010). A fragment includes the residue(s) to be inserted and counterparts of residues that flank the indel in the template. The default distribution of the ModeRNA software allows for inserting fragments up to 17 residues long (not counting the flanks). The choice of the maximum length was conditioned by the size of the library file (20 MB). The standard fragment library includes 128,169 fragments (*n*-grams) of RNA structure that are 2–19 residues long and have a continuous backbone. It has been derived from the representative set of 172 RNA tertiary structures in the RNADB2005 set (Richardson et al. 2008), which provides manually curated, nonredundant RNA structures from different families, including large structures, e.g., the ribosome, and is expected to cover all known types of local RNA structure. For modeling of longer insertions, a larger library covering fragments up to 100 nt long, derived from the same database, is available for download from the ModeRNA website.

For each indel, ModeRNA attempts to identify a backbone fragment with the appropriate length and superimposes its flanking residues onto the corresponding residues flanking the indel in the template structure so as to maximize its fit to the anchor and to minimize steric clashes with the rest of the molecule. The fragment search includes a prefiltering stage, where the geometry of the flanking residues is compared to all fragments of appropriate length from a library, and a fitting stage where the 50 most promising candidates are evaluated by inserting them into the model. The best-fitting candidate is retained. By default, the two residues flanking the insertion site are retained in the template versions, so their counterparts from the fragment are deleted.

If the gap cannot be closed by the above-mentioned procedure, e.g., if an extended fragment of the template is to be deleted and the resulting ends are too far from each other, ModeRNA will generate a model with an unsealed gap and generate a warning that the model is discontinuous. Such situations often occur when modeling is attempted using an incorrectly chosen template or in regions where the target–template alignment is erroneous.

5.3.5 *Refinement of Models Generated by ModeRNA*

Possible discontinuities in the backbone (resulting, e.g., from an imperfect match of fragment ends to the flanking regions of the template) are repaired using the full cyclic coordinate descent (FCCD) algorithm that connects two ends with a minimal number of operations (Boomsma and Hamelryck 2005). ModeRNA rebuilds coordinates of the RNA backbone atoms between two residues, aiming to restore the following native-like features, ordered according to the priority: (1) acceptable bond lengths, (2) absence of interatomic clashes, (3) acceptable bond angles, and (4) acceptable torsion angles. Acceptable values of bond lengths and angles have been taken from a statistical analysis of structures in our fragment library (described above). Acceptable torsion angles were directly taken from Richardson et al. (2008). To avoid clashes, 42 RNA suites defined by Richardson et al. (2008) are tried one after another as starting conformations, until the first clash-free loop closure is found. If all variants exhibit clashes, then the variant with the smallest number of clashes is selected. Subsequently, the positions of the most flexible P and O5′ atoms are optimized by a simple stochastic search algorithm trying to satisfy angle and dihedral constraints. For generating coordinates at various stages of the procedure, the NeRF algorithm used in ROSETTA (Parsons et al. 2005) has been implemented. In case the entire procedure fails to close the backbone, details about the kind of distortion for the residues flanking the problematic site are reported.

For more extensive remodeling and searching for structures that are close to the global energy minimum, the user is expected to use other specialized software. ModeRNA contains a script to use the external molecular dynamics package MMTK (Hinsen 2000) for model optimization. It can be applied to perform conjugate gradient energy minimization using the AMBER force field to refine the model. The optimization may be restricted to particular regions of the model, in order to lower calculation time. Alternatively, any other molecular dynamics program or statistical potential for RNA can be used. ModeRNA is also compatible with the Adun package (Johnston et al. 2005) for molecular simulations.

5.4 SimRNA, a New Method for Template-Free RNA Structure Modeling

The development of SimRNA has been inspired by the success of the template-free protein modeling methods CABS (Kolinski 2004) and REFINER (Boniecki et al. 2003). It uses a simplified (coarse-grained) model of the RNA structure, samples the conformational space using the Monte Carlo simulated annealing approach, and evaluates the energy of conformations using a statistical potential, derived from analysis of experimentally solved RNA structures. SimRNA can model the folding of RNA molecules comprising single or multiple chains and can predict the structure based on sequence information alone, although it can also utilize starting structures provided by the user and additional distance restraints.

SimRNA is implemented in C++. It is available for UNIX systems, and it is distributed only in the executable version (the current prototype can be obtained from the authors upon request). SimRNA requires considerably larger computational resources compared to ModeRNA, and therefore, the length of sequences to be modeled de novo, without any restraints, is limited. The folding simulation of an RNA molecule of approximately 50 residues on a single processor ~2 GHz takes about 5–10 h.

5.4.1 Coarse-Grained Representation of RNA in SimRNA

SimRNA represents the nucleotide chain using only three atoms per nucleotide residue. The backbone is represented by atoms P of the phosphate group and C4′ of the ribose moiety, whereas the base is represented by just one nitrogen atom of the glycosidic bond (N9 for purines or N1 for pyrimidines). The remaining atoms are neglected. Such a simplistic representation allows to retain the main characteristics of the RNA molecule such as base pairing and stacking, and a spiral shape of the backbone in helices, while it significantly lowers the computational cost for conformational transitions and energy calculation.

Our approach relies on the previously proposed concept that the RNA backbone conformation can be described through two effective virtual bonds P-C4′ and C4′-P (Olson and Flory 1972). This is possible because the RNA backbone is rotameric and both the P-O5′-C5′-C4′ bonds and the C4′-C3′-O3′-P bonds tend to be approximately planar (Duarte and Pyle 1998; Murray et al. 2003). The atom representing the base has been chosen arbitrarily. Its position with respect to the backbone is established by defining a local coordinate system that depends on the backbone conformation with its origin on the C4′ atom.

5.4.2 Statistical Energy Function in SimRNA

Dealing with a reduced representation requires the energy function to capture the essential characteristics of the entire nucleotide chain that encompass also the atoms that are not explicitly modeled. This task can be accomplished by employing statistical potentials derived from frequency distributions of geometrical parameters observed in experimentally determined RNA structures. Terms of the SimRNA energy function were generated using reverse Boltzmann statistics (Sippl 1993):

$$E(p) = -\ln\left[f_{\text{observed}}(p)/f_{\text{expected}}(p)\right], \tag{5.1}$$

where $E(p)$ is the energy of a geometrical parameter p, $f_{\text{observed}}(p)$ is the frequency at which this is observed in a certain bin in the dataset, and $f_{\text{expected}}(p)$ is a reference frequency value assuming an unbiased distribution in all bins.

In SimRNA, short-range energy terms control the lengths of virtual bonds along the backbone (P-C4′ and C4′-P), flat angles (P-C4′-P and C4′-P-C4′), and torsion angles (P-C4′-P-C4′ and C4′-P-C4′-P). All short-range energy terms are sequence independent. The values of adjacent torsion angles are correlated (Duarte and Pyle 1998); therefore, this term was treated as a function of two variables, while terms corresponding to bond lengths and angles were defined as independent functions.

For long-range base–base interactions, we employed a similar approach to the one used in FARNA (Das and Baker 2007). For each type of a standard nucleotide (A, C, G, U), we collected information about its spatial neighbors in the representative set of 172 RNA tertiary structures from the RNADB2005 set (Richardson et al. 2008). We defined a second local coordinate system based on the conformation of the N-C4′-P unit with the origin on the C4′ atom and calculated the frequency of occurrence of the nitrogen atom of the interacting base, independently for each combination of base types. The spatial distributions were mapped onto 3D grids and converted into 3D histograms independently for each combination of interacting bases. This procedure allowed to calculate the relative preferences for all types of nucleotide–nucleotide interactions, including stacking and canonical (Watson–Crick) as well as noncanonical base pairing.

SimRNA models the total energy of RNA as the linear combination of three short-range terms that assess the local geometry of the chain and a single term that assesses all interactions between nucleotide residues. In doing so, we implicitly assume that the total energy of the system can be partitioned as the sum of independent contributions of the following terms:

$$E_{\text{total}} = E_{\text{bonds}} + E_{\text{flat\_angles}} + E_{\text{torsion\_angles}} + E_{\text{pair\_interactions}}. \quad (5.2)$$

5.4.3 Conformational Sampling in SimRNA

For searching the conformational space, we employed Monte Carlo dynamics. The simulation is controlled by an asymmetric Metropolis method (Metropolis and Ulam 1949) that accepts or rejects new conformations depending on the energy change associated with the conformational change. The chance of a move being accepted is also related to the "temperature" of the system. The probability of acceptance of the trial move is:

$$1 \text{ when } \Delta E \text{ is } \leq 0 \text{ or } e^{-\Delta E \beta} \text{ when } \Delta E \text{ is } > 0, \quad (5.3)$$

where ΔE is the energy change associated with the conformational change and β^{-1} is the reduced temperature (T^*). Thus, energetically favorable changes are always accepted, while energetically unfavorable changes may be accepted depending on the temperature. In order to find the global energy minimum, the conformational sampling is subjected to simulated annealing (SA), which involves a gradual decrease of the temperature of a system. The dependency is such that conformations are allowed to change almost randomly when T^* is large, but the conformational freedom is progressively restricted to low-energy variants as the system is cooled down. A properly parameterized simulation is expected to guide the system into a global energy minimum. The allowance for moves that increase the energy allows for crossing energy barriers between low-energy conformations and prevents the system from becoming stuck in local energy minima.

in the protein 3D structure modeling field. The development of useful methods for protein structure prediction has been driven by the benchmarking experiments, in which blind predictions are objectively compared to the experimentally solved structures. In the protein structure prediction community, there are periodic evaluation experiments that rigorously test the accuracy of prediction methods, e.g., CASP (biannually; http://www.predictioncenter.org/casp9/) and Livebench (continuously; http://meta.bioinfo.pl/livebench.pl). The ability to objectively assess the structure prediction methods, their relative performance, as well as the typical accuracy of predictions using an established set of measures (Moult et al. 2009) has proven indispensable for progress in this field of research.

The assessment of model accuracy requires reliable and meaningful metrics for comparisons between the models and the experimentally determined structures used as a "gold standard". One of the measures used commonly for comparison of macromolecular models is the root-mean-square deviation (RMSD) between pairs of equivalent atoms in the optimally superimposed structures. Typically, only backbone atoms are considered, e.g., $C\alpha$ in protein structures or P in RNA structures, but RMSD can be also calculated for any (or all) atoms. However, RMSD is not a perfect measure. A small perturbation in just one part of the structure (e.g., a hinge movement of two domains) can create a large RMSD suggesting that the two structures are very different overall. To take into account both local and global structural similarities, several metrics have been developed. The global distance test (GDT_TS) score (Zemla 2003) and the template matching (TM) score (Zhang and Skolnick 2004b) are examples of metrics developed for comparison of protein structures that have been generally accepted in the protein structure prediction field and used by assessors in the CASP experiment; they can be also applied to compare RNA structures and measure the accuracy of RNA models.

The GDT_TS score is defined as the average coverage (fraction of superimposed residues) of one structure by another in superpositions carried out with four different distance thresholds: 1, 2, 4, and 8 Å. The exact per-residue deviation values are ignored (e.g., residues with deviations ranging from 4.1 to 8.0 Å from native have identical contributions to the score). The GDT_TS score as well as the RMSD and many other metrics of structural similarity are dependent on the molecule size: if randomly selected molecules of the same size are compared, the score deteriorates with the molecule size. To eliminate the dependence on protein size, Levitt and Gerstein converted the structure similarity score into the P-value, i.e., a statistical significance score, based on the statistics of random structure comparisons (Levitt and Gerstein 1998). The TM score extends the approaches used in the Levitt–Gerstein score and in the GDT_TS score and attempts to eliminate the dependence on protein size by taking into account the radii of gyration of compared structures (Zhang and Skolnick 2004b). The value of the TM score always lies in range (0, 1), with better templates having higher TM scores. Recently, Hajdin et al. have analyzed the dependence of the structure similarity on the molecule size in small RNAs ($<$161 nt length) with relatively complex tertiary structures. They found that the compactness of folded RNA molecules is slightly lower than for proteins with the same mass. Based on their analysis, they defined an expression relating RMSD with the *P*-value that describes prediction significance (Hajdin et al. 2010).

Measures of structural similarity developed for protein models are not always ideal for RNA structures. They may capture the general 3D shape, local deviations of the structure, intradomain deformation, or interdomain deviations but are agnostic about important features that are unique to RNA, i.e., the base-pairing and base-stacking patterns. Parisien et al. developed an RNA 3D structure comparison measure called the deformation index (DI), which evaluates and indicates the deviations between two RNA 3D structures with both RMSDs and base interactions (stacking and pairing) (Parisien et al. 2009). They also developed another measure called a deformation profile (DP) that highlights dissimilarities between structures at the residue level for both intradomain and interdomain interactions. The DP score can be also used for proteins.

The number of crystal and NMR structures solved for RNA molecules that are sufficiently large for meaningful analysis is probably still too small to provide a sufficient number of targets for CASP-like intense modeling over a few months every year. While "CASP for RNA" has not fully developed yet, there have been several initiatives aiming at objective assessment of different methods and approaches for RNA modeling. One question is how these methods compare to each other when run in a fully automated mode, and how well they perform in the hands of different users. First, in the fall of 2010, Eric Westhof and Neocles Leontis organized a CASP-like RNA prediction challenge, with just three targets for a few groups of human predictors. At the time of the writing of this article, the results remained confidential, as the experimentally solved structures of the targets have not been published yet. Second, organizers of the CASP experiment expressed interest in including RNA structures as a possible new type of predictions, perhaps in CASP-10 (to be organized in 2012). Third, in the meantime, we have started a project similar to Livebench (again, an inspiration from the field of protein structural bioinformatics), which aims to become an objective benchmark of fully automated methods for RNA structure prediction. The CompaRNA web server (http://comparna.amu.edu.pl, T.P., K.R., Łukasz Kozłowski, Ewa Tkalińska, J.M.B., manuscript in preparation) provides a continuous benchmark for stand-alone and web server methods. Currently, it addresses only fully automated methods for RNA 2D structure prediction, but we intend to extend it to include methods for RNA 3D structure prediction that will become available as public web servers and/or local installations that can be run in a fully automated mode with default parameters and do not require large computing resources. This approach excludes expert-based modeling and methods that are not yet fully automated or require high-performance computing; hence, it is complementary to CASP-like modeling by human experts. We are convinced that these (and perhaps other) efforts will significantly stimulate the progress in the RNA structure prediction field.

5.6 Note

In the section devoted to ModeRNA software, this article includes extended passages from a research article "ModeRNA: A tool for comparative modeling of RNA 3D structure," (Rother et al. 2011b) © M.R., K.R., T.P., J.M.B. 2010. After this chapter

Chapter 6
The RNA Folding Problems: Different Levels of sRNA Structure Prediction

Fredrick Sijenyi, Pirro Saro, Zheng Ouyang, Kelly Damm-Ganamet, Marcus Wood, Jun Jiang, and John SantaLucia Jr.

Abstract RNA 3D structure prediction is analogous to the protein folding problem, particularly the astronomical size of the conformational search space and the challenge of appropriately scoring native versus decoy alternatives. However, RNA presents important differences compared to proteins, notably the existence of a low-energy secondary structure intermediate on the pathway to tertiary folding. The availability of a secondary structure facilitates de novo prediction using assembly of fragments. RNA mutants and close homologs are readily predicted with high accuracy using homology modeling. Evolutionarily distant RNAs often require a combination of homology and de novo modeling approaches. The greatest challenges to RNA structure prediction are posed by multihelix loops, certain types of pseudoknots, and multidomain packing. There are also a variety of partial folding problems for RNA and opportunities for whole database structure prediction. Herein we describe a unified suite of programs called "*RNA123*" for the analysis and prediction of RNA structure.

F. Sijenyi • P. Saro • Z. Ouyang • K. Damm-Ganamet
DNA Software, Inc., 334 E. Washington St., Ann Arbor, MI 48104, USA
e-mail: Fred@dnasoftware.com; Pirro@dnasoftware.com; Zheng@dnasoftware.com; Kelly@dnasoftware.com

M. Wood • J. Jiang
Department of Chemistry, Wayne State University, Detroit, MI 48202, USA
e-mail: mwoo@chem.wayne.edu; Jonathan@chem.wayne.edu

J. SantaLucia Jr. (✉)
DNA Software, Inc., 334 E. Washington St., Ann Arbor, MI 48104, USA

Department of Chemistry, Wayne State University, Detroit, MI 48202, USA
e-mail: John@dnasoftware.com

N. Leontis and E. Westhof (eds.), *RNA 3D Structure Analysis and Prediction*, Nucleic Acids and Molecular Biology 27, DOI 10.1007/978-3-642-25740-7_6,

6.1 Introduction

RNA performs diverse functions in the cell, including catalysis of protein synthesis, catalysis of RNA processing (e.g., cleavage, splicing, editing, chemical modification, silencing, and degradation), binding of small-molecule ligands, interactions with proteins, and regulation of gene expression at the transcription and translation levels. More than half of the human genome is transcribed into noncoding RNAs such as small interfering RNAs (siRNA) and microRNAs (miRNA) that participate in gene regulation, and yet, structures are available for only a small fraction of noncoding RNAs. Messenger RNAs (mRNA) also contain interesting secondary structures within the coding regions, introns, and in the 5′ and 3′ untranslated regions (UTR); additional functional RNAs include ribozymes, riboswitches, and in vitro selected aptamers. Many RNAs must form complex three-dimensional (3D) structures to carry out their functions. Thus, knowledge of the tertiary structure of these RNAs is essential to unraveling their roles in the cell.

Current experimental methods for three-dimensional structure determination are difficult and slow compared to the pace of discovery of interesting RNA sequences from genome sequencing projects. For example, the PDB currently (as of February 2010) contains only 1,730 RNA-containing structures of which 759 are RNA-only structures, but there are more than three million RNA sequences in the RefSeq and fRNAdb (Mituyama et al. 2009) databases and greater than 1.1 million RNAs in the curated Rfam database (Gardner et al. 2009). In the future, the number of functional RNAs discovered is expected to grow exponentially. Of this number, most are relatively simple structures or repetitive examples of similar folds. Thus, there is a compelling need to develop tools for 3D structure prediction based only on the sequence and any available experimental constraint information. Although several approaches and associated software tools have been developed to address this need (Das and Baker 2007; Ding et al. 2008; Jonikas et al. 2009; Lu and Olson 2003; Maier et al. 1999; Massire and Westhof 1999; Mueller and Brimacombe 1997; Parisien and Major 2008; Tan et al. 2006), general solutions with high accuracy are yet to be achieved.

Two of the major hurdles in RNA structure prediction continue to be the formulation of an accurate free-energy function and a conformational searching or sampling methodology that is capable of locating the energy minima (Das and Baker 2008). Consequently, most work has focused on the prediction of small RNAs with fewer than 100 nucleotides (nts) due to the computational complexity associated with conformational sampling for larger RNAs. Our goal is to increase the accuracy and size limit of predicted RNA structures. We have developed the software package *RNA123*, which contains a suite of tools for analyzing RNA structures, performing structure-based sequence alignments, secondary structure prediction, and 3D homology modeling of RNA structures (and protein complexes) as large as the bacterial ribosome. A prototype module in *RNA123* for de novo prediction of RNAs with single-domain structures consisting of helices, bulges, internal loops, and hairpin loops has also been completed. Addition of functionality in the de novo module for predicting multihelix loops and multidomain structures is underway but will not be discussed in this contribution.

6.2 Comparison of Protein and Nucleic Acid Folding Problems

Tremendous progress has been made on the "protein folding problem" in the past 30 years (Ben-David et al. 2009; Blum et al. 2010; Pandit et al. 2010; Raman et al. 2010). Nonetheless, it is still very challenging to accurately predict protein structure from sequence information alone, particularly for protein sequences longer than 100 amino acids, and rational protein design is still in its infancy (Kaufmann et al. 2010). Some aspects of the folding problem are easier to solve for RNA than for proteins (Tinoco and Bustamante 1999), while others are harder (Shapiro et al. 2007). Considering a protein and an RNA structure that have the same number of residues, the conformational search space for RNA is vastly larger than that for proteins due to the fact that each RNA residue has six backbone dihedrals, compared to only two in proteins. However, RNA only has four different residues, all of which contain a heterocyclic aromatic base, while proteins have 20 different amino acids with diverse chemical functionality (i.e., apolar, charged, sulfhydryl, aromatic, etc.) and diverse conformational freedom (e.g., glycine, proline, alanine, and lysine have different numbers of side-chain dihedrals and functional groups, and thus, they have very different conformational preferences). The limited RNA alphabet allows for easy determination of the library of preferred dimer conformations (Das and Baker 2007; Murray et al. 2005). In addition, RNA has strong pairing rules (G–C and A–U), while there are no such rules for proteins. The strong pairing rules result in a well-defined hierarchy of folding in RNA in which most of the folding free-energy change is due to secondary structure formation (Draper 2008; Jaeger et al. 1990; Mathews and Turner 2006; Tinoco and Bustamante 1999; Turner and Mathews 2010; Turner et al. 1988). Thus, the neglect of tertiary interactions is a reasonable first approximation for nucleic acids, making accurate prediction of secondary structure generally possible (described below). This contrasts with proteins where secondary structures are much weaker and multiple secondary structures are transiently sampled en route to the folded conformation (Alm et al. 2002; Duan and Kollman 1998; Karanicolas and Brooks 2003; Krivov and Karplus 2004). The strong pairing rules of nucleic acids also result in well-defined secondary structure boundaries (i.e., the beginnings and ends of helices) that are readily predicted by comparative sequence analysis even when the primary sequence similarity is low (which is not the case for proteins). The α-helical and β-strand elements of proteins exhibit significant variable bending, which is difficult to predict. In contrast, double helices in folded RNAs are fairly rigid and usually in A-form conformation, and structural flexibility is achieved through the connecting single-stranded loops (e.g., bulges, non-Watson–Crick base pairs, internal loops, multihelix loops, and exterior loops that connect domains). In addition, secondary structure in proteins involves local symmetry among residues that are close in the sequence (i.e., approximately twofold helical symmetry for β-strand and 3.6-fold helical symmetry for α-helices), while in RNA, secondary structure involves double helices that involve hydrogen bonds (H-bonds) between residues that are far apart in sequence.

Thus, the notion of secondary structure for RNA is related to protein supersecondary structure topology (e.g., the antiparallel β-sheet), which describes interactions among secondary structure elements. This qualitative difference in secondary structure makes it possible to deconstruct a large RNA fold into a set of smaller fragments that are coupled and subject to certain topological and steric constraints.

The nature of tertiary packing in RNA and globular proteins is also quite different. In proteins, secondary structure elements (α-helices and β-sheets) are stabilized by H-bonds between backbone atoms while side chains radiate outwards. Additionally, globular protein tertiary folding largely excludes water and is stabilized by weak cumulative interactions including London dispersion, hydrophobic effect, electrostatics, and H-bonds. By contrast, RNA secondary structure is stabilized by H-bond and stacking interactions (i.e., London dispersion interactions) on the inside of the double helix and by a cloud of nonspecific solvent and counterion interactions on the outside of the duplex. RNA tertiary structure is stabilized by a high degree of solvation and complex networks of H-bond and stacking interactions of loop residues from different parts of the sequence. Additional stabilization is contributed by specific interactions of hydrated magnesium ions particularly at locations of high phosphate density. RNA stacking is enthalpically driven and does not exhibit the classical hydrophobic effect (Bloomfield et al. 2000); all of the common bases A, C, G, and U are polar, though the modified nucleotides can potentially form small pockets of hydrophobic interactions. The lack of hydrophobic effect moieties in RNA significantly simplifies RNA packing compared to proteins. Note, however, that ligand binding to RNA (e.g., drugs, substrates, and proteins) can provide nonpolar surface area that is buried upon complex formation. This creates a hydrophobic component to such interactions that cannot be neglected. Interestingly, in RNA, there appears to be no concept analogous to "fold classification" (e.g., TIM barrel, Greek key, and beta-sandwich) (though there it is possible that RNA coaxial stacking and side-by-side helical packing may be amenable to fold classification). Thus, the threading approach that sometimes applies to evolutionarily unrelated protein folds cannot be applied to whole RNAs that are not evolutionarily related but does apply to individual loop motifs (i.e., many different sequences of unrelated function can have similar loop folds). As such, this principle is utilized extensively in our de novo modeling approach. However, the concept of threading for evolutionarily related RNAs (i.e., two RNAs that have a close common ancestor) forms the basis of our homology modeling approach.

6.3 Structure Prediction of RNA

Macromolecular tertiary structure prediction can be accomplished by four different approaches: folding pathway simulation, conformational sampling, fragment assembly, or threading (including homology modeling). RNA has a huge conformation space of ~3^{7N} (where N is the number of nucleotides, 7 is the total number of

RNA backbone and base dihedrals per residue, and 3 is the assumed number of conformational minima for each dihedral angle). This vast conformational space renders pathway and sampling methods computationally intractable for large N. The folding process for RNA is complex with numerous intermediates, but the formation of secondary structure occurs very quickly, often during transcription (Herschlag 2009; Koculi et al. 2006; Misra et al. 2003). The existence of a distinct low-energy secondary structure intermediate on the folding pathway for most RNAs makes RNA particularly suited to an approach that combines a dynamic programming algorithm (DPA) for secondary structure prediction and a 3D motif assembly algorithm for construction of tertiary structures. Secondary structure base-pairing constraints can be inferred from chemical probing experiments or from phylogenetic covariation analysis (Gutell et al. 2002) or from free-energy minimization-based secondary structure prediction algorithms (Mathews and Turner 2006; SantaLucia and Hicks 2004). Challenges to accomplishing accurate tertiary structure prediction using secondary structure restraints in this pathway are discussed at length below. The threading approach forms the basis for homology modeling, which is well suited to RNA structure prediction whenever the structure of a close homolog is available.

6.3.1 Types of RNA Structure Prediction

Figure 6.1 shows the different types of RNA structure prediction arranged by difficulty and the amount of experimental data included to aid the prediction. The simplest type of modeling involves analyzing an experimentally determined 3D structure to identify and correct regions that contain modeling errors; we refer to this as structure "conditioning," "regularization," or "preparation" (Davis et al. 2004). Somewhat harder is to complete a partially solved structure by performing de novo structure prediction on the unknown portions of the structure subject to the constraints of the known parts of the structure. More difficult still is to complete a full de novo prediction. The most challenging of all is de novo prediction of protein–RNA complexes (i.e., quaternary structure). Homology modeling, on the other hand, can be performed if a 3D template structure is available that has a high level of sequence or secondary structure similarity with the query. If the template has a low level of homology with the query (i.e., evolutionarily distantly related RNAs), then a combination of homology modeling of conserved portions of the secondary structure and de novo prediction of diverged regions can be performed. Generally, once the structure of a single RNA of a given type is known, then, in principle, homology modeling can be used to predict the 3D structures of an entire sequence database of that RNA type. For example, there are currently five known 3D structures for 5S rRNA, while GenBank contains >60,000 examples of 5S rRNA from different organisms (though a significant fraction of these sequences are currently incorrectly annotated). Thus, it is now possible to use *RNA123* to predict atomic 3D models for all known 5S rRNA sequences.

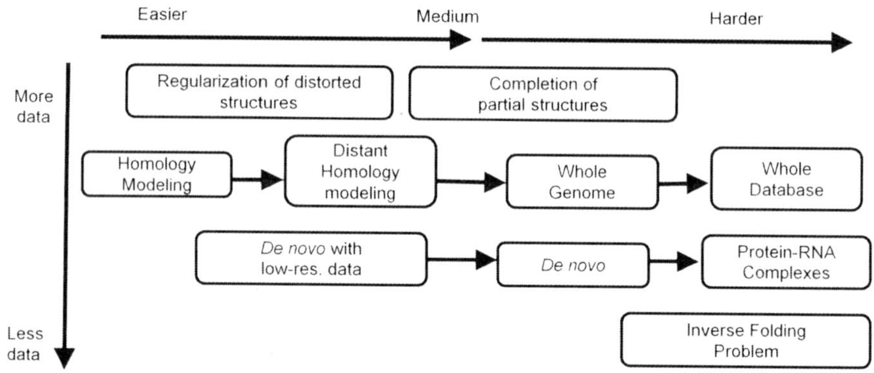

Fig. 6.1 Different levels of RNA folding problems arranged by difficulty (*horizontal axis*) and amount of experimental data used as restraints (*vertical axis*)

We have begun using *RNA123* to construct complete homology models for entire bacterial 30S ribosomes (including the proteins). Even for such large complexes as the ribosome (as described in Sect. 1.7.4), the homology models have global RMSDs <4 Å when compared to the published crystal structures, contain most of the tertiary H-bonds, and are free of steric clashes or bond gaps. Note that a significant portion of the observed global RMSD for homology models is due to different functional states of different crystal forms rather than actual errors in modeling. The *RNA123*-generated homology models often have near crystal structure quality as judged by global RMSDs less than 2 Å. In some instances, the homology models from *RNA123* appear to be better than published crystal structures because the models from *RNA123* correct modeling errors (described in Sect. 1.7.2) in the template and also complete regions of the structures that are disordered in the crystals. It will be more challenging, however, to homology model eukaryotic ribosomes because they contain large variable insertion regions, though the recently published low-resolution structure of the yeast ribosome will be helpful (Taylor et al. 2009). For some of these inserts, the correct secondary structures are not even known.

Achieving accurate de novo structure prediction is more challenging than homology modeling because much less information is available to guide the prediction and much more conformational sampling is required. There are a variety of experimental techniques that provide information that can improve the accuracy of the predictions accompanied by a decrease in the amount of conformational sampling required. The experimental methods include FRET distances, psoralen or UV cross-linking (Jaeger et al. 1993), hydroxyl radical footprinting (Sclavi et al. 1997), chemical modification reactivity, cryo-EM electron density (Lasker et al. 2009), and small-angle X-ray scattering (SAXS) (Forster et al. 2008; Lamb et al. 2008). Lastly, there is the inverse folding problem in which the goal is to predict a sequence that can fold into a designed 3D structure. This has been achieved for a few proteins (Dahiyat and Mayo 1997; Godzik et al. 1993; Hellinga 1997; Kuhlman et al. 2003; Looger et al. 2003), DNAs (Sherman and Seeman 2006), and RNAs (Nasalean et al. 2006).

6.3.2 Secondary Structure Prediction

There are several methods of obtaining restraints and constraints to aid secondary structure prediction ranging from phylogenetic covariation studies (Gutell et al. 2002), experimental techniques such as SHAPE analysis (Merino et al. 2005), chemical probing experiments (Jaeger et al. 1993), and free-energy minimization methods using nearest-neighbor thermodynamic data (Mathews et al. 2004; Turner et al. 1988). For an RNA of length N, there are approximately 1.86^N possible secondary structures (Zuker and Sankoff 1984). This fact makes the brute force search for the global optimum intractable for $N > 50$. Fortunately, the discrete nature of base pairing makes it possible to apply dynamic programming algorithms (DPA) to RNA folding. DPAs are very efficient; the global minimum and suboptimal structures are guaranteed to be found with calculation time proportional to N^3 and memory proportional to N^2, which is computationally tractable for $N < 10{,}000$ with desktop computers (Zuker 1989). The current parameterization of DPAs in programs such as MFold, RNAstructure, and Visual OMP result in approximately 73% or 90% correct secondary structure prediction of RNA (Mathews et al. 1999) and DNA (SantaLucia and Hicks 2004), respectively. The accuracy of secondary structure prediction can be dramatically improved with the addition of experimental restraint information from chemical probing (Mathews et al. 2004), SHAPE analysis (Deigan et al. 2009), and microarray hybridization (Kierzek et al. 2006).

The DPA in *RNA123* has been engineered for fast performance and minimized memory usage, includes pairing and nonpairing restraints, allows for setting different temperatures and solution conditions, has thermodynamic parameters for different strand types (DNA, RNA, 2′-*O*-methyl-RNA, PNA, phosphorothioates), and accounts for many modified nucleotides (LNA, inosine, riboT, diaminopurine, etc.). Such modified nucleotides are widely found in natural RNAs and are used in biotechnology applications of RNA including siRNA and antisense oligonucleotide design. In addition, the DPA in *RNA123* contains a novel algorithm for including "fuzzy restraints" in which a nucleotide is constrained to pair within a specified range of nucleotides, allowing for a rough secondary structure fold to be imposed on a given sequence and bias the prediction to be consistent with those restraints. Even without experimental restraints, when the suboptimal RNA secondary structures that are within a 2% energy window are examined, one of the suboptimal structures is typically >90% correct (Mathews et al. 1999). However, currently, there is no way to distinguish the correct from incorrect structures, although work has been done to specify the most reliable regions of the prediction (Zuker 1989). In comparison, current protein secondary structure predictions claim about 80%

accuracy (Jones 1999). The accuracy for RNA secondary structure prediction is more impressive than for proteins, however, because the secondary structure of RNA involves double helices with interactions that are far apart in sequence, whereas protein secondary structure (i.e., including β-strands but not including β-sheets) is purely local. The main drawback of using DPAs, however, is that RNA residues are represented by letters rather than 3D atomic structures. In addition, incorrect secondary structure predictions are generally the result of approximate and incomplete thermodynamic rules (i.e., neglect of the sequence dependence of the free-energy changes of loop motifs), neglect of the stabilizing effects of tertiary interactions (including "pseudoknots," coaxial stacking, tertiary H-bonds, and stacking), neglect of protein interactions, kinetically trapped folding (i.e., the functional structure is not the free-energy minimum structure), and neglect of metal ion and solvent interactions. We hypothesize that representation of the full atomic detail and force field energy of 3D models corresponding to suboptimal secondary structures will result in a more accurate energy ranking of the alternative secondary structures. An underlying assumption of this hypothesis is that the correct secondary structure is supplemented by a network of non-Watson–Crick hydrogen bonds and stacking and tertiary interactions (selected by evolution) that stabilize the functional structure. Incorrect secondary structures, however, cannot form such a network of stabilizing interactions and may also have steric clashes in 3D. An important question, however, is whether it is possible to model tertiary structure accurately enough to discover the native network of stabilizing non-Watson–Crick and tertiary interactions that would improve the ranking of secondary structures.

6.3.3 Tertiary Structure Prediction

As noted above, the prediction of a tertiary structure from a corresponding secondary structure is theoretically more difficult for RNA than for proteins due to the vast size of the conformational landscape available to RNA with its six backbone dihedrals per residue. The approximate number of backbone conformations for RNA is 3^{6N}. This rough estimate is verified by the conformational entropy change for the formation of Watson–Crick base pairs from random coils in both DNA and RNA (SantaLucia 1998; Xia et al. 1998). If half of the residues are known to be in A-form geometry (i.e., the secondary structure is given), then the dependence becomes approximately 3^{3N}, which is still astronomical for large N, and worse than the combinatorial explosion for proteins of about 3^{2N} (only considering phi and psi angles). These considerations suggest that nucleic acid backbones are more flexible than proteins, which implies that the tertiary folding problem is more difficult for RNA than proteins. Importantly, however, RNA secondary structure provides additional long-range (i.e., far apart in sequence) 3D constraints from Watson–Crick pairs. Full-atom classical molecular dynamics (MD) simulations are incapable of widely searching the conformation space due to limitations on computational resources. As a result of the limited searching, classical MD simulations rarely find the native structure when starting with an extended or random structure (Ding et al. 2008; Duan and Kollman 1998). The implementation of coarse-grained methods, however, can dramatically improve the sampling for a given amount of CPU time, albeit at a cost in accuracy (Das and Baker 2007; Jonikas et al. 2009; Tan et al. 2006).

6.3.4 Current Software for Tertiary Structure Predictions of Nucleic Acids

Sparked by the intense interest in RNA 3D structural modeling, a host of computational tools have been developed in recent years. The field has come a long way given that the first large all-atom RNA structure prediction was published in 1990 by Michel and Westhof (Michel and Westhof 1990). They predicted the structure of a group I intron by manual modeling based on available experimental biochemical and genetic data (Michel and Westhof 1990). Later, the Westhof group developed the program *MANIP*, which allows an expert user to manually link together 3D fragments and place them into a desired location (Massire and Westhof 1999). *MANIP* was used to create a model of ribonuclease P (Massire et al. 1998; Tsai et al. 2003). Recently, the Westhof group published a program named ASSEMBLE that has an intuitive GUI that aids in the manipulation and modeling of 3D structures of RNA (Jossinet et al. 2010). *MC-SYM* predicts the full atomic structures of small RNAs with fewer than 100 nts using nucleotide cyclic motifs (Parisien and Major 2008). These tools iteratively mix and match base pairs (both WC and non-WC) and dimer rotamers from a structure database using force field energy to rank candidate structures (Gautheret et al. 1993; Major et al. 1991) Sklenar's group has developed *JUMNA* to perform conformational searches in small hairpin loops (Maier et al. 1999). Macke and Case developed the *Nucleic Acid Builder* (*NAB*) tool (Macke 1998), which allows users to link together motifs to create rough structures suitable for AMBER refinement. ERNA-3D was written by Mueller and Brimacombe to model the ribosome (Mueller and Brimacombe 1997; Mueller et al. 2000) and has been used to generate a structure of the signal recognition particle receptor (SRP) RNA (Zweib and Muller 1997). However, *ERNA-3D* also requires manual intervention to generate structural models. Olson's group has developed the software *3DNA* to assemble double helical structures given helical parameters and also model single-stranded structures and complex motifs found in nucleic acids (Lu and Olson 2003). *FARNA* was developed by the Baker group, which developed *ROSETTA*, a leading program for protein folding, and is one of the more recent automated methods that can predict structures of small RNA fragments with complex folds stabilized by base triples and pseudoknots (Das and Baker 2007; Jonikas et al. 2009). *FARNA* employs a knowledge-based approach that uses trinucleotide torsion angle libraries in modeling; however, it is limited to prediction of small RNAs (~30 nts) due to severe computational demands for conformational searches. *FARNA's* prediction performance, although respectable, reveals the need for improvement in RNA structure prediction algorithms.

Other recent prediction tools have focused on enhancing the conformational searches by incorporating a coarse-grained molecular representation and conformational searching before full atomic computations to allow the modeling of large RNAs in a reasonable computation time. *YUP*, developed by Tan et al. (2006), is a flexible molecular mechanics framework that can incorporate coarse-grained and full atomic models and associated energy potentials and has been used to model RNA, DNA, and protein structures. *DMD*, developed in 2008, is also a coarse-grained molecular dynamics tool that utilizes an energy function to account for base-pairing and base-stacking interactions terms (Ding et al. 2008). *NAST* is the latest of the recent fully automated coarse-grained tools. Developed by Altman and coworkers in 2009, *NAST* employs an RNA-specific knowledge-based potential within a coarse-grained molecular dynamics engine to generate candidate structures (Jonikas et al. 2009). *NAST* has been successfully used to model yeast phenylalanine tRNA (76 nts), the P4–P6 domain of the *T. thermophila* group I intron (Cate et al. 1996; Murphy and Cech 1993) (158 nts), and to model missing loops in the *Azoarcus* and *Twort* ribozyme crystal structures.

6.4 *RNA123* Software for 3D Structure Prediction

Existing software packages do not contain force fields that are sufficient for optimization of nucleic acids that contain modified nucleotides, gaps from deletions, overlaps from nucleotide insertions, or model building errors and are inefficient at optimizing highly distorted geometries. For example, AMBER and CHARMM often fail to optimize gaps and atom overlaps generated in preliminary models because of the huge energy penalties from long covalent bond lengths and close van der Waals contacts. Both gaps and overlaps can cause numerical convergence problems or break chemical bonds, which terminates molecular dynamics simulations and energy minimization algorithms. In all fairness, however, classical simulations are designed to model *real* chemical systems, whereas starting structures in homology modeling and de novo prediction contain gaps and overlaps that are artificial. We have created a new force field called NA_FF (nucleic acid force field) that is optimized for RNA but also allows for inclusion of proteins and small-molecule ligands. The force field takes into account the charged phosphate backbone of RNA, planar preference for hydrogen bonding to aromatic bases (Chen et al. 2004), *gauche* effect of the gamma dihedral (Perez et al. 2007), preferred non-Watson–Crick H-bonding interactions (Leontis and Westhof 2001), optimized atomic partial charges [including modified nucleotides (Aduri et al. 2007)], optimized weighting of van der Waals and electrostatic interactions, novel pseudopotential for gaps, and a variety of structural restraints. *RNA123* also contains a general hierarchical strategy for conformational optimization (i.e., energy minimization) that works in torsion angle space and is computationally efficient for all classes of biopolymers (DNA, RNA, proteins, and carbohydrates). This algorithm is called DSTA (discrete sampling of torsion angles). The DSTA search is multidimensional and utilizes a novel method for modeling the local potential energy surface and finding an analytical minimum. As a measure of capability, the DSTA algorithm can optimize the conformation of an entire 16S rRNA (~1,540 nucleotides) homology model in about 2 h on a single core of a 2.0-GHz Centrino Duo laptop computer. However, the DSTA algorithm is limited to local conformational optimization and is not capable of searching vastly different conformational folds to find global minima. Methods for global conformational searching in the de novo module of *RNA123* are currently under development. The protein community has utilized structure decoys to optimize the balance of force field terms so that scoring functions can discriminate native from decoy folds (Jagielska et al. 2008). For RNA, however, there is a need to develop such a database of high-quality decoy structures to enable the testing and evaluation of scoring functions.

6.4.1 RNA123 *Visualization GUI*

RNA123 has a "smart" graphical user interface (GUI) that automatically accepts commonly used coordinate file formats (old and new PDB, AMBER, Xplor, mol2, mmCIF, etc.) and converts the file into standard PDB format on the fly once the coordinate file is loaded. The *RNA123* GUI, shown in Fig. 6.2, automatically analyzes the 3D coordinates and computes the secondary structure (including Watson–Crick and non-Watson–Crick paired bases, and also lists distorted pairs) and identifies all of the tertiary contacts (pseudoknots, base triples and quartets, and other tertiary H-bonds, and tertiary stacking interactions). These pairs are then classified and annotated by type based on the nomenclature of Leontis and Westhof (2001). The tertiary structure visualization component of *RNA123* is derived from RasMol (Sayle and Milner-White 1995). The program was redesigned and encapsulated as a reusable software component that can be seamlessly used by a Windows application or embedded into Web pages. Optimized coding, scripting language, and enhanced features for RNA/DNA structures allow researchers to utilize *RNA123* to visualize and manipulate macromolecules on personal computers. Based on this visualization component, *RNA123* also offers intuitive tools for editing structure, evaluating structure superimposition, and tracking conformation changes during the structural modeling process.

6.4.2 Structure Conditioning

Accurate 3D coordinates are a prerequisite for RNA structures to be used as templates in homology modeling and also for the creation of a reliable motif library or use in de novo structure prediction. The majority of the structures deposited in

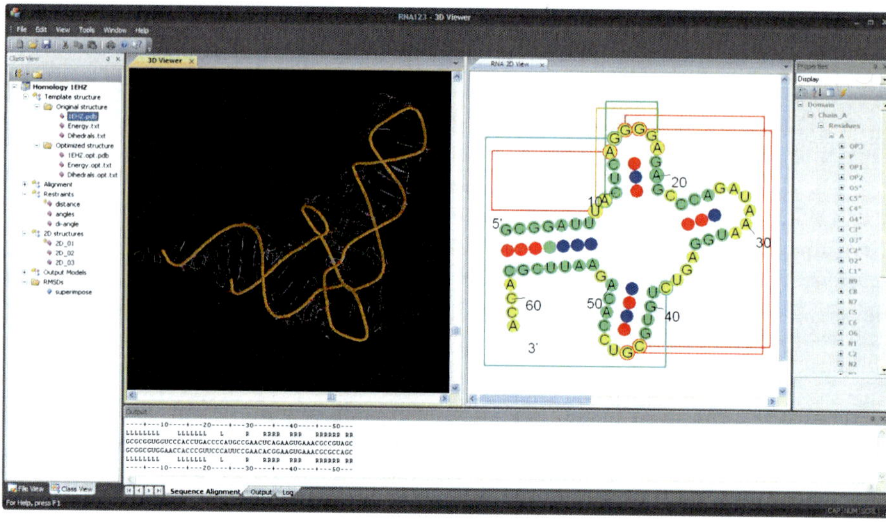

Fig. 6.2 Snapshot of *RNA123* GUI showing secondary structure, pseudoknots, and tertiary structure of the tRNAphe from *yeast* (PDB ID: 1EHZ)

the PDB (Protein Data Bank), however, have modeling errors such as nonstandard bond lengths or bond angles, residues with steric clashes, and incorrect base conformation where the chi dihedral is *syn* when instead it should be *anti*, distorted base pairs, missing atoms, missing residues, stereochemical errors, etc. Many of these errors are noted in the remarks section of the PDB entry (generated by the Procheck, NUCheck, SFCheck, and other programs) (Laskowski et al. 1993; Vaguine et al. 1999; Wheeler et al. 2007) but nonetheless are not corrected and cause problems for force field-based methods. A significant source of these modeling errors in X-ray crystal structures is a consequence of the neglect of hydrogen atoms in the modeling process. The later addition of hydrogen to heavy-atom-only structures often reveals steric clashes that are inconsistent with known van der Waals radii (Word et al. 1999). In addition, annotation errors, such as nonstandard naming of atoms or residues, or changing the order of atoms can also cause problems for many force fields.

For evaluating protein structures, Richardson's lab has developed methods for detecting errors in the side-chain conformations of asparagine, glutamine, and histidine (Word et al. 1999) as well as remodeling of structures with rare/high-energy rotamers (Murray et al. 2005). For RNA, similar modeling errors are apparent in a majority of the X-ray structures deposited in the PDB that are modeled using only heavy atoms. Richardson's lab has developed online tools for the analysis and correction of structural errors in nucleic acids (Davis et al. 2004). Such errors are within the Luzzati error of the structure determination, but they can cause problems for modeling. For instance, if one naively uses a crystal structure that contains errors as a source of folding motifs for RNA, then upon threading the original sequence into the backbone structure, a poor force field energy is obtained for the correct fold. Energy minimization can help to reduce this problem, but it is by no means a general solution because the structures are often trapped in local minima and thus remain in a poor energetic conformation. Similarly, NMR structures also often exhibit poor global geometry, especially if the structural refinement was carried out without residual dipolar couplings (Zhou et al. 2000). The application of simulated annealing as part of the modeling process for NMR structures can result in distorted geometries for Watson–Crick and non-Watson–Crick base pairs particularly for residues with sparse experimental restraints. In addition, NMR structures can also exhibit poor backbone modeling for residues, and ^{31}P chemical shifts are not available to restrain alpha and zeta dihedrals (Gorenstein 1984), where proton-phosphorus J-couplings are not available for restraining the beta and epsilon dihedrals (Lankhorst et al. 1984), or where gamma dihedrals are not restrained due to severe spectral overlap that precludes measurement of H4′-H5′/H5″ J-couplings.

To correct the structural errors mentioned above, *RNA123* contains an automated feature for identifying and fixing the majority of modeling and nomenclature errors in PDB structures. The process of correcting these structures using this *RNA123* feature is referred to as "structure conditioning," and it is an essential part of obtaining accurate de novo structure predictions using the fragment and motif assembly approach. During structure conditioning, residues with nonstandard bond lengths or bond angles are replaced with standard residues and geometry optimized with the DSTA algorithm. In addition, the *RNA123* conditioning algorithm detects steric clashes and residues with rare conformers (such as a syn geometry base) and attempts to remodel such structures with an alternative lower energy conformer, if such a structure is possible. We used this algorithm on the entire PDB to create a database of 453 conditioned RNA structures (out of 1,730 total in the PDB as of February, 2010).

6.4.3 Development of an Extensive Motif Library

Once "conditioned" structures are obtained, a complete library of coordinates for all possible RNA motifs (base pairs—both WC and non-WC, hairpin loops, bulges, internal loops, and multihelix loops) of various lengths and sequences is generated for de novo modeling using the fragment and motif assembly approach. Figure 6.3 illustrates the overall process for constructing the motif database in *RNA123*. The motif identification and extraction algorithm analyzes the 3D coordinates of the given "conditioned" structures and automatically determines the secondary structures of the RNAs and all the motifs present in the structures (hairpin loops, bulges, internal loops (including mismatches), and multihelix loops). Then the algorithm stores the backbone coordinates of those motifs in a library, including annotations such as the PDB_ID, base dihedrals, nucleotide sequence, and population of the motif. To avoid redundancy, each new candidate motif is superimposed on those already in the library; if the candidate motif is not present in the library

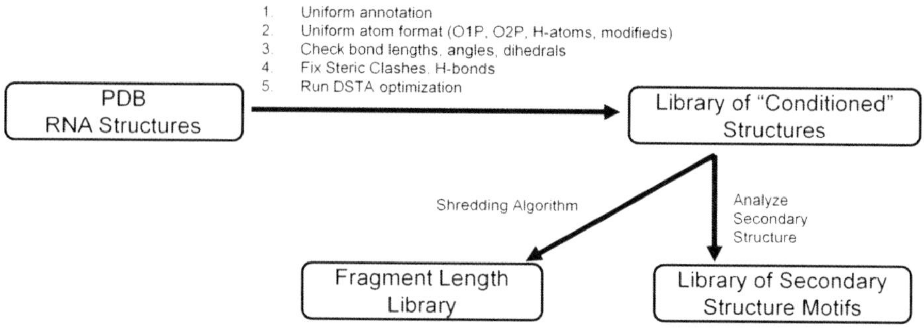

Fig. 6.3 Constructing the motif and fragment databases for de novo structure prediction. The library of PDB structures is first conditioned to fix modeling and other errors. The conditioned library of structures is then used to extract fragments of different lengths and motifs including bulges, internal loops, hairpin loops, and multibranched loops

Table 6.1 Summary of the contents of the current motif and fragment database in *RNA123*

Hairpins[a]		Bulges		Internal loops		Fragments		Multiloops	
Length	Number	Length	Number	Length	Number	Length	Number	Branches	Number
3	18	1	64	1 × 1	30	1	11	3	155
4	109	2	23	1 × 2	34	2	212	4	56
5	62	3	17	1 × 3	28	3	1,241	5	25
6	64	4	3	1 × 4	13	4	2,545	6	9
7	49	5	2	1 × 5	2	5	3,771	7	4
8	35	6	4	1 × 6	4	6	4,964	9	1
9	28	11	1	2 × 2	14	7	6,223	10	2
10	22	–	–	2 × 3	27	8	7,420	–	–
11	14	–	–	2 × 4	14	9	8,539	–	–
12	8	–	–	2 × 5	3	10	9,548	–	–
13	10	–	–	2 × 7	1	–	–	–	–
14	5	–	–	3 × 3	34	–	–	–	–
15	7	–	–	3 × 4	56	–	–	–	–
16	3	–	–	3 × 5	7	–	–	–	–
17	6	–	–	3 × 6	8	–	–	–	–
19	3	–	–	3 × 7	4	–	–	–	–

[a]This database was generated from 453 conditioned RNA structures from the PDB. The database contains more entries for larger internal loops up to size 16 × 17, but they are not listed in the table due to limited space. An RMSD cutoff of 1 Å is used to identify distinct motifs

(i.e., all RMSDs >1.0 Å), it is added. Additionally, the algorithm parses the structures into fragments of given lengths and stores these in a separate fragment library (e.g., a library of all unique dimers, trimers, and tetramers). The current library is presented in Table 6.1. An important point for de novo prediction is the need for a *complete* database of *all possible* structural folds for each type and size of motif (e.g., all possible examples of hairpin loops of length 8). Our current motif library (Table 6.1) has 35 entries for hairpin loops of length 8. This is likely an incomplete representation of all the structures that can form hairpin loops of length 8. The same

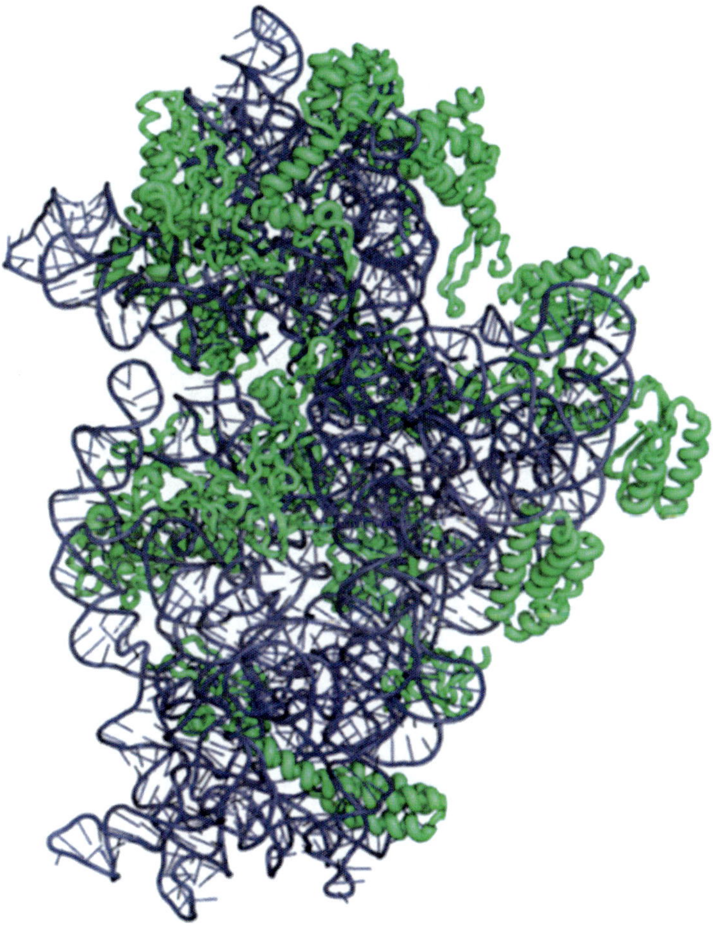

Fig. 6.5 All-atom homology model of the *P. aeruginosa* 30S ribosomal subunit generated by *RNA123*. rRNA is shown in *blue*. rProteins are shown in *green*. Only backbones are shown for clarity

RNA123 has also been used to accurately homology model a variety of other RNAs such as tRNAs, 5S rRNAs, a riboswitch, and an RNase P (Fig. 6.6). A cross-validation study using the three available bacterial 5S rRNAs (from *E. coli*, *T. thermophilus*, and *D. radiodurans*) to homology model each other in all permutations (e.g., using *E. coli* 5S rRNA to predict *T. thermophilus* 5S rRNA and vice versa) resulted in homology models that have <2 Å RMSD on average between the model and its representative crystal structure. A validation study of the whole 16S rRNA of *E. coli* and *T. thermophilus* has been similarly successful (See Tables 6.2 and 6.3).

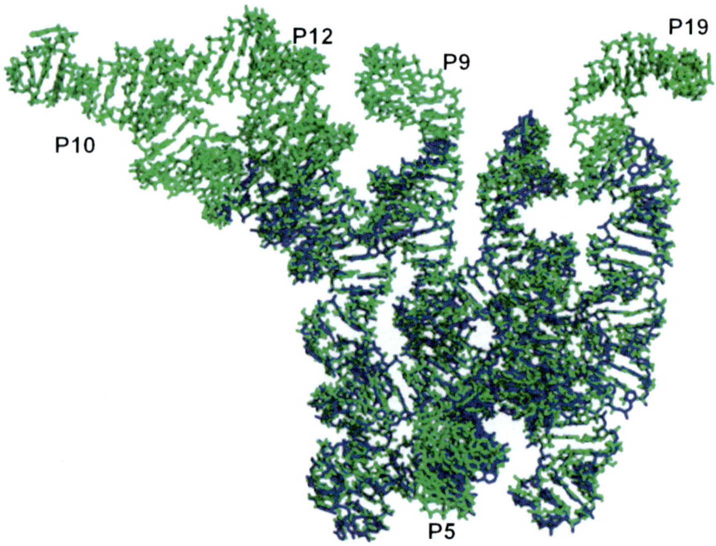

Fig. 6.6 Superposition of the complete model of the *B. stearothermophilus* RNase P structure (*in green*) modeled using *RNA123* and the X-ray crystal structure (PDB: 2A64, *in blue*). The labeled regions (P5, P9, P10, P15, and P19) were not resolved in the crystal structure and thus were modeled using *RNA123*. P10 and P12 helices were predicted using *RNA123* homology modeling utilizing the specificity domain from *B. subtilis* (PDB: 1NBS) as the template. Helices P9, P15, and P19 were predicted using the de novo module of *RNA123*

Table 6.2 Homology modeling results for the four domains of *E. coli* 16S rRNA

16S rRNA domain	Alignment score[c] (%)	No. of nucleotides[b]	RMSD between homology model and 2AVY crystal structure[a] (Å)
5′ Domain	82.0	560	3.91
Central domain	86.4	351	1.84
3′ Major domain	90.7	482	5.90
3′ Minor domain	79.5	134	3.67
Averages	84.7	–	3.83

[a]The RMSD of the homology models are compared to the 2AVY crystal structure
[b]The *T. thermophilus* structure (2J00) was used as the homology modeling template
[c]The alignment score is from the SBSA algorithm

Modeling of other RNAs such as 23S rRNA, group I and group II introns, riboswitches, siRNAs, and microRNAs is possible in principle provided that a homologous crystal or NMR structure is available. However, there are many interesting RNA sequences that have no closely solved homologous structures available (e.g., the deluge of noncoding RNAs). In addition, the existing homology modeling module in *RNA123* cannot currently predict widely divergent RNAs that have large insertions, particularly involving changes in complex multihelix loops. For example, consider the ribosome field where there are now structures available

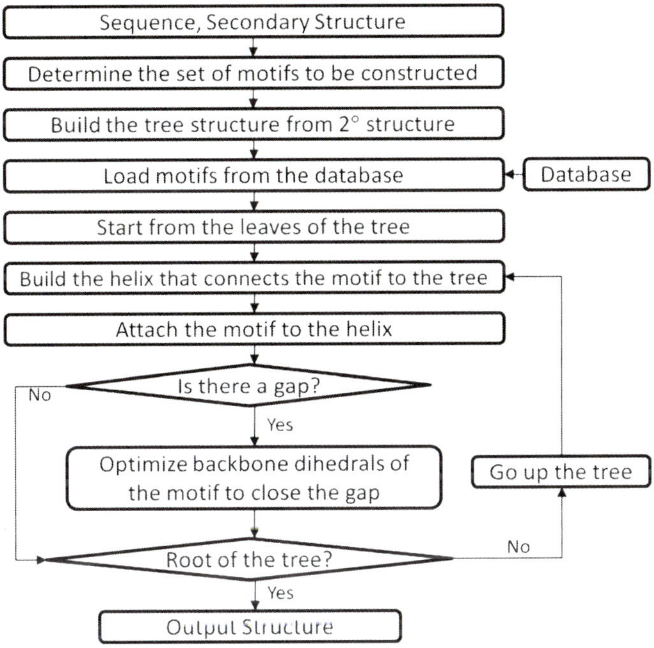

Fig. 6.8 Flowchart of the *BUILDER* algorithm

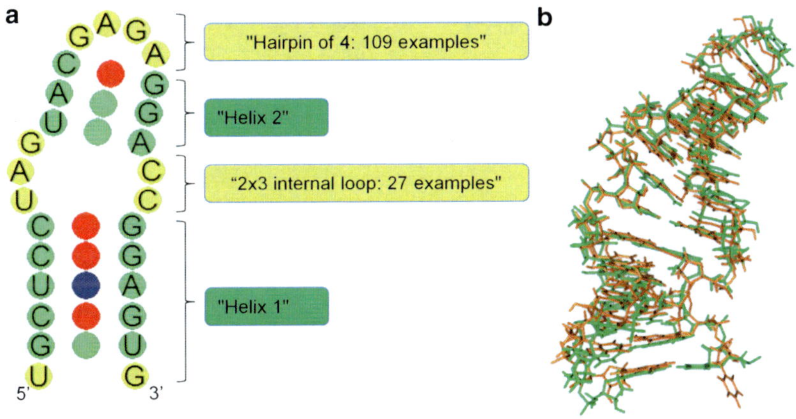

Fig. 6.9 The *BUILDER* strategy and results for a simple secondary structure. (**a**) The secondary structure motifs are labeled, and the number of examples for each motif type and size in our library are given. (**b**) Superimposition of one of the predicted 3D structures (in *green*, RMSD = 1.26 Å) and the crystal structure PDB ID: 1Q9A (in *orange*)

steps and was used to generate the structure in Fig. 6.9. The *BUILDER*, however, requires added functionality to encompass our full vision. While the current *BUILDER* strategy is effective for simple single-domain unbranched RNA structures (i.e., those with only helices, bulges, base pairs, internal loops, and a single hairpin loop), it is naïve about three crucial points: (1) it relies on only the motifs that are in the motif library, which is currently limited to those extracted from known structures in the PDB, (2) it cannot build multihelix loops, and (3) it neglects interactions between motifs that occur in complex structures such as multihelix loops, pseudoknots, and multiple domains. Work on *RNA123* is currently underway to address each of these issues.

6.4.7 The Combinatorial Explosion Problem for Multihelix Loops

Perhaps the most difficult aspect of the RNA folding problem is dealing with multihelix loops, although multidomain packing and pseudoknots are also challenging. There are currently only 252 unique multihelix loops in our motif database, which represents most of what is available in the PDB (Table 6.1). Shapiro's group has recently published a database of multihelix junctions (Bindewald et al. 2008). However, these 252 multihelix loops are a tiny fraction (less than one millionth) of the number possible. There are billions of permutations of multiple helices connected by different length single-strand linkers. As an example, consider a six helix junction with six linking single-stranded regions. If the single strands are allowed to vary in length between 0 and 9 nucleotides each, then there are exactly one million permutations, and each of those permutations can form multiple structures for different sequences. Multihelix loops with up to ten branches are found in bacterial 23S ribosomal RNAs, and even larger multihelix loops are present in some eukaryotic ribosomal RNAs. When a candidate RNA sequence contains a multihelix loop that is found in the database, then it can be assembled with *BUILDER* in the same fashion as other motifs (like hairpin loops, bulges, and internal loops). A multihelix loop can be considered a circular motif if the closing base pairs are considered to be connected. *RNA123* contains a novel algorithm to align a candidate multihelix loop in the query with circular permutations of those present in the motif library. This ensures that the multihelix loop is assembled in the proper orientation by the *BUILDER algorithm*. However, since the database vastly underrepresents the possible multihelix loops, there is a high probability that the database does not contain a correct structure for a new sequence. Thus, *RNA123* also assembles many trial multihelix loop structures from fragments (from the fragment library) subject to certain topological and steric constraints. This is an area that is currently under development within *RNA123*.

6.4.8 Case Study of the De Novo Algorithm in RNA123

To test the 3D structure prediction functionality of our preliminary de novo tool, we predicted the 3D structure of a 27 mer (PDB ID: 1Q9A), corresponding to the sarcin/ricin domain from *E. coli* 23S rRNA crystal structure solved at 1.04 Å resolution (Correll et al. 2003). In this study, we first used the dynamic programming algorithm in *RNA123* to predict 20 secondary structures (optimal and suboptimal folds). Next, using the predicted secondary structures, the de novo *BUILDER* algorithm assembled the structure from the base pairs, internal loops, and hairpin loops in its motif library (but not including any motifs from 1Q9A). Figure 6.9a illustrates the fragment approach that *BUILDER* sequentially follows by constructing: (1) helix 1, (2) UU mismatch, (3) 1 × 2 internal loop, (4) helix 2, and (5) the hairpin loop. For each motif, the actual sequence was threaded into all of the candidate conformers (each of which are energy minimized using DSTA), and the one conformer with the lowest NA_FF energy was used to build the structure. The computation took ~20 min to run using a standard laptop computer. Structures generated from the top three optimal folds had the best structures with an RMSD of 1.26, 1.39, and 1.14 Å, respectively, compared to the crystal structure (Fig. 6.9b). This result inspires confidence that the *RNA123* force field can correctly score native folds over decoys and also that *BUILDER* can assemble complete structures. We are currently implementing algorithms for predicting multibranched loop structures and multiple domains, which will enable predictions of larger RNAs.

To date, the CASP (critical assessment of structure predictions) competition has only assessed protein-based modeling tools; however, the RNA community is currently discussing such a competition for RNA predictions. The goal of the CASP competition is to assess the current state of the structure prediction field and bring to light areas where improvement is needed. As such, it will be the ultimate test for structure prediction tools from the community and described throughout this book.

6.5 Conclusion

RNA123 is a software package developed primarily for secondary and tertiary structure prediction and analysis of RNA (though it can handle DNA and protein as well). Several algorithms have been developed in *RNA123*; main components include a platform for de novo prediction and homology modeling of RNAs. Other functionalities include structure-based sequence alignment and a host of RNA structure visualization and analysis tools. *RNA123* is a product from DNA Software, Inc., and is available by licensing to academic (with discounted pricing), nonprofit, and industrial users (see http://www.dnasoftware.com for more information). *RNA123* is written in C++ and currently runs on Windows operating systems with implementations for Linux and Web access under development.

Acknowledgments We thank Norm Watkins for managing NIH grant R44GM085889 and Astrid Tuin for preparing Fig. 6.8. This work was supported by NIH grants R01-GM073179 (P.I. John SantaLucia), U01-AI061192 (P.I. Philip Cunningham), and R44GM085889 (P.I. Norman E. Watkins, Jr., and Fredrick Sijenyi).

References

Aduri R, Psciuk BT, Saro P, Taniga H, Schlegel HB, SantaLucia J Jr (2007) AMBER force field parameters for the naturally occurring modified nucleosides in RNA. J Chem Theor Comput 3:1464–1475

Alm E, Morozov AV, Kortemme T, Baker D (2002) Simple physical models connect theory and experiment in protein folding kinetics. J Mol Biol 322:463–476

Ben-David M, Noivirt-Brik O, Paz A, Prilusky J, Sussman JL, Levy Y (2009) Assessment of CASP8 structure predictions for template free targets. Proteins 77(Suppl 9):50–65

Bindewald E, Hayes R, Yingling YG, Kasprzak W, Shapiro BA (2008) RNAJunction: a database of RNA junctions and kissing loops for three-dimensional structural analysis and nanodesign. Nucleic Acids Res 36:D392–397

Bloomfield VA, Crothers DM, Tinoco IJ (2000) Nucleic acids: structures, properties and functions. University Science, Sausalito, CA

Blum B, Jordan MI, Baker D (2010) Feature space resampling for protein conformational search. Proteins 78:1583–1593

Cate JH, Gooding AR, Podell E, Zhou K, Golden BL, Kundrot CE, Cech TR, Doudna JA (1996) Crystal structure of a group I ribozyme domain: principles of RNA packing. Science 273:1678

Chen Y, Kortemme T, Robertson T, Baker D, Varani G (2004) A new hydrogen-bonding potential for the design of protein-RNA interactions predicts specific contacts and discriminates decoys. Nucleic Acids Res 32:5147–5162

Correll CC, Beneken J, Plantinga MJ, Lubbers M, Chan YL (2003) The common and the distinctive features of the bulged-G motif based on a 1.04 A resolution RNA structure. Nucleic Acids Res 31:6806–6818

Dahiyat BI, Mayo SL (1997) De novo protein design: fully automated sequence selection. Science 278:82–87

Das R, Baker D (2007) Automated de novo prediction of native-like RNA tertiary structures. Proc Natl Acad Sci U S A 104:14664–14669

Das R, Baker D (2008) Macromolecular modeling with rosetta. Annu Rev Biochem 77:363–382

Davis IW, Murray LW, Richardson JS, Richardson DC (2004) MOLPROBITY: structure validation and all-atom contact analysis for nucleic acids and their complexes. Nucleic Acids Res 32:W615–619

Deigan KE, Li TW, Mathews DH, Weeks KM (2009) Accurate SHAPE-directed RNA structure determination. Proc Natl Acad Sci U S A 106:97–102

Ding F, Sharma S, Chalasani P, Demidov VV, Broude NE, Dokholyan NV (2008) Ab initio RNA folding by discrete molecular dynamics: from structure prediction to folding mechanisms. RNA 14:1164–1173

Draper DE (2008) RNA folding: thermodynamic and molecular descriptions of the roles of ions. Biophys J 95:5489–5495

Duan Y, Kollman PA (1998) Pathways to a protein folding intermediate observed in a 1-microsecond simulation in aqueous solution. Science 282:740–744

Forster F, Webb B, Krukenberg KA, Tsuruta H, Agard DA, Sali A (2008) Integration of small-angle X-ray scattering data into structural modeling of proteins and their assemblies. J Mol Biol 382:1089–1106

Gardner PP, Daub J, Tate JG, Nawrocki EP, Kolbe DL, Lindgreen S, Wilkinson AC, Finn RD, Griffiths-Jones S, Eddy SR et al (2009) Rfam: updates to the RNA families database. Nucleic Acids Res 37:D136–D140

Gautheret D, Major F, Cedergren R (1993) Modeling the three-dimensional structure of RNA using discrete nucleotide conformational sets. J Mol Biol 229:1049–1064

Godzik A, Kolinski A, Skolnick J (1993) *De novo* and inverse folding predictions of protein structure and dynamics. J Comput Aided Mol Des 7:397–438

Gorenstein DG (1984) Phosphorus-31 NMR: principles and applications. Academic, Orlando

Gutell RR, Lee JC, Cannone JJ (2002) The accuracy of ribosomal RNA comparative structure models. Curr Opin Struct Biol 12:301–310

Hellinga HW (1997) Rational protein design: combining theory and experiment. Proc Natl Acad Sci U S A 94:10015–10017

Herschlag D (2009) Biophysical, chemical, and functional probes of RNA structure, interactions and folding: Part A. Preface. Methods Enzymol 468:xv

Jaeger JA, Zuker M, Turner DH (1990) Melting and chemical modification of a cyclized self-splicing group I intron: similarity of structures in 1M Na+, in 10 mM Mg2+, and in the presence of substrate. Biochemistry 29:10147–10158

Jaeger JA, SantaLucia J Jr, Tinoco I Jr (1993) Determination of RNA structure and thermodynamics. Annu Rev Biochem 62:255–287

Jagielska A, Wroblewska L, Skolnick J (2008) Protein model refinement using an optimized physics-based all-atom force field. Proc Natl Acad Sci U S A 105:8268–8273

Jones DT (1999) Protein secondary structure prediction based on position-specific scoring matrices. J Mol Biol 292:195–202

Jonikas MA, Radmer RJ, Laederach A, Das R, Pearlman S, Herschlag D, Altman RB (2009) Coarse-grained modeling of large RNA molecules with knowledge-based potentials and structural filters. RNA 15:189–199

Jossinet F, Ludwig TE, Westhof E (2010) Assemble: an interactive graphical tool to analyze and build RNA architectures at the 2D and 3D levels. Bioinformatics 26:2057–2059

Karanicolas J, Brooks CL (2003) Improved Go-like models demonstrate the robustness of protein folding mechanisms toward non-native interactions. J Mol Biol 334:309–325

Kaufmann KW, Lemmon GH, Deluca SL, Sheehan JH, Meiler J (2010) Practically useful: what the Rosetta protein modeling suite can do for you. Biochemistry 49:2987–2998

Kierzek E, Kierzek R, Turner DH, Catrina IE (2006) Facilitating RNA structure prediction with microarrays. Biochemistry 45:581–593

Koculi E, Thirumalai D, Woodson SA (2006) Counterion charge density determines the position and plasticity of RNA folding transition states. J Mol Biol 359:446–454

Krivov SV, Karplus M (2004) Hidden complexity of free energy surfaces for peptide (protein) folding. Proc Natl Acad Sci U S A 101:14766–14770

Kuhlman B, Dantas G, Ireton GC, Varani G, Stoddard BL, Baker D (2003) Design of a novel globular protein fold with atomic-level accuracy. Science 302:1364–1368

Lamb J, Kwok L, Qiu X, Andresen K, Park HY, Pollack L (2008) Reconstructing three-dimensional shape envelopes from time-resolved small-angle X-ray scattering data. J Appl Crystallogr 41:1046–1052

Lankhorst PP, Haasnoot CA, Erkelens C, Altona C (1984) Carbon-13 NMR in conformational analysis of nucleic acid fragments. 3. The magnitude of torsional angle epsilon in d(TpA) from CCOP and HCOP NMR coupling constants. Nucleic Acids Res 12:5419–5428

Lasker K, Topf M, Sali A, Wolfson HJ (2009) Inferential optimization for simultaneous fitting of multiple components into a CryoEM map of their assembly. J Mol Biol 388:180–194

Laskowski RA, MacArthur MW, Moss DS, Thornton JM (1993) PROCHECK: a program to check the stereochemical quality of protein structures. J Appl Crystallogr 26:283–291

Leontis NB, Westhof E (2001) Geometric nomenclature and classification of RNA base pairs. RNA 7:499–512

Leontis NB, Stombaugh J, Westhof E (2002) The non-Watson-Crick base pairs and their associated isostericity matrices. Nucleic Acids Res 30:3497–3531

Looger LL, Dwyer MA, Smith JJ, Hellinga HW (2003) Computational design of receptor and sensor proteins with novel functions. Nature 423:185–190

Lu X-J, Olson KW (2003) 3DNA: a software package for the analysis, rebuilding and visualization of three-dimensional nucleic acid structures. Nucleic Acids Res 31:5108–5121

Macke TJ, Case DA (1998) Modeling unusual nucleic acid structures. ACS Symp Ser 682:379–393

Maier A, Sklenar H, Kratky HF, Renner A, Schuster P (1999) Force field based conformational analysis of RNA structural motifs: GNRA tetraloops and their pyrimidine relatives. Eur Biophys J Biophys Lett 28:564–573

Major F, Turcotte M, Gautheret D, Lapalme G, Fillion E, Cedergren R (1991) The combination of symbolic and numerical computation for three-dimensional modeling of RNA. Science 253:1255–1260

Massire C, Westhof E (1999) MANIP: An interactive tool for modeling RNA. J Mol Graphics Modeling 16:197–205

Massire C, Jaeger L, Westhof E (1998) Derivation of the three-dimensional architecture of bacterial ribonuclease P RNAs from comparative sequence analysis. J Mol Biol 279:773–793

Mathews DH, Turner DH (2006) Prediction of RNA secondary structure by free energy minimization. Curr Opin Struct Biol 16:270–278

Mathews DH, Sabina J, Zuker M, Turner DH (1999) Expanded sequence dependence of thermodynamic parameters improves prediction of RNA secondary structure. J Mol Biol 288:911–940

Mathews DH, Disney MD, Childs JL, Schroeder SJ, Zuker M, Turner DH (2004) Incorporating chemical modification constraints into a dynamic programming algorithm for prediction of RNA secondary structure. Proc Natl Acad Sci U S A 101:7287–7292

Merino EJ, Wilkinson KA, Coughlan JL, Weeks KM (2005) RNA structure analysis at single nucleotide resolution by selective 2'-hydroxyl acylation and primer extension (SHAPE). J Am Chem Soc 127:4223–4231

Michel F, Westhof E (1990) Modelling of the three-dimensional architecture of group I catalytic introns based on comparative sequence analysis. J Mol Biol 216:585–610

Misra VK, Shiman R, Draper DE (2003) A thermodynamic framework for the magnesium-dependent folding of RNA. Biopolymers 69:118–136

Mituyama T, Yamada K, Hattori E, Okida H, Ono Y, Terai G, Yoshizawa A, Komori T, Asai K (2009) The functional RNA Database 3.0: databases to support mining and annotation of functional RNAs. Nucleic Acids Res 37:D89–92

Mueller F, Brimacombe R (1997) A new model for the three-dimensional folding of *Escherichia coli* 16 S ribosomal RNA. I. Fitting the RNA to a 3D electron microscopic map at 20 A. J Mol Biol 271:524–544

Mueller F, Sommer I, Baranov P, Matadeen R, Stoldt M, Wohnert J, Gorlach M, van Heel M, Brimacombe R (2000) The 3D arrangement of the 23 S and 5 S rRNA in the *Escherichia coli* 50 S ribosomal subunit based on a cryo-electron microscopic reconstruction at 7.5 A resolution. J Mol Biol 298:35–59

Murphy FL, Cech TR (1993) An independently folding domain of RNA tertiary structure within the Tetrahymena ribozyme. Biochemistry 32:5291–5300

Murray LJ, Richardson JS, Arendall WB, Richardson DC (2005) RNA backbone rotamers–finding your way in seven dimensions. Biochem Soc Trans 33:485–487

Nasalean L, Baudrey S, Leontis NB, Jaeger L (2006) Controlling RNA self-assembly to form filaments. Nucleic Acids Res 34:1381–1392

Needleman SB, Wunsch CD (1970) A general method applicable to the search for similarities in amino acid sequence of two proteins. J Mol Biol 48:443–453

Pandit SB, Brylinski M, Zhou H, Gao M, Arakaki AK, Skolnick J (2010) PSiFR: an integrated resource for prediction of protein structure and function. Bioinformatics 26:687–688

Parisien M, Major F (2008) The MC-Fold and MC-Sym pipeline infers RNA structure from sequence data. Nature 452:51–55

Perez A, Marchan I, Svozil D, Sponer J, Cheatham TE 3rd, Laughton CA, Orozco M (2007) Refinement of the AMBER force field for nucleic acids: improving the description of alpha/gamma conformers. Biophys J 92:3817–3829

Raman S, Vernon R, Thompson J, Tyka M, Sadreyev R, Pei J, Kim D, Kellogg E, DiMaio F, Lange O et al (2010) Structure prediction for CASP8 with all-atom refinement using Rosetta. Proteins 77:89–99

SantaLucia J Jr (1998) A unified view of polymer, dumbbell, and oligonucleotide DNA nearest-neighbor thermodynamics. Proc Natl Acad Sci USA 95:1460–1465

SantaLucia J Jr, Hicks D (2004) The thermodynamics of DNA structural motifs. Annu Rev Biophys Biomol Struct 33:413–438

Sayle RA, Milner-White EJ (1995) RASMOL: biomolecular graphics for all. Trends Biochem Sci 20:374

Sclavi B, Woodson S, Sullivan M, Chance MR, Brenowitz M (1997) Time-resolved synchrotron X-ray "footprinting", a new approach to the study of nucleic acid structure and function: application to protein-DNA interactions and RNA folding. J Mol Biol 266:144–159

Shapiro BA, Yingling YG, Kasprzak W, Bindewald E (2007) Bridging the gap in RNA structure prediction. Curr Opin Struct Biol 17:157–165

Sherman WB, Seeman NC (2006) Design of minimally strained nucleic Acid nanotubes. Biophys J 90:4546–4557

Tan RKZ, Petrov AS, Harvey SC (2006) YUP: a molecular simulation program for coarse-grained and multiscaled models. J Chem Theor Comput 2:529–540

Taylor DJ, Devkota B, Huang AD, Topf M, Narayanan E, Sali A, Harvey SC, Frank J (2009) Comprehensive molecular structure of the eukaryotic ribosome. Structure 17:1591–1604

Tinoco I Jr, Bustamante C (1999) How RNA folds. J Mol Biol 293:271–281

Tsai H-Y, Masquida B, Biswas R, Westhof E, Gopalan V (2003) Molecular modeling of the three-dimensional structure of the bacterial RNase P holoenzyme. J Mol Biol 325:661–675

Turner DH, Mathews DH (2010) NNDB: the nearest neighbor parameter database for predicting stability of nucleic acid secondary structure. Nucleic Acids Res 38:D280–282

Turner DH, Sugimoto N, Freier SM (1988) RNA structure prediction. Annu Rev Biophys Biophys Chem 17:167–192

Vaguine AA, Richelle J, Wodak SJ (1999) SFCHECK: a unified set of procedures for evaluating the quality of macromolecular structure-factor data and their agreement with the atomic model. Acta Crystallogr D Biol Crystallogr 55:191–205

Wheeler DL, Barrett T, Benson DA, Bryant SH, Canese K, Chetvernin V, Church DM, DiCuccio M, Edgar R, Federhen S et al (2007) Database resources of the national center for biotechnology information. Nucleic Acids Res 35:D5–12

Word JM, Lovell SC, Richardson JS, Richardson DC (1999) Asparagine and glutamine: using hydrogen atom contacts in the choice of side-chain amide orientation. J Mol Biol 285:1735–1747

Xia T, SantaLucia J Jr, Burkard ME, Kierzek R, Schroeder SJ, Jiao X, Cox C, Turner DH (1998) Thermodynamic parameters for an expanded nearest-neighbor model for formation of RNA duplexes with Watson-Crick base pairs. Biochemistry 37:14719–14735

Zhou H, Vermeulen A, Jucker FM, Pardi A (2000) Incorporating residual dipolar couplings into the NMR solution structure determination of nucleic acids. Biopolymers 52:168–180

Zuker M (1989) On finding all suboptimal foldings of an RNA molecule. Science 244:48–52

Zuker M, Sankoff D (1984) RNA secondary structures and their prediction. Bull Math Biol 46:591–621

Zweib C, Muller F (1997) Three-dimensional comparative modeling of RNA. Nucleic Acids Symp Ser 36:69–71

Chapter 7
Computational Prediction and Modeling Aid in the Discovery of a Conformational Switch Controlling Replication and Translation in a Plus-Strand RNA Virus

Wojciech K. Kasprzak and Bruce A. Shapiro

Abstract This chapter presents computational tools used to predict the secondary structure and model the 3D structure of a novel translation enhancer element found within the 3′-UTR of the Turnip crinkle virus (TCV). Our Massively Parallel Genetic Algorithm program (MPGAfold) was used to predict the secondary structure, including one H-type pseudoknot, of the translation enhancer element. The results were confirmed and augmented by experiments. The combined secondary structure information was used to create a 3D model of the enhancer element with the aid of our program RNA2D3D. The 3D structure resembles that of tRNA, while its secondary structure is different from canonical tRNAs. It is the first such element found within a 3′-UTR, and it is a part of a conformational switch involved in the control of translation and transcription. It is possible that similar mechanisms may exist in other eukaryotic genomes.

7.1 Introduction

This chapter describes the computational tools we have developed and applied to elucidate the structure and mechanism of a translational enhancer element in the 3′-UTR of the Turnip crinkle virus (TCV). The study was conducted in close cooperation between computational and experimental groups, and involved prediction of the secondary structure of the 3′-UTR of TCV with the aid of our stochastic secondary structure prediction Massively Parallel Genetic Algorithm (MPGAfold)

W.K. Kasprzak
Basic Science Program, SAIC-Frederick, Inc., NCI Frederick, Frederick, MD 21702, USA
e-mail: kasprzaw@mail.nih.gov

B.A. Shapiro (✉)
Center for Cancer Research Nanobiology Program, National Cancer Institute Frederick, Frederick, MD 21702, USA
e-mail: shapirbr@mail.nih.gov

N. Leontis and E. Westhof (eds.), *RNA 3D Structure Analysis and Prediction*,
Nucleic Acids and Molecular Biology 27, DOI 10.1007/978-3-642-25740-7_7,

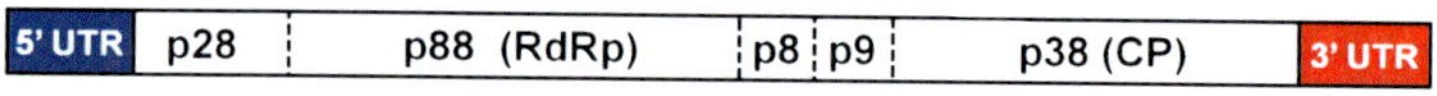

Fig. 7.1 TCV genomic layout showing the relative positions of the five overlapping ORFs coding for proteins, as well as the flanking 5′ and 3′ untranslated regions (UTRs). The translational enhancer element is found within the 3′-UTR

(Shapiro and Navetta 1994; Shapiro and Wu 1996, 1997; Shapiro et al. 2001b; Wu and Shapiro 1999). The predicted secondary structure, including a novel H-type pseudoknot, was combined with an experimentally elucidated pseudoknot, and this composite 2D structure model was used as the input information to our interactive, geometry-driven 2D to 3D modeling program RNA2D3D (Martinez et al. 2008). A 3D shape resembling a tRNA emerged in modeling and suggested a possible functional role for the new tRNA-shaped structure (TSS) as a ribosome binder. Experiments verified this hypothesis and, in addition, led to discovering a functional structural switch between translation and replication of TCV. The structural stability of the TSS was evaluated by subjecting it to molecular dynamics (MD) simulations, which indicated areas of particular flexibility in one of its elements. The TSS model and the MD data showed good agreements with the solution of the TSS structure in solvent, based on data from small angle X-ray scattering and residual diploar coupling (SAXS/RDC) (Wang et al. 2009; Zuo et al. 2010).

TCV is a *Carmovirus* from the family Tombusviridae. It has a single short (4,045 nt) RNA plus strand genome and is well suited for the study of the switching between the alternative mechanisms of translation and replication. Figure 7.1 depicts the genomic layout of TCV. Its five partially overlapping open reading frames (ORFs) encode proteins required for replication (p28 and a readthrough p88, which encodes the RNA-dependent RNA polymerase, RdRp), cell-to-cell movement of the virus (p8 and p9) and encapsidation (coat protein p38 or "CP").

In the prevailing eukaryotic translation mechanism, a 5′-capped and 3′-polyadenylated mRNA template is brought together to form a circular structure with the help of initiation factor proteins (eIF) and a poly(A)-binding protein. Recruitment of the small (40S) and the large (60S) ribosomal subunits leads to the full ribosome assembly and translation (Merrick 2004; Preiss and Hentze 2003). Alternative mechanisms to recruit ribosomes via internal ribosome sequences (IRES) have been found for RNA viruses missing the 5′-cap (Dreher and Miller 2006; Fechter et al. 2001; Fraser and Doudna 2007; Hellen and Sarnow 2001; Lancaster et al. 2006). For many plant viruses lacking the 5′-cap, translation initiation mechanisms involve cap-independent translational elements (CITEs) located in their 3′-UTRs (Miller et al. 2007). TCV is among the viruses which are neither 5′-capped nor polyadenylated. Prior experimental data indicated the existence of an unidentified element within the 3′-UTR which enhances translation in synergy with the 5′-UTR. However, the structure and mechanism of this element were not known (Qu and Morris 2000; Yoshii et al. 2004).

Computational structure prediction, in-line probing, and mutagenesis of the RNA were used to predict and verify its secondary structure. The RNA secondary structure of the enhancer element was predicted by our MPGAfold to contain one pseudoknot (Ψ_3 in Figs. 7.3 and 7.4), the existence of which was experimentally verified (McCormack et al. 2008; Zhang et al. 2006). In addition, pseudoknot Ψ_2, shown in Fig. 7.4, was identified experimentally (Zhang et al. 2006). Pseudoknot Ψ_1 pairs the 3′ side of the large symmetric loop (LSL) in H5 with the 3′ terminal residues past the Pr stem-loop. The above-mentioned structures play roles in translation and transcription (McCormack and Simon 2004; McCormack et al. 2008; Sun and Simon 2006; Wang et al. 2009). The self-contained and functional subdomain corresponding to the sequence between the pseudoknots Ψ_3 and Ψ_2, shown in Fig. 7.4, was selected for 3D modeling (McCormack et al. 2008).

Using as input the secondary structure pairing information of the central region of the 3′-UTR between pseudoknots Ψ_3 and Ψ_2, our modeling program, RNA2D3D, very rapidly produced a 3D structure resembling a tRNA (TSS). This result was unexpected since the elucidated secondary structure, including the two pseudoknots, did not have the typical cloverleaf shape of the canonical tRNA secondary structure. The shape of the 3D structure immediately suggested the hypothesis that the TSS could function as a translational enhancer by potentially binding ribosomes or ribosomal subunits.

Further refinement of the model followed, and, in tandem, ribosome binding experiments were performed on various 3′-UTR fragment sizes containing the TSS, as well as some additional surrounding sequence context. The ribosome binding experiments were successful and revealed that the 60S subunit binds better to the TSS than the full 80S does. In addition, the 43S ribosomal subunit was shown to bind to the 5′-UTR of TCV. These results suggested a model for translation initiation whereby the 5′-bound 43S ribosomal unit in combination with the 3′-bound 60S ribosomal unit aid in cyclizing the 5′ and 3′ ends of TCV, thereby initiating translation. In addition, the experimental results indicated that the TSS element binds strongly to the P site of the ribosome, but with lower affinity than canonical tRNAs, an important property required for successful completion of the translation initiation process (Stupina et al. 2008). Further studies showed that the translation product necessary for the replication (by RdRp) causes reversible structural changes in the 3′-UTR region, involving elements of the TSS structure (Yuan et al. 2009). Thus, the TSS structure is a part of a conformational switch controlling the functional use of the genomic template for translation or for replication.

7.2 Computational Prediction of the TCV TSS

Several software packages developed by our group were used to elucidate the structure of the tRNA-shaped translation enhancer element (TSS) found in the 3′-UTR of TCV. These programs, MPGAfold, StructureLab and RNA2D3D, will now be described in some detail with specific illustrations of how they were used to determine the tRNA-shaped motif. Together with many other programs, all the tools mentioned in this chapter are publicly available on our Web site: http://www.ccrnp.ncifcrf.gov/~bshapiro/sl.

7.2.1 MPGAfold

It has been reported that the folding of RNA in many circumstances is hierarchically determined (Tinoco and Bustamante 1999). That is, the secondary structure of the RNA forms first and this is followed, usually in the presence of magnesium ions, by collapse to a state containing tertiary interactions. This is the model of folding we are assuming for this chapter, although there is evidence for nonhierarchical transient interactions and that folding and tertiary interactions may form cooperatively as an RNA molecule is transcribed (Cruz and Westhof 2009; Ding et al. 2008; Heilman-Miller and Woodson 2003; Wilkinson et al. 2005). Using this hierarchical assumption, we first generate an RNA secondary structure given a sequence, which then serves as the starting point for predicting its three-dimensional structure. Several algorithms exist for predicting RNA secondary structure. These algorithms fall into two broad categories, those which are deterministic, e.g., dynamic programming based, and those which are stochastic, e.g., genetic algorithm based. The algorithm that was applied in this case was MPGAfold, a massively parallel genetic algorithm for secondary structure prediction that was developed by our group (Shapiro and Navetta 1994; Shapiro and Wu 1996, 1997; Shapiro et al. 2001b; Wu and Shapiro 1999).

MPGAfold is based on the principles of genetic algorithms (GA), as originally described by Holland (1992). GAs can be used as optimization procedures to search large solution spaces for results that are "best" or near optimal. In the case of RNA secondary structure determination, we are optimizing an objective function that estimates the free energy of the folded RNA secondary structure. In other words, we are looking for an RNA fold such that the free energy of the folded structure is optimal or near the optimal free energy possible for the given sequence and a given set of energy rules. The strength of the algorithm lies in its Boltzman-like characteristics, preferring the most probable solutions over the strictly lowest energy ones (Wu and Shapiro 1999). As such, the algorithm must be run repeatedly (usually 20–25 times) to determine the consensus solution of multiple runs. In MPGAfold, the alphabet consists of all possible contiguous fully base-paired stems (stem pool), the fundamental building elements of the algorithm that are derived (precomputed) from the given sequence. These stems may be shortened by a conflict-driven peel-back operator (see below). Also, added to the alphabet are motifs that consist of two stems that together can form bulge loops or internal loops of sizes 1×1, 1×2, and 2×2 (i.e., loops with combinations of one or two nucleotides in their $5'$ and/or $3'$ sides). The basic GA operators of mutation, recombination, and selection are applied in sequence to generate new RNA secondary structures that reside on a rectangular (or square) grid representing a population that is a power of two in size (see Fig. 7.2). Typically grid sizes range from as low as 2 K to as high as 128 K. The size range is normally chosen as a function of the size of the sequence being folded. Longer sequences usually require a broader range of sizes. Each population element in the grid, which consists of an RNA secondary structure, can "see" its eight neighbors (N, S, E, W, NE, NW, SE, and SW). The grid is toroidally wrapped so that all population elements have eight neighbors. Thus, for example, the West neighbor of a population element on the left edge of the

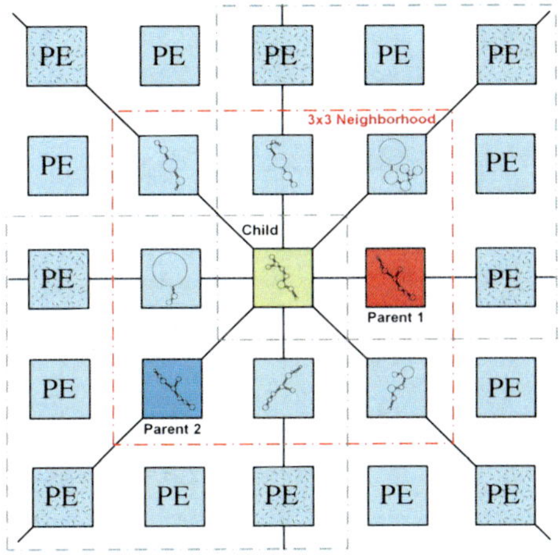

Fig. 7.2 Schematic representation of the rectangular MPGAfold population layout. Population sizes are powers of two and typically run between 4 and 128 K. The central element of each 3 × 3 neighborhood is replaced at each generation in parallel

grid will be an element on the right edge of the grid, etc. The algorithm is implemented to run on a parallel cluster computer, distributing the population across a set of power of 2 processors. MPGAfold's speed scales almost linearly with the addition of more processors. Figure 7.2 illustrates a portion of the grid layout of the population elements

The three GA operators are mutation, recombination, and selection:

1. Mutation—Stems are drawn from the stem pool at random. As a precursor to each recombination step (see below), two child structures are initialized with some stems mutated in from the stem pool.
2. Recombination—Each population element and its nine nearest neighbors (see above) are placed in a sorted array of nine elements. The sorting is done from the lowest energy structure to the highest. Two "parent" structures are chosen with a biased sampling from the array (low energy structures have a higher probability of being chosen than high energy structures). Stems from the parents are distributed to the two child structures (which will already contain some stems previously added by the mutation operator). Both mutation and crossover operators incorporate a probablilistic conflict-driven peel back mechanism to resolve potential conflicts (overlapping base pairs) between stems being added to a structure and the ones already a part of it. Thus, instead of being completely rejected, a conflicting stem can be peeled back (i.e., shortened, or have some base pairs removed) in order to fit into the existing structure. The use of this mechanism improves the resolution of the algorithm, allowing structures to contain single base pair stems, while at the same time permitting the drawing from a precomputed stem pool consisting of only maximum-sized contiguous stems.
3. Selection—An objective function calculates the free energy of the two child structures. The child structure that has the lower free energy then replaces the element that is in the center of the 3 × 3 mesh window.

These three GA operators are iteratively applied in parallel across the entire population grid. Thus, after each GA cycle, a population of 16 K may contain 16 K new elements (secondary structures). An annealing mutation operator is applied that gradually lowers the mutation rate. As a consequence of this, after multiple iterations, the population stabilizes resulting in termination of the algorithm. Several kinds of solution structures can be output from MPGAfold. Typically, either the population-wide consensus structure is output as a solution or the lowest (best) energy structure. The population-wide consensus structure corresponds to the most frequent entry in a histogram of free energy values calculated for an evolving population at each generation. Also, since MPGAfold is a stochastic algorithm, it is run multiple times (20–25 times) in order to obtain the consensus folding patterns from the final or intermediate outcomes of multiple runs for the given sequence.

It is often useful to run MPGAfold using different population sizes for a given sequence (4–128 K, for example) because the algorithm has the demonstrated property of terminating with what may be significant intermediate folds at lower population sizes (Gee et al. 2006; Kasprzak et al. 2005; Linnstaedt et al. 2006, 2009; Shapiro et al. 2001a). These intermediate folds are usually seen as transient structures that exist in larger population runs.

7.2.1.1 H-Type Pseudoknots

Because MPGAfold uses stems as the alphabet for the algorithm, it is relatively easy to generate pseudoknotted structures. However, because of the limited knowledge regarding the energies of complex pseudoknots, the algorithm is currently restricted to generating only H-type pseudoknots, i.e., base pair interactions between a simple hairpin loop structure and a flanking single stranded region, for which the free energies can be calculated and used in the objective function. In principle, however, MPGAfold is capable of considering more complex structures. MPGAfold does not implement any 3D constraints on pseudoknot formation at this time. However, a scheme of geometric constrains recently implemented in our CyloFold algorithm (and Web server) is being considered for inclusion into MPGAfold (Bindewald et al. 2010).

7.2.1.2 Miscellaneous Information Regarding MPGAfold

The MPGAfold program suite provides other facilities proven to be very valuable for predicting and interpreting the results obtained from the algorithm (Shapiro et al. 2001b, 2006):

1. Co-transcriptional folding—Since RNA is normally transcribed in the 5′- to 3′-direction, the structure into which an RNA folds can depend on the transcription (elongation) process, resulting in different secondary structures than would be obtained by folding the entire sequence at once. MPGAfold has the ability to explore the secondary structures that form while the sequence strand elongates. The rate of elongation can also be changed to study its impact on the maturation of a predicted structure (Linnstaedt et al. 2009; Shapiro et al. 2001a). One has to keep in mind that the rate of elongation parameter was not designed to mimic known experimental transcription rates. It controls how fast bases are made available for consideration at

the 3′-end per generation of the algorithm and thus, how fast stems with the increasing 3′-end interactions can be used by the mutation and recombination operators. It does not explicitly control how quickly structures propagate throughout the evolving population or become the population-wide consensus structures.
2. Folding visualization—MPGAfold can be run in conjunction with a Java-based interactive front-end visualizer. Color-coded two-dimensional maps of the population grid can be viewed and manipulated while the algorithm is running. Population energy distributions, pseudoknots, and the existence of predefined stems can be monitored in real time. In addition, when MPGAfold is run with the Visualizer and StructureLab (see below), drawings of individual RNA secondary structures from the population can be displayed (Shapiro et al. 2006).

Since MPGAfold is stochastic in nature, with no two runs for the same sequence following exactly the same folding intermediate conformations, it is impossible to describe its performance as a function of the input sequence length. Sequences of the same length may display very different folding characteristics, i.e., long convergence (high number of generations) to the final answers, or very rapid folds. Finding such differences, especially when the stable intermediates may be biologically functional, is the main advantage of MPGAfold (Linnstaedt et al. 2006, 2009; Shapiro et al. 2001a). In the studies cited here the biologically significant intermediates, as well as the final answers, reached the status of population-wide consensus structures. By comparison, the algorithms sampling the DPA solution spaces for the same sequences either missed the structures of interest or assigned extremely low probabilities to them. On the other hand, MPGAfold cannot compete with the DPA-based programs in terms of execution times. To give the reader a rough idea of its performance, a 368-nt long fragment of HIV-1 (nl43) 5′-UTR subjected to 20 runs at the 16 K population level takes 10:01 min on an 8 processors of an Intel Nehalem 2.8 GHz machine. The algorithm scales approximately linearly with the changes in the population size. Dependence on the number of processors is not perfectly linear and depends on the efficiency of communication between processors (Shapiro et al. 2001b). For example, almost linear scaling can occur when the algorithm is run on symmetric multiprocessors with uniform memory access. MPGAfold can be obtained by contacting us directly.

7.2.2 *StructureLab*

StructureLab is a graphical data mining program that permits interactive exploration of databases of RNA secondary structures (Kasprzak and Shapiro 1999; Shapiro and Kasprzak 1996; Shapiro et al. 2006). These structures may be derived from dynamic programming algorithms such as Mfold (Mathews et al. 1999; Zuker 2003), RNAstructure (Mathews et al. 1999, 2004), the Vienna package RNAfold (Hofacker 2003; Hofacker et al. 1994) or from MPGAfold described above. Also, see reviews (Mathews and Turner 2006; Shapiro et al. 2007). The ability to explore multiple structures is very useful for gaining a perspective on the diversity of structures obtained from the various folding algorithms. For a complete description of the functions available in StructureLab, the reader is referred to Kasprzak and Shapiro (1999) and Shapiro et al. (2006). StemTrace is an important and useful tool within StructureLab that was employed in this TCV study, and so it will be described here.

StemTrace produces an interactive two-dimensional plot that represents the stems that are found in RNA secondary structures. Each position along the *X*-axis represents an RNA secondary structure. Each position along the *Y*-axis represents a unique stem found in one or more of the plotted structures. By unique stem we mean a unique triplet of values describing the 5′-start position of a stem, its 3′-stop position, and its size (i.e., number of base pairs). The position of a stem along the *Y*-axis can determine either by the order in which it is generated (i.e., the first appearance in the predicted structures), or by a user-selected sort criterion, for example, the increasing 5′-start positions of the stems. Thus a vertical line drawn from a specific position along the *X*-axis will intersect all the points representing stems present in an RNA secondary structure. Frequently a StemTrace plot displays horizontal bands appearing at specific *Y*-positions. These bands indicate stems that re-occur as one moves along the *X*-axis (i.e., the stems appear in multiple structures). The bands are color coded to represent the frequency of occurrence of individual stems within the displayed set of structures. StemTrace can be used in a variety of ways to depict the emergence of a single RNA secondary structure (in one MPGAfold run, for example) or to permit the comparison of thousands of RNA secondary structures derived from a single sequence or from several sequences in a family. Four of the more commonly used arrangements are described below.

1. StemTrace can display the evolution of an RNA secondary structure generated by a single MPGAfold run. In this case, stems plotted for the lower *X*-axis values will together correspond to relatively immature RNA secondary structures in the early stages of development. As one proceeds along the *X*-axis, the RNA secondary structure "matures" (gains more stems and achieves a lower free energy). Typically by the end of an MPGAfold run, the structures plotted by StemTrace become constant indicating that the population of solutions converges to a single, dominant, and stable structure.
2. StemTrace can display the final (converged) structures from multiple runs of MPGAfold. In this case a StemTrace could represent, for example, 20 structures that are generated in 20 independent MPGAfold runs of a given RNA sequence.
3. StemTrace can display the final (converged) structures for a family of related sequences. In this representation, the solutions for each sequence are depicted as a block of contiguous runs along the *X*-axis (e.g., 20). The *Y*-axis position for stems from different sequences of the family can be adjusted to account for sequence insertions and deletions, thereby maintaining the proper *Y*-axis ordinate for the stems.
4. StemTrace can display the output from a dynamic programming algorithm such as Mfold, RNAstructure or RNAfold. Here, for example, the optimally folded structure can occupy the left-most position along the *X*-axis with suboptimal solutions occupying succeeding higher positions along the *X*-axis.

StemTrace can be used interactively by placing the cursor at any position on the graph and by clicking different mouse buttons to retrieve and display individual stem information or full structural information and, optionally, to draw the corresponding secondary structure by automatically passing this information to another tool from StructureLab.

7.3 Prediction and Analysis of the Secondary Structure of the TCV tRNA-Shaped Domain

We applied MPGAfold to the region containing the TCV translational enhancer (the last 216 nucleotides from the 3′-UTR; 3,839–4,054 in the TCV genome). MPGAfold was run 20 times to obtain a consensus structure (as defined above and illustrated in Fig. 7.3). It was known from previous experiments that a base pair interaction existed between the 3′-most four bases and the four nucleotides corresponding to the central four positions in the 3′ side of the H5 large symmetric loop (LSL). Because it was believed that this interaction, Ψ_1, was not important for translation and was part of a switch induced by RdRp, this interaction was removed from consideration in subsequent runs (McCormack et al. 2008; Zhang et al. 2006). Figure 7.3 depicts the StemTrace plot for 20 runs of MPGAfold and shows the dominant structure obtained from the plot. It should be noted that this plot is sorted, with the stems along the *Y*-axis depicted in increasing 5′ order. The StemTrace and drawing color codes reflect the

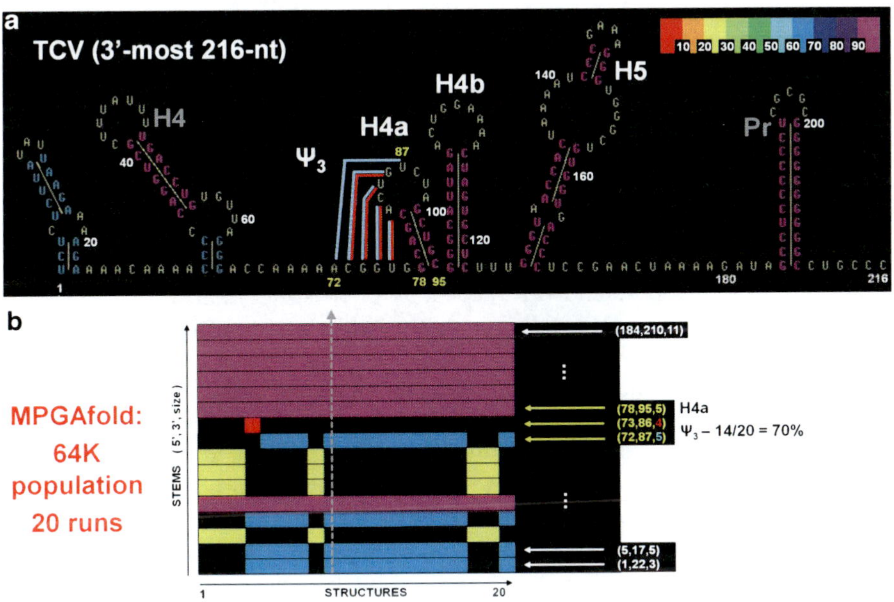

Fig. 7.3 Secondary structure depiction and StemTrace plot from StructureLab of MPGAfold predicted structures. (**a**) Color-coded dominant secondary structure derived from the StemTrace plot indicated by the *vertical dotted line* in the StemTrace plot shown in (**b**). The color-coding scale is displayed in the *upper-right corner*. (**b**) All the stems shown in the dominant structure are either *purple* or *blue* indicating that they appear at least 70% of the time in 20 MPGAfold runs

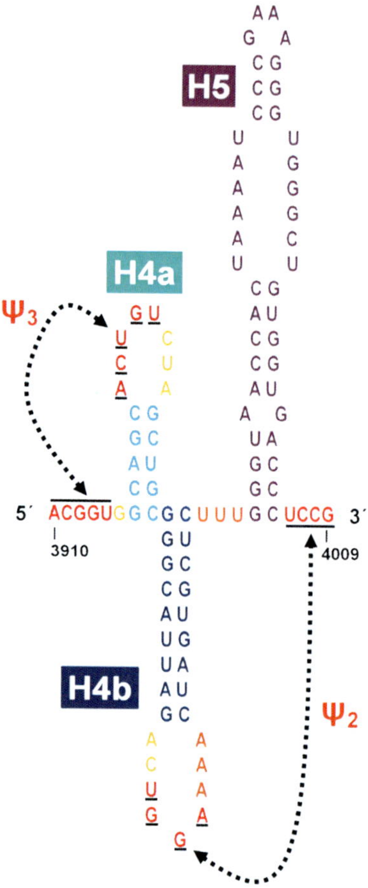

Fig. 7.4 Secondary structure representation of TCV enhancer element. Two key pseudoknots Ψ_2 and Ψ_3 are shown. (Copyright © American Society for Microbiology, Journal of Virology, 82(17), 2008, pp. 8706–8720, doi:10.1128/JVI.00416-08)

frequency of occurrence of the individual stems. The magenta stems appear in all 20 runs. The blue stems appear in 13 or 14 out of the 20 runs. One of them, the stem that is involved in the Ψ_3 pseudoknot, consists of 5 base pairs in 13 runs (light blue in Fig. 7.3) and 4 base pairs in 1 run (red in Fig. 7.3). Experiments support existence of this pseudoknot and indicate its role in ribosome binding (McCormack et al. 2008; Stupina et al. 2008). In addition, the adenylates upstream of Ψ_3 appear to stabilize it (Yuan et al. 2009). Another pseudoknot was found experimentally that involves the loop bases in stem H4b and the bases downstream of stem loop H5. The initial region of interest resides between base 72 and 171. Stem Pr is part of a promoter structure and stem H4 is important for RdRp binding and is also part of a structural switch to be discussed later. Figure 7.4 shows the secondary structure of the translational enhancer element with the two pseudoknots. In-line probing experiments and site-directed mutagenesis studies that included compensatory base pair mutations verified the structure that is shown (McCormack et al. 2008; Zhang et al. 2006).

7.4 Computational Determination of the 3D Structure of the TCV Translational Enhancer Element

7.4.1 RNA2D3D

To understand how the secondary structure depicted in Fig. 7.4 facilitates translation, we modeled the 3D structure of this element. The program RNA2D3D accepts primary RNA sequences and their secondary structure representations as input (Martinez et al. 2008). The secondary structure description may include pseudoknots. The program then rapidly generates a first-pass, three-dimensional model from the secondary structure. Because of its speed, RNA2D3D can be used interactively to quickly explore alternative 3D conformations. Further refinements of the models are usually necessary due to the limitations of the idealized geometry approach (see below), but in many cases the first pass model will give significant clues as to how to proceed. Some of the capabilities of this program will be described in detail in this section, along with illustrative examples using the secondary structure model indicated in Fig. 7.4. In brief, the program facilitates manipulation of the whole molecule or selected subparts. With the secondary and 3D structures displayed side by side, as shown in Figs. 7.5 and 7.6, one can interactively select a region of interest in either of the depictions, whichever is easier to work with, and take actions affecting both the 2D and 3D models. Addition or deletion (opening) of individual base pairs is possible, as well as coaxial stacking and "compactification" of paired regions (see below). One can interactively edit bond angles, as well as "clean up" the model via calls to energy minimization and short molecular dynamics runs. Incorporation of known motifs from a database, such as PDB for example, is also possible. Some aspects of RNA-based nano-scale structure design are also serviced by the RNA2D3D tools. Quick exploration of alternatives is aided by the capability to save and retrieve multiple models. The program features rapid interactive, geometrically driven exploration with heuristic stacking of stems and pseudoknots in a 3D RNA structure to facilitate manual modeling by a user having sufficient expertise in modeling to identify modular motifs and structure fragments to substitute in the preliminary, automatically generated model. No capabilities are provided for automatic, knowledge-based augmentation of the initial model.

RNA2D3D first generates a secondary structure representation from the provided base pairing data. Pseudoknots are automatically arranged as coaxially stacked stems. The secondary structure drawing acts as a starting point for the 3D rendering. The 2D structure representation can be thought of as a graph whose nodes represent base positions and arcs represent backbone covalent bond connections and noncovalent hydrogen bond base pair interactions. The graph contains two bond length components. One is the connection from one base to the next along the backbone. This is defined here as the distance between two successive phosphates. The other important distance is the one between the two phosphates that constitute the nucleotides involved in a base pair. Loop regions in the drawing are represented by

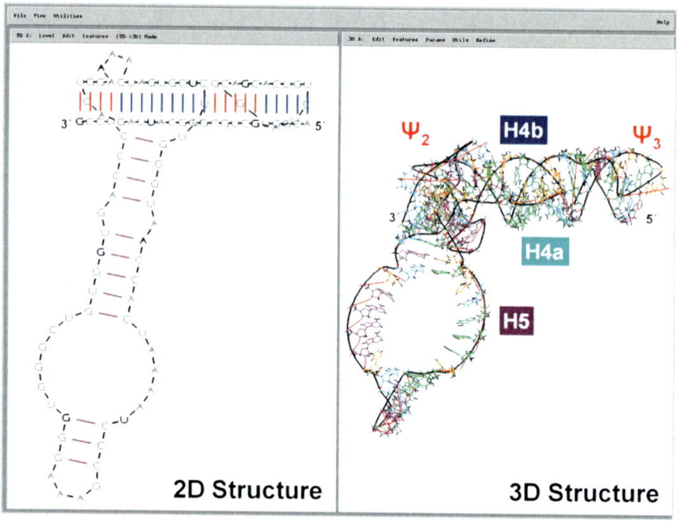

Fig. 7.5 Depiction of the secondary structure layout (*left panel*) and its initial 3D rendering (*right*). The pseudoknot layout is shown horizontally in both panels. The pseudoknots consisting of stems H4a and H4b coaxially stack on each other, while stem H5 is near perpendicular. The *right panel* also shows the initial winding of the A-form helices and the embedding of the nucleotides in the stems as provided by the secondary structure information. Labels shown correspond to the labels in Fig. 7.4

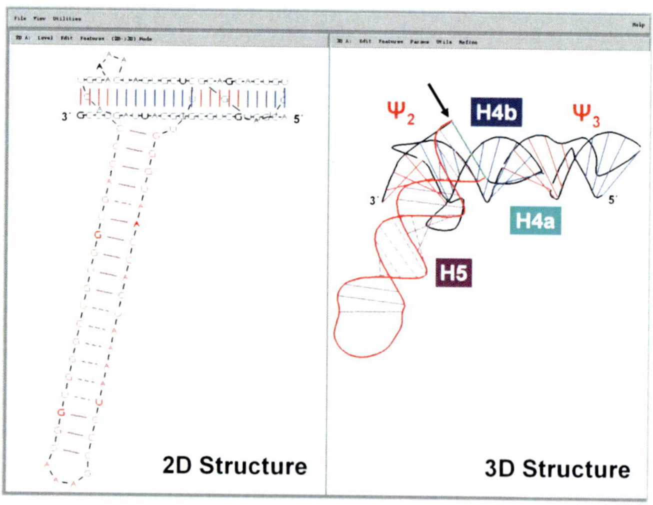

Fig. 7.6 Depiction of the "compactification" operation. The *left panel* shows how the original internal loops have now been incorporated in a stem with induced base pairing indicated by the *dotted lines* within the H5 stem-loop. The *right panel* shows the 3D rendering of this operation. The entire H5 is highlighted in *red* in both panels, which indicates that it has been interactively selected for further manipulation. The segment highlighted in *red* can be translated and/or rotated around the axis shown as a *dashed green line* between the 5′ and 3′-most positions of the selected segment (indicated by the *black arrow* for added clarity). Many other editing options exist (see the text)

circular arc segments that are interrupted by protruding base-paired stems. Using the well-known geometries of individual nucleotides and base pairs, i.e., A, C, G, and U as well as A–U, G–C, and G–U, an embedding is performed that places these objects, as rigid bodies, at their corresponding positions in the secondary structure-based graph [for details, please refer to (Martinez et al. 2008)].

Once the embedding is completed, base-paired stems are "wound " (transformed) into standard A-form double stranded helices, which are commonly found in many experimentally determined RNA structures (Duarte and Pyle 1998; Wadley et al. 2007). The individual bases that are found in loops are placed such that their plane is oriented in a direction that bisects the line that joins the base position in the secondary structure drawing that precedes and succeeds the base being embedded. Figure 7.5 illustrates this first step. During the 3D transformation, the loop regions along with their bounding base pairs are treated as rigid bodies. This simplified and incorrect geometry of the loop regions is the first pass approximation, to be dealt with by the modeler later with the aid of structure editing tools, substitutions with experimental PDB structures and energy minimization of the single stranded regions, structure fragments, or the full structure model. The process of "winding" is done recursively, emanating out, for example, from a multibranch loop, transforming stems in order as the algorithm proceeds. It should be noted that dangling ends, i.e., single strands that do not contain any base pairs and are not part of loops, also follow an A-form geometry. The procedure described above is very fast and can be accomplished in less than a second for a structure that contains over 1,600 nucleotides. Figure 7.5 illustrates the process just described for the TCV enhancer domain shown in Fig. 7.4. The left panel shows the rendering of the secondary structure with the two pseudoknots, Ψ_2 and Ψ_3, and the right panel shows the initial 3D modeling results after embedding and "winding." Pseudoknots Ψ_2 and Ψ_3, which are composed of stems H4b and H4a respectively, coaxially stack on each other, while stem H5 is roughly perpendicular to these pseudoknots. At this point in the modeling protocol, the model is improved by applying interactively driven modifications. After examining the initial 3D rendering of TCV, it becomes fairly obvious that the existence of the large internal loop appears to be somewhat artifactual. RNA2D3D contains a tool called compactification which extends helical stems into loop regions which may or may not involve canonical interactions. The result of this type of operation applied to TCV can be seen in Fig. 7.6. The dashed lines indicate these induced base pairs.

Upon closer examination of the structure, it was found that the junction between the "vertical" and "horizontal" sections contained steric conflicts and unrealistic backbone angles. Since interactive model editing can be accomplished quite rapidly with RNA2D3D, the H5 arm was rotated −75° with respect to the horizontal stack of stems, immediately resulting in a T-shaped structure. In addition, a series of minor translations were performed. These operations significantly relieved the apparent stresses and collisions present in the first depiction. H5 was selected for further manipulation via translations and rotations around the axes or points of reference chosen by the user, such as the axis between the 5′ and 3′-most positions of the selected segment, pointed to by the black arrow in Fig. 7.6. Other minor

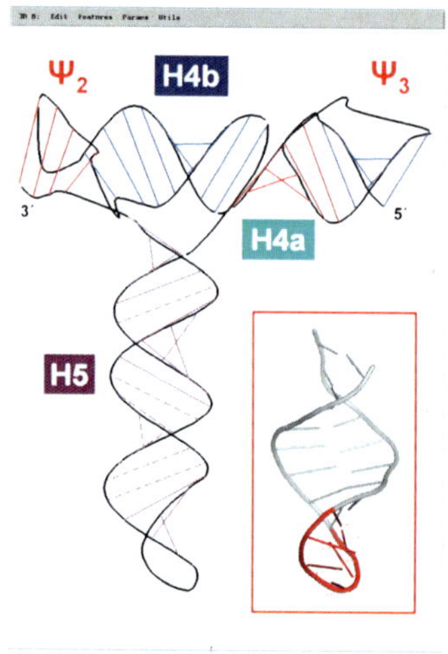

Fig. 7.7 Interactive refinement of the 3D TCV model. The 3D drawing shows the result of rotating the H5 stem by $-75°$ around the closing base pair 5′–3′ axis (*green dashed line* in Fig. 7.6) and translating the stem with respect to the horizontal pseudoknot motif. These operations and several other minor adjustments were performed to alleviate steric conflicts. In addition, a GNRA tetraloop from the PDB database was used to replace the equivalent structure in the RNA2D3D-generated model at the bottom of H5. The *red inset window* shows the PBD entry 1F9L with the fragment used highlighted in *red*. Finally, small adjustments to the 5′ and 3′ pseudoknots were made and molecular mechanics was employed to clean up the model shown

adjustments (rotations and translations) were applied to several small fragments of the structure.

The model was further refined by editing in a GNRA tetraloop from the PDB database (PDB ID:1F9L) to the end of the H5 stem loop structure, as shown in Fig. 7.7. This illustrates the philosophy of utilizing known motifs in the modeling process whenever possible, which can be accomplished with a "Replacement tool." As RNA2D3D can be used to operate on two models in parallel, it is possible to read in a reference PDB structure, such as the 1F9L, into one of these model spaces, and then select in both of them the corresponding subsets (the program can operate at the level of the whole molecule, branches, and user-defined subsets). Once the corresponding subsets have been defined, one can replace the 3D coordinates of the modeled subset with the data from the reference structure. Ideally, an overlap of two matching nucleotides within a paired-up helix should be used to assure the correct alignment of the reference subset with the modeled structure. However, in the case of 1F9L we could use only the base pair closing the tetra loop due to sequence discrepancy with the TCV. The results were satisfactory and nearly identical to a more extensive match performed with the aid of other tools, which, however, required far more tedious substitution procedures. For all practical purposes one can define a structure subset to be an up to $n - 1$-long fragment within a structure of n nucleotides, allowing for a lot of flexibility in mixing the elements of known structures or alternative models via 3D coordinate substitutions.

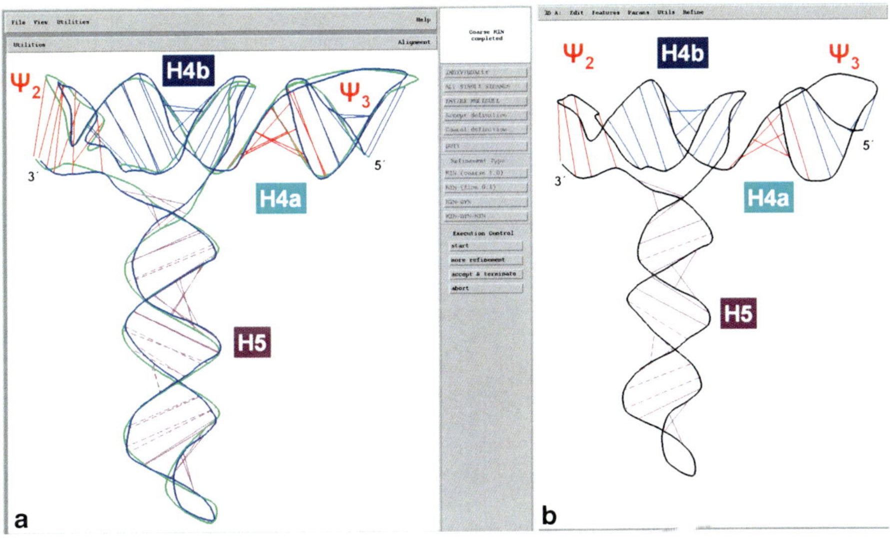

Fig. 7.8 Refinement of the 3D TCV model with the help of molecular mechanics (minimization) and dynamics. (**a**) The *left panel* shows, in *green*, the results of single strand refinements for the starting structure shown in Fig. 7.7. The results of global minimization of the entire structure are depicted in *blue*. This was followed by a short molecular dynamics run (not shown). The *gray, right-hand side panel* shows the many refinement options that can be applied to the entire structure or its user-defined subsets. (**b**) Modeling state after another round of minor interactive edits to the 5′ and 3′ pseudoknot regions followed by molecular mechanics and short molecular dynamics applied to the full structure

A next set of steps used the "refinement" tools that are part of RNA2D3D. This set of tools essentially connects RNA2D3D to the "Tinker" package which applies molecular mechanics and dynamics to the generated 3D coordinates (Ponder 2006). Energy minimization was performed with the Tinker's MINIMIZE module, with Na+ ions placed along the backbone to neutralize phosphate group charges. First, RNA2D3D steps through each single stranded segment (including those that may be part of the compactification process) minimizing its energy based only on its local context. The results of this step are depicted as the green-colored structure in Fig. 7.8a. After this, further single strand minimizations may be performed or one can interactively pick a segment and minimize it, allowing only the picked segment to move, but in this operation, the minimization process sees its global context. This step, however, was not needed in the refinement of TCV. Finally, a global minimization was applied, the results of which are illustrated in Fig. 7.8a as the blue-colored structure. This was followed by a 1 picosecond molecular dynamics run, which was then followed by a global minimization. The results of this refinement phase did not alter the model significantly and are not depicted, to keep Fig. 7.8a as readable as possible. These last steps were applied to clean up bond length issues and/or steric clashes. Another round of interactive editing, mostly with respect to the two pseudoknot regions, followed by dynamics and minimization runs was performed to produce the result illustrated in Fig. 7.8b. Since the shape of the TCV structure that

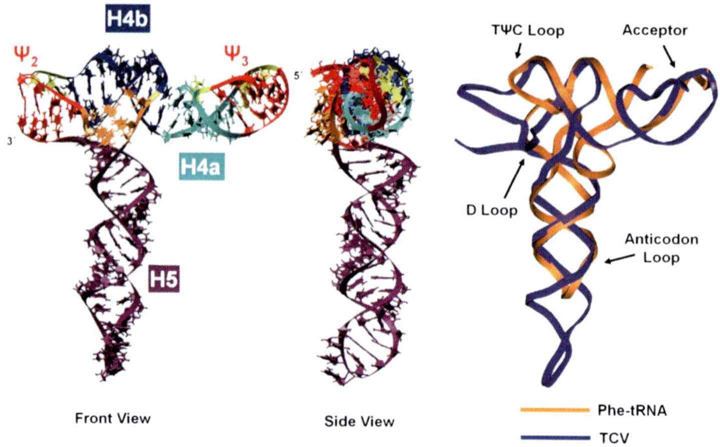

Fig. 7.9 Depiction of the front and side (5′ end toward the viewer) views of the TCV model and an overlay of the model with the phenylalanine tRNA. The striking resemblance to the Phe tRNA is readily apparent. Copyright © American Society for Microbiology, Journal of Virology, 82(17), 2008, pp. 8706–8720, doi:10.1128/JVI.00416-08)

emerged from the modeling looked strikingly similar to a tRNA, the final few minor changes in the relative positions of the structure elements were guided by the phenylalanine tRNA PDB structure, as shown in Fig. 7.9. The sequence of steps presented above is not a rigid protocol, but an outline of options. After each step, one should examine the results and accept or reject them and perform more manual refinements as needed. It has to be stressed that the use of molecular mechanics and short molecular dynamics runs is intended as a "touch-up" procedure, to be followed by longer MD runs with help of tools independent of RNA2D3D.

In general, editing options offer a lot of flexibility by providing the 5′–3′ axis and another one linking the middle point of the 5′–3′ axis (MP) with the center of mass of the selected segment (COM). In addition, pivot points located at the 5′, 3′, MP, and COM locations can be selected, and rotations and translations in the Cartesian space with the origin of the X, Y, Z axes located at these points are available to the user. Selected segments can be grouped together for further manipulations as a rigid body. The combination of the choices of segments and groups to be worked on with the many points of reference around which one can translate and rotate the selected fragments of the molecule allows for a flexible and rapid generation of alternative conformations. Combined with an option to store the current state, as a so-called 3DM file, one can pursue multiple modeling choices with a quick option to return to the last satisfactory state, in case one reaches a dead-end branch in a tree of choices. While stem stacking was accomplished automatically in the TCV modeling, this is an option that the system user can control interactively and add or remove stacking based on his or her knowledge or modeling insight. Another editing feature of the program, which adds to its overall flexibility, is the ability to add or remove base pairs to the model produced from the initial secondary structure input. In cases in which one does not have a reference structure to be substituted for terminal loops, an option of shaping such loops by extending the A-form helix geometry into a loop from its 3′ end up to a user-selected point within the loop is provided. Such loop shaping, combined with base-pairing options makes it possible to approximate kissing loop interactions that may bring together not only distant parts of a large model but also multiple molecules (RNA chains), which may be used as building blocks for nano-scale RNA assemblies. This can be accomplished interactively or via topology descriptor files. Thanks to these features, RNA2D3D has been proven capable of modeling nanostructures such as tectosquares and tectosquare meshes in which building blocks connected by kissing loop interactions can form programmable shapes (Chworos et al. 2004; Jaeger and Chworos 2006; Martinez et al. 2008).

7.4.2 *The Enhancer Element Is tRNA-Like in Appearance and Function*

Once the three-dimensional model was realized, it became apparent that the shape of the defined element bears a striking resemblance to tRNA, as shown in the overlay in Fig. 7.9 between phenylalanine tRNA and the TCV element, aligning the Ψ_3 pseudoknot with the amino acceptor stem of the tRNA. This further reinforced our idea that this element acts as a translational enhancer by recruiting ribosomes. Experiments were performed which verified this hypothesis. These experiments found that the 60S ribosomal subunit bound somewhat more preferentially to the element than the 80S. It was also found that the 40S subunit bound to the 5′-UTR of the virus, thus suggesting a potential role for cyclization of the virus whereby the 3′ and 5′ ends come together with the aid of the formation of the 80S ribosomal complex by the interaction of the 60S unit with the 40S unit.

7.4.3 *Molecular Dynamics Simulations of the TCV TSS Element and the H5 Stem Loop Structure*

In order to better understand the stability and the dynamic nature of the TSS enhancer element, molecular dynamics simulations (MD) were performed on the entire TSS element as well as the H5 alone. Amber 9 was used for the full TSS simulation, using the Cornell force field for RNA (Case et al. 2005; Wang et al. 2000). The entire TSS was run for 50.3 ns using the Particle Mesh Ewald method for determining electrostatics (Essmann et al. 1995). 99 Na+ counterions were added to neutralize the RNA. An additional 63 Na+ and Cl− pairs were added with 28,075 TIP3P water molecules, yielding a 0.1 mol/L relative salt concentration. The system of 87,666 atoms in total was maintained at the temperature of 300 K throughout the MD run.

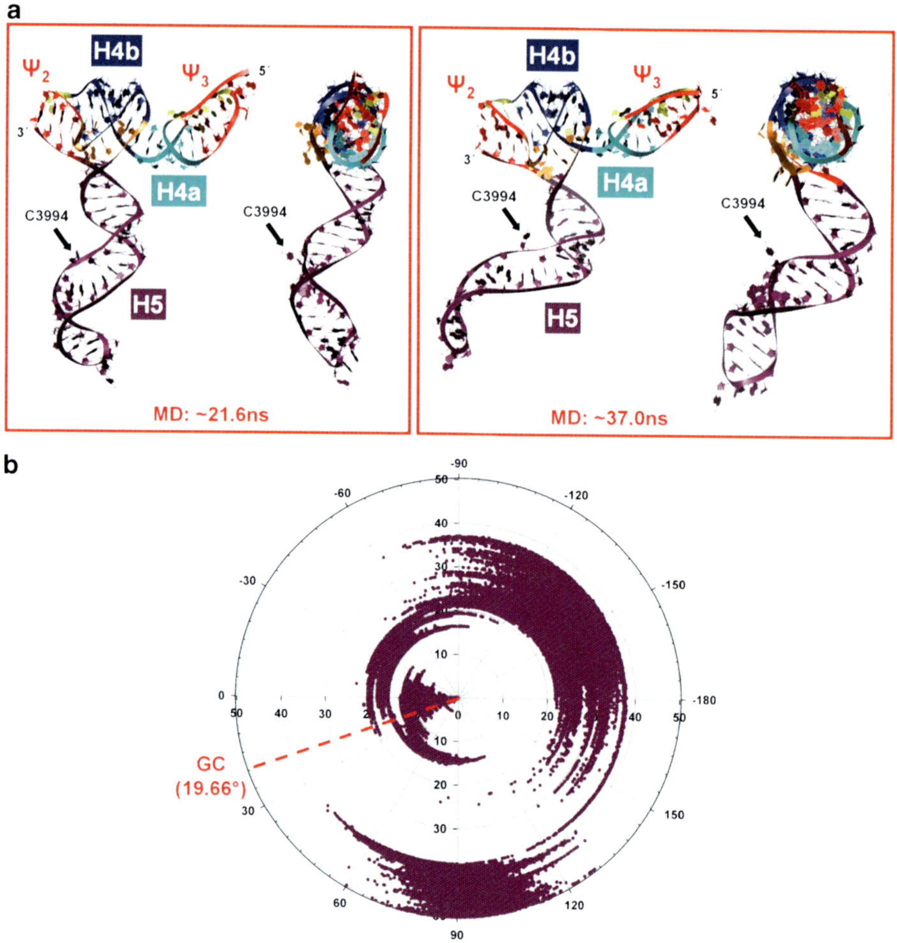

Fig. 7.10 Flexibility of the H5 stem-loop structure and the mobility of C3994. (**a**) Illustrates two conformations of H5 selected from the 50.3 ns trajectory. The position of base C3994 is shown. (**b**) A polar plot of the dihedral angle variations between G3993 and C3994 during the full MD trajectory. The dihedral angle was measured for atoms G3993(N1), G3993(C3′), C3994(C3′), C3994(N3). The *radial lines* correspond to the MD simulation time, starting from the center and increasing outwardly. The angles range from −180° to +180°. The *dotted red line* indicates a reference dihedral angle (measured for the same four atoms—refer to the text) for a neighboring G–C base pair found in an ideal A-form helix. [Panel (a): Copyright © American Society for Microbiology, Journal of Virology, 82(17), 2008, pp. 8706–8720, doi:10.1128/JVI.00416-08]

An interesting finding from the dynamic simulation is the relative flexibility of the H5 stem loop structure compared to the rest of the TSS element. The flexibility seems to be derived, at least in part, from the mobility of C3994 (full genome numbering) found in the large symmetric internal loop (LSL). The higher mobility observed for this base, compared to the flanking nucleotides, is consistent with results from in-line probing experiments (McCormack et al. 2008). Figure 7.10

illustrates two of the orientations of H5 observed during the 50.3 ns simulation. Also depicted is a plot of the dihedral angle of C3994 over the full trajectory showing that C3994 samples almost all possible angles and switches rapidly between stable states (i.e., persistent intervals of smaller angular variations). The neighboring relatively stable nucleotide G3993 serves as a reference point for the measurement, and the results capture nearly exclusively the motions of the C3994.

It is important to note that experiments have shown that the LSL within H5 also interacts with the 3′-most nucleotides of the TCV genomic template (Zhang et al. 2006). In another interaction, the H5 LSL and the hairpin loop in the H4 stem-loop structure (see Fig. 7.3) form pseudoknot Ψ_4, which inhibits ribosome binding (Stupina et al. 2008). All these elements appear to be part of a structural switch that changes conformations as a function of RdRp binding. In other words, two mutually exclusive conformations serve two mutually exclusive functions (translation and replication) (Yuan et al. 2009). The dynamic nature of H5 may therefore play a role in the functionality of the switch.

The dynamic properties of the H5 domain were explored further in a 63-ns long MD simulation of just the H5 stem-loop (42 nt) extracted from the model of the full TSS (100 nt). Amber 10 was used with the same Cornell force field for RNA, using the Particle Mesh Ewald (Case et al. 2008; Essmann et al. 1995, Wang et al. 2000). Forty one Na+ counterions were added to neutralize the RNA, and an additional 24 Na+ and Cl− pairs with 28,075 TIP3P water molecules were added to solvate the RNA to a 0.1 mol/L relative salt concentration. The system of 31,279 atoms was simulated at 300 K. One has to keep in mind that MD runs will not produce identical results, unless given identical starting conditions (parameters). Thus the full TSS and H5 MD simulations differed in some details or the amplitudes of movements, but they shared the same key large movement characteristics within the H5 domain. These included bending of the entire domain around the LSL, loss of ideal A-form helicity (uncoiling) in the central region (LSL), and intermittent coaxial contraction and elongation. While less pronounced than in the full TSS MD simulation, the rotational mobility of C3994 was also outstanding, leading to sharp local bends in the backbone and appearing to be the main cause of the full domain bending. Using the data from the entire trajectory and the most stable interval of the MD simulation (the last 18 ns), we measured the angle between the proximal and distal helices of the corresponding average structures to be 158° and 144°, respectively, showing good agreement with the latest experimental measurements (see below).

In the context of the MD results, it is not unreasonable to speculate that the measured lower affinity of the TCV translational enhancer for ribosomes in comparison to that of regular tRNAs may be related not only to the fact that the TSS is missing the anticodon loop, but also to the flexibility of the equivalent but longer H5 stem loop offering a less than perfect fit or a fit limited to a subset of dynamic states of TSS (Stupina et al. 2008).

7.5 TSS Structure in Solution

The global structure of the TSS in solvent was recently solved experimentally using a novel protocol combining data from SAXS/RDC (Wang et al. 2009; Zuo et al. 2010). The methodology allows one to elucidate relatively large RNA structures and is presented in detail in Chap. 16. In a nutshell, SAXS provides an envelope for the structure, while the RDC data allows one to determine orientations of helices within the structure. The combination of the two constraints leads to a globally correct structure with respect to the mutual orientations of the molecule's helical regions. The overall T-shape of the TCV translational enhancer was confirmed, and more data was obtained on the angles between the different domains of the TSS. Bending of the H5 domain was observed and measured to be $140 \pm 30°$, based on the low resolution envelope calculated from the SAXS data, and $159 \pm 2°$ based on the top simultaneous RDC fits (Zuo et al. 2010).

7.6 Current Model of the Functional Role of the TSS Element in Translation and Replication

Guided by the structure modeling, and grounded in the experimental results (McCormack et al. 2008; Stupina et al. 2008; Yuan et al. 2009; Zhang et al. 2006; Zuo et al. 2010), the following model of the functional role of the TSS is proposed. In the absence of the accumulated RdRp, the TCV genomic template maintains a stable conformation within its 3′-UTR which includes the TSS element capable of binding ribosomes. During the translation initiation process, the 60S ribosomal subunit binds to the TSS. How the 5′ and 3′ ends are brought together is still under investigation (Stupina et al. 2008; Yuan et al. 2009). However, it appears that the 43S subunit binds to the 5′-UTR and advances to the start codon, where it comes in contact with the large subunit bound to the 3′ TSS, thus affecting cyclization of the RNA template. It is also possible that the overall secondary structure of the genomic RNA brings the two ends in proximity to each other, helping to achieve this bridging. For the full ribosome assembly to be properly accomplished, the TSS has to be released from the P-site to allow the initiator tRNA to be properly positioned there. The observed preference in the binding affinity of the ribosome for the initiator tRNA over the TSS should facilitate this process. In the repeated process of translation, aided by the TSS's continual recruitment of ribosomal subunits, the RNA-dependent RNA polymerase (RdRp) from the p88 fragment of the genome (see Fig. 7.1) is produced, and its local concentration begins to rise. Above a threshold concentration the newly translated RdRp binds in the vicinity of the TSS and induces a conformational change within the translation enhancer region, disrupting interactions within the TSS and between the TSS and its flanking regions extending from H4 to the 3′ end (Yuan et al. 2009). The changed structure, the 3D picture of which remains to be solved, can no longer bind to the ribosomal subunit. Rather, transcription (replication) of the complementary strand of the TCV virus is initiated. Repeated transcriptions deplete the RdRp, since it is bound to the minus strand following the process and leaves the immediate vicinity of the plus strand template. Once the RdRp concentration falls below the threshold level required for binding to the translation enhancer region, the structure switches back to the more stable conformation with the TSS element in it, and the positive strand is once more open for translation initiation.

7.7 Summary

In summary, we presented a study of the translational enhancer element within the 3′-UTR of the TCV, stressing the importance and the integral role of the computational secondary structure prediction and 3D modeling in the study. Our MPGAfold program predicted a new pseudoknot as part of the secondary structure of the last 216 3′ nucleotides. 3D molecular modeling of the central region of the 3′-UTR enclosed between pseudoknots Ψ_2 and Ψ_3, performed with our program RNA2D3D, yielded a TSS starting from a secondary structure that differs from that of tRNA. The 3D model of the TSS suggested a translation enhancement mechanism related to ribosome binding. The model was tested by mutagenesis, in-line probing, and ribosome binding. Recently, the solution structure of the TSS was elucidated experimentally using a novel method combining SAXS and RDC data. Thus the study of the TCV 3′ translation enhancer demonstrated the first tRNA-shaped domain internal to a 3′-UTR, capable of ribosome binding and central to a larger structural switch controlling translation and replication of a viral genome. It is possible that similar structural elements will be found in other 3′-UTRs.

Acknowledgments We would like to thank the teams of Dr. Anne E. Simon of University of Maryland at College Park, MD, and Dr. Yun-Xing Wang of the National Cancer Institute in Frederick, MD, for the exciting and fruitful collaborations on this research project. We would like to acknowledge Dr. Yaroslava Yingling of North Carolina State University, Raleigh, NC, for her role in the TCV's TSS modeling. We would also like to thank Dr. Hugo Martinez for his work on enhancing the RNA2D3D capabilities and for fruitful discussions during preparation of this chapter.

This publication has been funded in part with Federal funds from the National Cancer Institute, National Institutes of Health, under Contract No. HHSN261200800001E. This research was supported in part by the Intramural Research Program of the NIH, National Cancer Institute, Center for Cancer Research. The content of this publication does not necessarily reflect the views or policies of the Department of Health and Human Services, nor does mention of trade names, commercial products, or organizations imply endorsement by the US Government.

References

Bindewald E, Kluth T, Shapiro BA (2010) CyloFold: secondary structure prediction including pseudoknots. Nucleic Acids Res 38:368–372

Case DA, Cheatham TE III, Darden T, Gohlke H, Luo R, Merz KM Jr, Onufriev A, Simmerling C, Wang B, Woods RJ (2005) The Amber biomolecular simulation programs. J Comput Chem 26:1668–1688

Case DA, Darden TA, Cheatham TE III, Simmerling CL, Wang J, Duke RE, Luo R, Crowley M, Walker RC, Zhang W, Merz KM, Wang B, Hayik S, Roitberg A, Seabra G, Kolossváry I, Wong KF, Paesani F, Vanicek J, Wu X, Brozell SR, Steinbrecher T, Gohlke H, Yang L, Tan C, Mongan J, Hornak V, Cui G, Mathews DH, Seetin MG, Sagui C, Babin V, Kollman PA (2008) AMBER10. University of California, San Francisco, CA

Chworos A, Severcan I, Koyfman AY, Weinkam P, Oroudjev E, Hansma HG, Jaeger L (2004) Building programmable jigsaw puzzles with RNA. Science 306:2068–2072

Cruz JA, Westhof E (2009) The dynamic landscapes of RNA architecture. Cell 136:604–609

Ding F, Sharma S, Chalasani P, Demidov VV, Broude NE, Dokholyan NV (2008) Ab initio RNA folding by discrete molecular dynamics: from structure prediction to folding mechanisms. RNA 14:1164–1173

Dreher TW, Miller WA (2006) Translational control in positive strand RNA plant viruses. Virology 344:185–197

Duarte CM, Pyle AM (1998) Stepping through an RNA structure: a novel approach to conformational analysis. J Mol Biol 284:1465–1478

Essmann U, Perera L, Berkowitz ML, Darden TA, Lee H, Pedersen LG (1995) A smooth particle mesh Ewald method. J Chem Phys 103:8577–8593

Fechter P, Rudinger-Thirion J, Florentz C, Giege R (2001) Novel features in the tRNA-like world of plant viral RNAs. Cell Mol Life Sci 58:1547–1561

Fraser CS, Doudna JA (2007) Structural and mechanistic insights into hepatitis C viral translation initiation. Nat Rev Microbiol 5:29–38

Gee AH, Kasprzak W, Shapiro BA (2006) Structural differentiation of the HIV-1 polyA signals. J Biomol Struct Dyn 23:417–428

Heilman-Miller SL, Woodson SA (2003) Effect of transcription on folding of the Tetrahymena ribozyme. RNA 9:722–733

Hellen CU, Sarnow P (2001) Internal ribosome entry sites in eukaryotic mRNA molecules. Genes Dev 15:1593–1612

Hofacker IL (2003) Vienna RNA secondary structure server. Nucleic Acids Res 31:3429–3431

Hofacker IL, Fontana W, Stadler PF, Bonhoeffer M, Tacker M, Schuster P (1994) Fast folding and comparison of RNA secondary structures. Monatshefte f Chemie 125:167–188

Holland JH (1992) Adaptation in natural and artificial systems: an introductory analysis with applications in biology, control, and artificial intelligence. MIT, Cambridge, MA

Jaeger L, Chworos A (2006) The architectonics of programmable RNA and DNA nanostructures. Curr Opin Struct Biol 16:531–543

Kasprzak W, Shapiro B (1999) Stem trace: an interactive visual tool for comparative RNA structure analysis. Bioinformatics 15:16–31

Kasprzak W, Bindewald E, Shapiro BA (2005) Structural polymorphism of the HIV-1 leader region explored by computational methods. Nucleic Acids Res 33:7151–7163

Lancaster AM, Jan E, Sarnow P (2006) Initiation factor-independent translation mediated by the hepatitis C virus internal ribosome entry site. RNA 12:894–902

Linnstaedt SD, Kasprzak WK, Shapiro BA, Casey JL (2006) The role of a metastable RNA secondary structure in hepatitis delta virus genotype III RNA editing. RNA 12:1521–1533

Linnstaedt SD, Kasprzak WK, Shapiro BA, Casey JL (2009) The fraction of RNA that folds into the correct branched secondary structure determines hepatitis delta virus type 3 RNA editing levels. RNA 15:1177–1187

Martinez HM, Maizel JV Jr, Shapiro BA (2008) RNA2D3D: a program for generating, viewing, and comparing 3-dimensional models of RNA. J Biomol Struct Dyn 25:669–684

Mathews DH, Turner DH (2006) Prediction of RNA secondary structure by free energy minimization. Curr Opin Struct Biol 16:270–278

Mathews DH, Sabina J, Zuker M, Turner DH (1999) Expanded sequence dependence of thermodynamic parameters improves prediction of RNA secondary structure. J Mol Biol 288:911–940

Mathews DH, Disney MD, Childs JL, Schroeder SJ, Zuker M, Turner DH (2004) Incorporating chemical modification constraints into a dynamic programming algorithm for prediction of RNA secondary structure. Proc Natl Acad Sci U S A 101:7287–7292
McCormack JC, Simon AE (2004) Biased hypermutagenesis associated with mutations in an untranslated hairpin of an RNA virus. J Virol 78:7813–7817
McCormack JC, Yuan X, Yingling YG, Kasprzak W, Zamora RE, Shapiro BA, Simon AE (2008) Structural domains within the 3′ untranslated region of Turnip crinkle virus. J Virol 82:8706–8720
Merrick WC (2004) Cap-dependent and cap-independent translation in eukaryotic systems. Gene 332:1–11
Miller WA, Wang Z, Treder K (2007) The amazing diversity of cap-independent translation elements in the 3′-untranslated regions of plant viral RNAs. Biochem Soc Trans 35:1629–1633
Ponder J (2006) Tinker – software tools for molecular design – version 4.2
Preiss T, Hentze MW (2003) Starting the protein synthesis machine: eukaryotic translation initiation. Bioessays 25:1201–1211
Qu F, Morris TJ (2000) Cap-independent translational enhancement of turnip crinkle virus genomic and subgenomic RNAs. J Virol 74:1085–1093
Shapiro BA, Kasprzak W (1996) STRUCTURELAB: a heterogeneous bioinformatics system for RNA structure analysis. J Mol Graph 14:194–205
Shapiro BA, Navetta J (1994) A massively parallel genetic algorithm for RNA secondary structure prediction. J Supercomput 8:195–207
Shapiro BA, Wu JC (1996) An annealing mutation operator in the genetic algorithms for RNA folding. CABIOS 12:171–180
Shapiro BA, Wu JC (1997) Predicting RNA H-type pseudoknots with the massively parallel genetic algorithm. Comput Appl Biosci 13:459–471
Shapiro BA, Bengali D, Kasprzak W, Wu JC (2001a) RNA folding pathway functional intermediates: their prediction and analysis. J Mol Biol 312:27–44
Shapiro BA, Wu JC, Bengali D, Potts MJ (2001b) The massively parallel genetic algorithm for RNA folding: MIMD implementation and population variation. Bioinformatics 17:137–148
Shapiro BA, Kasprzak W, Grunewald C, Aman J (2006) Graphical exploratory data analysis of RNA secondary structure dynamics predicted by the massively parallel genetic algorithm. J Mol Graph Model 25:514–531
Shapiro BA, Yingling YG, Kasprzak W, Bindewald E (2007) Bridging the gap in RNA structure prediction. Curr Opin Struct Biol 17:157–165
Stupina VA, Meskauskas A, McCormack JC, Yingling YG, Shapiro BA, Dinman JD, Simon AE (2008) The 3′ proximal translational enhancer of Turnip crinkle virus binds to 60S ribosomal subunits. RNA 14:2379–2393
Sun X, Simon AE (2006) A cis-replication element functions in both orientations to enhance replication of Turnip crinkle virus. Virology 352:39–51
Tinoco I Jr, Bustamante C (1999) How RNA folds. J Mol Biol 293:271–281
Wadley LM, Keating KS, Duarte CM, Pyle AM (2007) Evaluating and learning from RNA pseudotorsional space: quantitative validation of a reduced representation for RNA structure. J Mol Biol 372:942–957
Wang J, Cieplak P, Kollman PA (2000) How well does a restrained electrostatic potential (RESP) model perform in calculating conformational energies of organic and biological molecules? J Comput Chem 21:1049–1074
Wang J, Zuo X, Yu P, Xu H, Starich MR, Tiede DM, Shapiro BA, Schwieters CD, Wang YX (2009) A method for helical RNA global structure determination in solution using small-angle x-ray scattering and NMR measurements. J Mol Biol 393:717–734
Wilkinson KA, Merino EJ, Weeks KM (2005) RNA SHAPE chemistry reveals nonhierarchical interactions dominate equilibrium structural transitions in tRNA(Asp) transcripts. J Am Chem Soc 127:4659–4667

Wu J-C, Shapiro BA (1999) A Boltzmann filter improves the prediction of RNA folding pathways in a massively parallel genetic algorithm. J Biomol Struct Dyn 17:581–595

Yoshii M, Nishikiori M, Tomita K, Yoshioka N, Kozuka R, Naito S, Ishikawa M (2004) The Arabidopsis cucumovirus multiplication 1 and 2 loci encode translation initiation factors 4E and 4G. J Virol 78:6102–6111

Yuan X, Shi K, Meskauskas A, Simon AE (2009) The 3′ end of Turnip crinkle virus contains a highly interactive structure including a translational enhancer that is disrupted by binding to the RNA-dependent RNA polymerase. RNA 15:1849–1864

Zhang J, Zhang G, Guo R, Shapiro BA, Simon AE (2006) A pseudoknot in a preactive form of a viral RNA is part of a structural switch activating minus-strand synthesis. J Virol 80:9181–9191

Zuker M (2003) Mfold web server for nucleic acid folding and hybridization prediction. Nucleic Acids Res 31:3406–3415

Zuo X, Wang J, Yu P, Eyler D, Xu H, Starich MR, Tiede DM, Simon AE, Kasprzak W, Schwieters CD, Shapiro BA, Wang YX (2010) Solution structure of the cap-independent translational enhancer and ribosome-binding element in the 3′ UTR of turnip crinkle virus. Proc Natl Acad Sci USA 107:1385–1390

Chapter 8
Methods for Building and Refining 3D Models of RNA

Samuel C. Flores, Magdalena Jonikas, Christopher Bruns, Joy P. Ku, Jeanette Schmidt, and Russ B. Altman

Abstract Interest in RNA has grown tremendously in recent years as we uncover more and more roles for RNA in the cell. Investigation of RNA function is often hampered by the absence of even a tentative 3D structure which can guide experiments. Experimental structure determination is difficult because of RNA's large size, high charge and flexibility, propensity for kinetic trapping, and the lack of the distinctive surface features necessary for crystallization. Computational structure prediction is challenging for mostly the same reasons. In this work, we describe three methods which are used in different ways to predict the structure and dynamics of RNA. RNABuilder is an "erector set" for constructing RNA molecules based on experiments, hypotheses, or other information known to the user. NAST quickly produces ensembles of coarse-grained molecules based on the statistics of backbone conformation. Lastly, Zephyr uses the graphical processing unit rather than the CPU to speed up conventional molecular dynamics calculations.

8.1 Introduction

We are currently witnessing an explosion in the known roles of RNA, along with an increased recognition of its importance. While proteins perform most physiological functions in the cell, RNA plays a possibly dominant role in determining when and how much protein is produced, and even configures proteins through alternative splicing. An important aspect to understanding RNA function in these processes is the role of its 3D structure. Yet, structural information is lacking for many RNAs due to both experimental and theoretical challenges. Their size, long folding times, propensity for kinetic trapping, charge, flexibility, and lack of distinctive surface

S.C. Flores • M. Jonikas • C. Bruns • J.P. Ku • J. Schmidt • R.B. Altman (✉)
Department of Bioengineering, 318 Campus Drive, Clark Center S170, MC: 5444, Stanford, CA 94305-5444, USA
e-mail: samuelfloresc@gmail.com; russ.altman@stanford.edu

N. Leontis and E. Westhof (eds.), *RNA 3D Structure Analysis and Prediction*,
Nucleic Acids and Molecular Biology 27, DOI 10.1007/978-3-642-25740-7_8,

features (Ferre-D'Amare et al. 1998) make crystallography, as well as computational prediction, difficult.

RNAs range in size from a single residue (e.g., ATP) to thousands of residues (e.g., the ribosome). In between are many RNAs of biological interest, including tRNA with ~75 residues and various group I introns with hundreds of residues. Further, experimental folding times for RNAs tend to be too long for simulation. For example, the ~350 residue RNase P folds in ~400 s (Furtig et al. 2007) while current full atomic resolution molecular dynamics methods can simulate events on the microseconds scale. Depending on parameters used, such methods would take several thousand years to simulate 400 s. In addition, explicit-water and -ion simulations may be necessary to treat the effect of backbone charges (Bowman et al. 2008) further increasing the computational complexity.

In response to these challenges, a variety of approaches have been proposed to simplify the structure prediction problem. Some seek to reduce the computational cost by coarsening the granularity of the problem (Atilgan et al. 2001) modeling multiple atoms as a single unit. Others are using knowledge-based potentials (Ayton et al. 2007), forces that are empirically derived from experimental data, to drive their structure predictions, in contrast to physics-based potentials that are derived from theoretical principles of physics. These approaches, although computationally efficient, have some shortcomings.

Coarse-grained methods will inevitably have limited accuracy, and so in recent years a consensus has emerged that it is beneficial to work at more than one level of resolution, often simultaneously, in an approach known as multi-resolution modeling (MRM). MRM is an emerging paradigm for efficiently modeling large molecules. By treating them at coarse resolution when possible and at fine resolutions when necessary, we can benefit from the efficiency of the coarse-grained modeling and the accuracy of the atomic models.

MRM schemes are sometimes classified into serial and parallel, depending on whether the different granularities exist at separate times (serial) or simultaneously (parallel). Serial schemes often involve two sequential stages. The first stage is a parameterization stage, in which short-timescale molecular dynamics simulations, molecular knowledge, or thermodynamics are used to derive coarse-grained simulation parameters. This is followed by a dynamics stage, in which the derived simulation parameters are used to carry out a long-timescale coarse-grained simulation of the molecule of interest. In a parallel scheme, two or more models can run simultaneously at different levels of granularity, exchanging conformations from the simulations at different resolutions from time to time (resolution exchange); alternatively, different regions of a single model can have different granularities (Ayton et al. 2007).

Computational approaches can also be differentiated by the type of force field used: knowledge-based (KB) versus physics-based. A force field is generally considered to be physics-based if it induces the behavior of large molecules from the behavior of its smallest pieces, which are sufficiently well understood on a physical level. A KB force field works in the opposite direction; it deduces forces empirically from observations of the structure or behavior of large systems (Sippl 1993).

While both approaches enable modeling of the physical world and could potentially yield the same solution, there are necessarily trade-offs between the two. Since a KB approach models the behavior of a molecule based on parameters derived from observations of large systems, it may work even when the underlying physics is not well understood. A KB force field can also be designed to be computationally faster because a subset of atoms can be used instead of all atoms, as is required for a physics-based approach; also the interactions can be simpler in form and shorter in range. Lastly, conformational transitions may occur in less simulation time in the thus-simplified system. The biggest disadvantage of KB approaches is that they are only designed to reproduce the phenomena they were trained on and cannot predict previously unobserved phenomena. In contrast, physics-based force fields can, in principle, predict the correct thermodynamics, kinetics, and free energies of a novel system and are thus more widely applicable. Also, while a KB force field can be used to *reproduce* an observed behavior, a physics-based force field can be used to reveal *why* that behavior occurs, since it is inductive rather than deductive.

It is important to note that the distinction between KB and physics-based is not absolute but depends on the kind of investigation being performed. For example, a classical physics-based force field actually derives from observed aggregate behavior of quantum physical effects, and thus would appear to be knowledge-based from the point of view of a quantum physicist. Classical physics is unable to make predictions about novel quantum systems for which it is not "trained." What makes a force field KB is the use of it to perform investigations at the same level at which the knowledge was acquired. So if a force field was trained on RNA structures and is then used to produce an RNA structure, it is KB.

Below, we describe in detail two 3D RNA structure modeling tools we have developed, RNABuilder and Nucleic Acid Simulation Tool (NAST). These two examples illustrate the trade-offs of different approaches. RNABuilder is a KB, single-model parallel MRM scheme. Within a simulation, some fragments of the model are set to be rigid while others are flexible. RNABuilder allows users to incorporate information from disparate sources, such as coevolution, functional assays, single molecule experiments, homology, and secondary structure prediction, and gives users a large degree of flexibility in building a model. NAST is also KB, but it uses a serial MRM method. It represents RNA structures with a single particle per residue, deriving its coarse-grained force field from statistical analysis of known structures. A primary goal of NAST is to generate a plausible 3D structure in much less than 1 day. NAST has a complementary tool, Coarse to Atomic (C2A) for adding in full-atomic detail. Both NAST and RNABuilder are intended to model large molecules in a computationally efficient manner.

We will also discuss OpenMM Zephyr, which provides a graphical user interface for molecular dynamics simulations at atomic resolutions. Contrary to RNABuilder and NAST discussed above, OpenMM Zephyr's goal is not to introduce new methodologies but rather to bring existing advanced molecular dynamics capabilities to users with little experience with those techniques. Currently OpenMM Zephyr provides an interface to a modified version of GROMACS (van der Spoel et al. 2005; Hess et al. 2008), a versatile software package for performing molecular dynamics, i.e., simulations based on the Newtonian equations of motion for systems with hundreds to millions of particles (a physics-based approach). Though this is a computationally expensive approach, as explained earlier, GROMACS has built-in efficiencies and an emphasis on performance. In addition, by using the OpenMM library, the modified version of GROMACS has recently been implemented to run on the graphics processing units (GPUs) that come with most modern computers. Zephyr provides an interface to this accelerated version of GROMACS (Friedrichs et al. 2009). We will discuss how Zephyr complements existing 3D RNA modeling tools, allowing users to extend their research with full atomic dynamic simulations without extensive set-up times. NAST and its companion program C2A take advantage of Zephyr.

Significant time and resources have been spent developing and validating the above described methods. To ensure the scientific community benefits from these tools, we have invested substantially to make them broadly available and have provided easy-to-use interfaces. RNABuilder, NAST, C2A, and OpenMM Zephyr are all freely available on https://simtk.org, along with clear documentation and practical test cases, enabling other researchers to experiment with the tools independently. They are able to test the different approaches on a larger variety of problems, discovering their limitations and advantages, and importantly, build upon the existing methods in the true spirit of scientific research. It is a new, exciting chapter in our investigations of RNA, and with it, we hope comes a renewed openness in sharing those scientific results and methods.

8.2 RNABuilder: An Internal Coordinate Mechanics Approach for Multiresolution Molecular Modeling of RNA

8.2.1 Introduction

RNABuilder is a flexible, open-source tool for combining biological information from disparate sources to model the structure and dynamics of RNA via base-pairing interactions. It has been developed to address many of the challenges in predicting 3D RNA structures mentioned earlier, incorporating a number of elements to make it computationally efficient. First, RNABuilder is a parallel multiresolution method (Ayton et al. 2007), designed so that all or part of any molecule can be selectively rigidified for significant time savings. It uses a KB force field to translate available molecule-specific experimental information into forces which drive correct structure formation. Using a KB force field also allows RNABuilder to describe certain interactions, such as sterics, in a simpler form for additional computational savings.

With RNABuilder, we have been able to predict the structure of molecules by applying base-pairing interactions [including any of the types catalogued by Leontis et al. (2002)] obtained from coevolution, cross-linking, and functional assays.

RNABuilder can also align a flexible molecule of unknown structure onto rigid molecular fragments from homologs of known structure. Further, unlike many methods whose resource requirements explode with molecule size, RNABuilder's simulation time is proportional to the number of atoms, suggesting that significantly larger molecules can be treated economically. This opens the door to modeling the packaging of viral genomes, ribosomal translation (Tung et al. 2002), and other phenomena involving large RNA complexes.

Below, we describe RNABuilder in more detail and explain how we have applied it to model tRNA, the P4/P6 domain of the *Tetrahymena* ribozyme, and the entire *Azoarcus* ribozyme.

8.2.2 Using RNABuilder and Internal Coordinate Mechanics to Model Large Molecules

Any method designed to address the folding of large RNA molecules faces the problem of *scaling*, or how the time required to compute a fixed number of simulation steps increases with system size. Methods whose resource requirements are directly proportional to the number of bodies in the system are said to have *order-N* scaling (N being the number of bodies). In a full-atomic system, the number of bodies would be equal to the number of atoms; in a coarse-grained representation where, for instance, each residue is represented by a single particle, the number of bodies is equal to the number of residues.

Many methods have computational requirements which grow faster than this. For example, in the most naïve implementation of molecular dynamics, the scaling is order-N^2, since the nonbonded electrostatic and van der Waals interactions must be computed between all pairs of atoms (N atoms interacting with N atoms). Many larger molecules are well out of practical reach for such methods. An emerging paradigm for dealing with issues of computational economy is MRM, mentioned earlier, which seeks to treat molecules at multiple levels of granularity. RNABuilder follows this paradigm. It provides the flexibility to impose granularity by region, concentrating resources on the region of interest while spending little computing time on perhaps large but less important parts of the system.

RNABuilder's MRM capability comes from its use of the Simbody Internal Coordinate Mechanics library, which is freely available from http://simtk.org/home/simbody. Simbody works in internal coordinates, a framework in which different regions of a system can easily be treated with different degrees of flexibility. For molecules, bond lengths and angles are often fixed, and dihedral angles are the only geometric quantity that can change between the two bonded atoms. Bond dihedral angles can also be fixed, turning the bonded atoms into a single rigid body.

Simbody provides other features which are key to RNABuilder's modeling capabilities, including the basewise force field it uses. This force field pulls pairs of bases into specific configurations, in particular any of those documented by Leontis et al. (2002). Simbody supports this functionality through the concept of "frames," or attachment points, associated with each body. In RNABuilder, each base is a body, and these can be aligned through translations and rotations.

Simbody's formulation of internal coordinate mechanics computes kinematics in order-N time, with the consequence that RNABuilder can treat large molecules without disproportionate increase in cost. The user can also elect to rigidify certain regions of the molecule as mentioned to decrease the computational costs. Together these features, as well as others to be described, may enable the user to economically treat very large molecules. Below, we provide examples and benchmarking data that support this suggestion.

8.2.3 Enforcing Leontis–Westhof and Other Base Interactions

One of the key elements for generating a 3D structure with RNABuilder is the user-specified base interactions. In early work, Taketomi et al. (1975) showed that preferentially applying interactions observed in the native state (the Go model) led to folding and suggestive thermodynamic behavior of simplified proteins. In many cases, experimentalists may not know the 3D structure of the native state but may have extensive knowledge of base-pairing contacts. The knowledge may come from coevolution (Levitt 1969), functional assays (Tijerina et al. 2006), or other experiments which are much easier to perform than full 3D structure determination. This often incomplete or imperfect information can be used to drive folding to a predicted native state using RNABuilder. Similar forces can also be used to drive homology modeling ("threading"), another approach to structure prediction that incorporates information from known, similar 3D structures.

RNABuilder supports a number of different interactions: base pairs, as catalogued by Leontis et al. (2002), stacking interactions, and a "Superimpose" force used in threading. These interactions consist of forces and torques which pull the bases into the desired base-pairing orientation as mentioned earlier.

The task of parameterizing the RNABuilder force field is mostly that of determining the position and orientation of an attachment frame for each base. An attachment frame is a coordinate framework such that when it is aligned with the default (body) frame that RNABuilder defines for a base, the desired base-pairing geometry is achieved (Fig. 8.1). Parameters have been provided for over 240 base-pairing interactions, including all those classified by Leontis et al. (2002). This is sufficient for most users. For users who want to include additional interactions, we also provide an auxiliary program with the RNABuilder source code distribution, generate-base-pair-transform. This program extracts the attachment frame coordinates given the 3D structural coordinates of two residues engaged in the desired interaction, facilitating the description of customized interactions. The *attachment* and *body* frames are then pulled together by an interaction potential in translational and torsional space (Flores and Altman 2010).

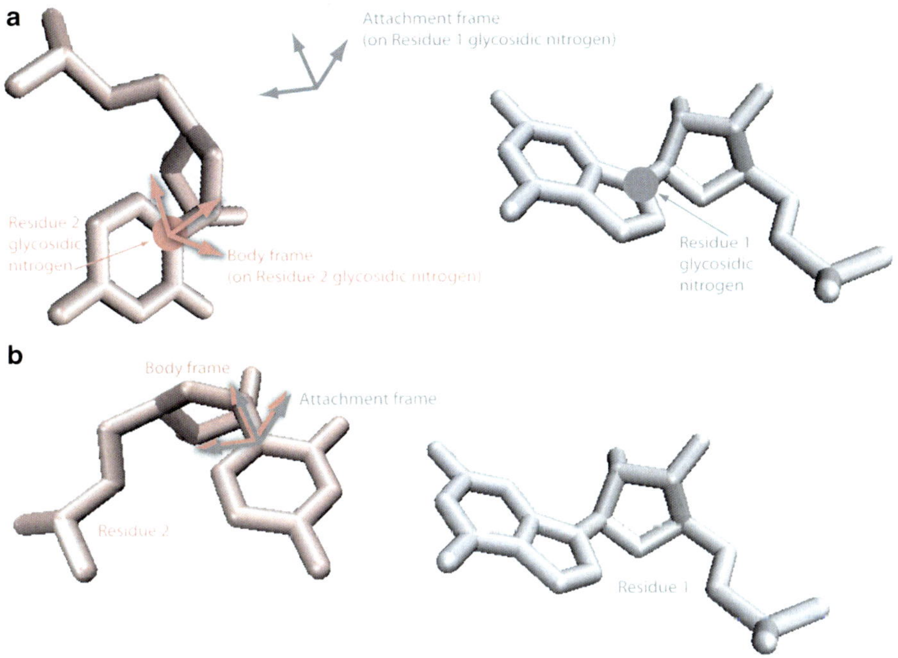

Fig. 8.1 Enforcing base pairs within RNABuilder. RNABuilder assigns a default *body frame* to each residue of a molecule. That frame is associated with the residue's glycosidic nitrogen. To describe a base-pairing interaction, an *attachment frame* is also defined for one of the two interacting bases. Base pairs are enforced by aligning the *attachment frame* of one residue with the *body frame* of the second residue. Illustration shows initial configuration (**a**) and equilibrated configuration (**b**)

8.2.4 Approximate Treatment of Sterics

Enforcing base-pairing geometry is not sufficient, however. In most practical cases, not all base pairs will be known, and further, since such interactions act only on bases, the backbone may take on incorrect conformations. To partially address these issues, we use Simbody's collision-detecting spheres to approximately recover the effects of steric exclusion. These use a fast collision detection algorithm, which does not calculate interactions between spheres until they are close enough to make contact. This treatment is a substitute for calculating the pairwise van der Waals and electrostatic interactions between atoms, which account for steric exclusion and other phenomena in MD simulations but are more costly computationally.

Simbody's contact spheres detect collision by tracking a bounding box around each sphere and computing interactions only when two such boxes overlap. RNABuilder allows the user to apply these spheres to every residue or to any stretch of residues according to various schemes: just to the phosphorus, C4*, and glycosidic nitrogen or to each heavy atom in each residue. The stiffness and radii of the spheres can be adjusted by the user. The default parameters have been optimized to produce duplexes similar to ones generated using James Stroud's popular make-na server (Stroud 2010), which is based on Nucleic Acid Builder (Macke and Case 1998), a computer language for performing various nucleic acid manipulations. Make-na generates regular helices of idealized geometry.

8.2.5 Building RNA Structures from Experimental Data

As we showed above, a single base pair of any of the supported types can be enforced. One could imagine that by specifying the interaction between each base and the next, the structure of an entire motif can be enforced. Below, we show that we can model various RNA structures, even without knowing all of the base-pairing contacts.

Using RNABuilder, we modeled a tRNA structure using base-pairing and stacking contacts obtained from experiments which were, or could have been, performed without reference to the structural coordinates. There were 85 such contacts (Flores and Altman 2010), including base-stacking interactions in helices, which RNABuilder applies automatically. We count each base pair or stacking interaction as a single contact. In the last stage of the simulation, the molecule ranged from 8.1 to 11.1 Å RMSD, averaging 9.6 Å RMSD, with respect to the crystallographically observed structure. To show how the predicted structure can improve as more base-pairing information becomes available, we then inspected the crystallographically obtained 3D structure and added or corrected the contacts as they were observed there. The additions and corrections resulted in a new total of 87 contacts (including automatically applied helical stacking interactions) and lowered the RMSD to an average of 6.1 Å (ranging from 4.4 to 8.4 Å RMSD) in the final stage of the simulation.

Similarly, we were able to generate a model of the bigger 160-residue P4/P6 domain of the *Tetrahymena* ribozyme (Cate et al. 1996). For this molecule, we manually applied 87 base-pairing and stacking contacts known by various noncrystallographic means much as before (Flores and Altman 2010), while RNABuilder automatically applied additional stacking interactions between consecutive bases in helices, for a total of 166 base-pairing and stacking contacts. The applied contacts were extracted from phylogenetic, UV cross-linking, dimethyl sulfate footprinting assays, NMR, and other noncrystallographic experiments. Motifs such as the tetraloop receptor are well characterized in multiple molecules and can be generated by applying a tightly defined set of noncanonical base-pairing and stacking interactions (Flores and Altman 2010). We ran for 8 ns of simulation time. In the latter 4 ns, the RMSD with respect to the crystal structure ranged from 8.7 to 11.3 Å, with an average of 10.0 Å (Fig. 8.2). However, the crystal structure lacks L2 which is in contact with P5c; therefore, the orientation of P5c in the crystal is not what it would be in the full-length ribozyme. The RMSD computed in a similar run without P5c

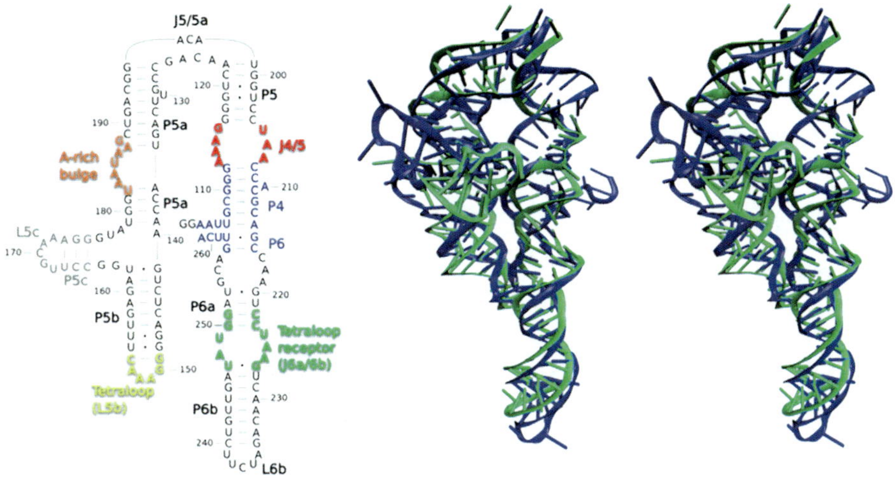

Fig. 8.2 Building P4/P6. Secondary structural and other contacts (*left*) obtained without reference to the crystallographically observed structural coordinates were enforced to generate 3D structure (*right*, in stereo). RNABuilder generated a model (*right*, *blue*) which compares well with the crystallographically observed structural coordinates (*green*). Agreement between model and observed structure averaged 10.0 Å RMSD in the latter half of the simulation

was about 8.6 Å. Prior work on this system yielded lower accuracy (Jonikas et al. 2009a, b) and/or required considerable computational expertise (Michel and Westhof 1990). When we enforced all contacts observed by crystal gazing (a total of 185), the RMSD in the last stage of the simulation decreased to an average of 9.6 Å, with range from 9.4 to 10.0 Å (Flores and Altman 2010).

8.2.6 Solving Structure by Multiple Template Homology Modeling

We have described how RNABuilder can use readily available experimental information in the absence of 3D coordinates to predict structure. In some cases, the molecule may be too large to be practically solved by that method. In such cases, although the structure of the molecule is unknown, various domains or fragments of the molecule may bear structural similarity to molecules of known structure. These circumstances call for homology modeling ("threading").

In a novel approach to threading, we rigidify the templates, or fragments of known structure, and use RNABuilder's "Superimpose" force to pull bases from the molecule of unknown structure, or model, into alignment with corresponding bases on the templates (Flores et al. 2010). Small regions of the model which have no suitable template are built up using base interaction forces as before. While other threading algorithms exist, the algorithm we present here is unique in that it is one

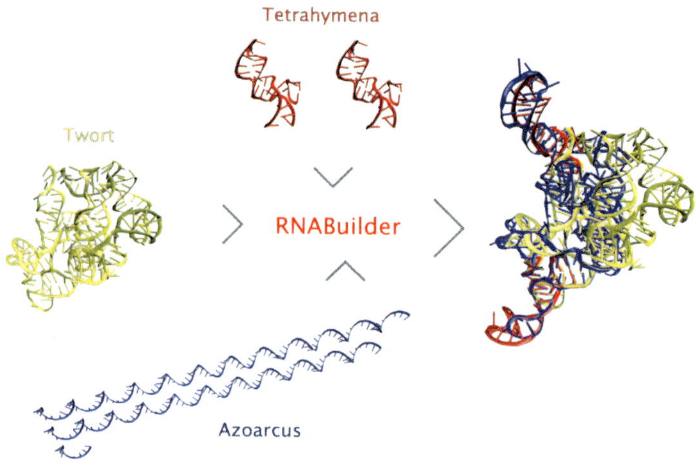

Fig. 8.3 Threading *Azoarcus* ribozyme onto templates from *Twort* and *Tetrahymena*. A flexible *Azoarcus* ribozyme of presumed unknown structure (*blue*) was aligned onto the construct composed of the core region from *Twort* (*yellow*) and tetraloop-receptor fragments from *Tetrahymena* (*red*). The threaded *Azoarcus* ribozyme agreed with the crystallographically observed 1–4.6 Å RMSD (Flores et al. 2010)

of the few that has been applied to RNA (Tung et al. 2002), and it easily threads one or more molecules to multiple templates.

We were able to use RNABuilder to solve the structure of the ~200-residue *Azoarcus* ribozyme by threading it to templates from *Tetrahymena* and *Twort* (Fig. 8.3). The resulting structure had a 4.6 Å RMSD[27] when compared with the crystal structure, especially remarkable when considering that the sequence identity between the model and the templates is <50%, much less than what previous methods appear to have required. A similar structure was previously obtained by fragment assembly methods (Rangan et al. 2003). Further, the three ribozymes are connected with multiple tertiary contacts, which is difficult to model with fragment assembly methods, such as Tung et al. (2002).

8.2.7 *Scaling*

We have described how RNABuilder can be applied to molecules of up to 200 residues. A natural question is "How much larger can I go?". The answer depends on the specifics of the problem, namely, how much of the structure is known, how much of it can be threaded to known structures, and how much of it matters at all. The intuition and insight of the user in applying flexibility, sterics, and forces can often turn a prohibitively difficult problem into a relatively simple one. In this section, we show that there is no intrinsic barrier to the modeling of very large molecules.

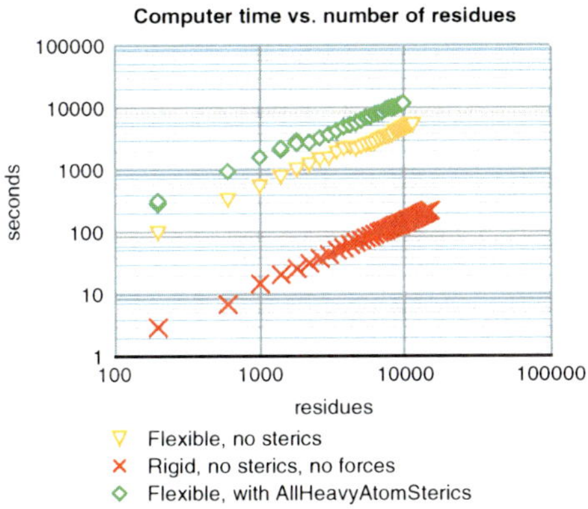

Fig. 8.4 Scaling of computer time with number of RNA residues. The computational cost increases linearly with the number of residues for flexible RNA chains without sterics. Residues with sterics applied incur an increase in cost (not necessarily linear). For rigid chains, the cost is linear because currently the atomic coordinates are updated at every time step, even within rigid regions; nonetheless, the overall cost is considerably lower than for flexible regions. Performance varies with system and simulation parameters. The user can choose the flexibility and sterics by region for efficient modeling

The discovery of recursive methods for solving the equations of motion in internal coordinates in O(n) time (linear in the number of bodies) made the latter method practical for dynamics. Our benchmarking study verifies that the computational cost of computing the dynamics for a fixed period of time is linear with the number of residues (Fig. 8.4); thus, no explosive increase in computational cost might prohibit modeling very large structures. In addition, we show that rigidifying molecules drastically reduces expense, indicating that rigidification is a viable thrift measure when parts of the system are known, converged, or uninteresting. The benchmark created chains of more than 10,000 residues before exhausting the 4 GB of memory addressable in a 32-bit system, suggesting very large complexes can be modeled using modest equipment.

8.2.8 Advantages and Limitations of Modeling with RNABuilder

RNABuilder offers a versatile, computationally efficient, scalable means of modeling 3D RNA structures. In taking advantage of the features provided by the multibody dynamics engine Simbody, RNABuilder is able to reduce the computational cost of modeling 3D RNA structures, as compared with molecular dynamics codes where cost increases as constraints are added. Unlike many other force fields, such as those used in MD, the RNABuilder force field only acts between bases which the user knows to interact in certain ways. This means that RNABuilder is less general than MD; on the other hand, it saves computer time and more importantly it enables the user to impose contacts known from experiments, theory, or insight.

RNABuilder is perhaps most similar in goals to Nucleic Acid Builder (Macke and Case 1998), but has the advantage of being able to model large scale conformational changes in a short period of time. In Nucleic Acid Builder, the resulting structure is achieved using a classical MD force field, AMBER; therefore, just as in classical MD simulations, although small scale rearrangements can easily be made, substantial domain motions are out of reach because of the computer time requirement. For that reason, with Nucleic Acid Builder the user must have a fairly clear idea of the final 3D structure before building a model. On the other hand, with RNABuilder the time required to fold an extended chain into its final configuration while simultaneously enforcing base-pairing contacts may be quite affordable, and thus minimal to no 3D information may be needed.

Whereas by some measures MD is the most computationally expensive modeling technique, very coarse-grained modeling tools like NAST (described in the next section) are among the least expensive. Fast tools of the latter class enable many simulations to be done at low cost, but often accuracy is limited, as is the number and type of constraints that can be applied. RNABuilder takes the middle road, offering greater diversity in the number and types of constraints available, as compared with most coarse-grained tools, and while not as fast as them, it does offer considerable speed ups when compared with MD.

RNABuilder is a dynamical code and thus should be distinguished from fragment assembly methods such as MC-Sym (Major et al. 1993; Parisien and Major 2008). These fragment assembly methods sample small RNA fragments from a database of known 3D structures and, as implied by the name, assembles them to build complete molecules. At the small scale, this approach generates structures that have physically reasonable configurations. However, if the molecule is large and has tertiary contacts, the probability of finding just the right fragments that will close those contacts is small. RNABuilder does not rely on databases; rather, it works by applying forces to bases, leaving the backbone flexible, so it is capable of building structures that have never been observed experimentally. Furthermore, if two or more applied base pairs are not quite mutually compatible, the bases may equilibrate to a compromise structure rather than failing altogether.

It should be clear that each approach has advantages for different applications. RNABuilder, with its use of internal coordinate mechanics and dynamics, provides users with a flexible, computationally efficient method that is suitable and practical for modeling large molecules without much prior 3D structural knowledge.

8.2.9 Downloading RNABuilder

RNABuilder can be freely downloaded from https://simtk.org/home/rnatoolbox.

8.3 Coarse-Grained Molecular Modeling with the NAST

8.3.1 Introduction

The NAST is a KB coarse-grained modeling tool for RNA 3D structures. NAST primarily addresses the challenge of generating plausible coarse 3D models of RNA structure within a short amount of time (<<1 day). NAST can be applied to RNA molecules with two or more known (or predicted) helical regions. It assumes that the known secondary structure of the molecule adopts the geometry of observed RNA helices (in crystal structures) and asks how these helices pack together to form a 3D structure. The result is an ensemble of 3D structures at the nucleotide resolution that resemble observed RNA geometries at that same resolution while satisfying the user-specified primary sequence, secondary structure, and tertiary contacts.

NAST simplifies the computationally expensive problem of molecular dynamics in two ways (1) Reducing the complexity of the molecule to a single point per residue, thereby significantly decreasing the number of pairwise interactions that need to be calculated, and (2) using a relatively simple energy function based on observed nucleotide-level RNA geometries in the Protein Data Bank (PDB). This energy function implicitly treats a wide variety of phenomena, including the effect of backbone and counter-ion charges. The simplifications make NAST a good tool for modeling large RNA structures quickly. Although the resulting model is coarse-grained, full atomic detail can be added into the model, which can then be energy minimized with tools such as OpenMM Zephyr (discussed later in this chapter) and refined to satisfy particular atomic-level interactions with tools such as RNABuilder. In this section, we will provide a brief description of NAST and its implementation and show several uses of NAST, including the prediction of structures and the generation of a diverse ensemble of conformations.

8.3.2 Coarse-Graining and the NAST Energy Function

NAST uses a coarse-graining scheme where each residue is represented by the position of its C3′ atom. This approach significantly reduces the complexity of the molecule, decreases the amount of time needed to achieve significant conformational change, and essentially treats the polymer as a "chain of beads."

The NAST energy function is based on nucleotide-level geometries observed in ribosomal RNA crystal structures. The only geometric relationships included in the energy function are distance, angle, and dihedral geometries for C3′ to C3′ atoms in sequential residues, and nonbonded distances for all pairs of C3′ atoms. The observed value distributions are empirically fit and each term's contribution to the energy function is determined using the Boltzmann relationship between probability distribution and energy. The nonbonded interactions are modeled with only the repulsive term of the Lennard-Jones potential. Since the energy function is based on observations in ribosomal RNA crystal structures, the resulting NAST models will satisfy observed RNA geometry at the local residue level, while remaining flexible enough to explore conformational space and satisfy user-supplied constraints.

The user provides the secondary structure, which defines additional distances, angles, and dihedrals needed to constrain an ideal A-form helix. These regions are strongly constrained to maintain their ideal helical geometries, allowing NAST to focus on exploring the possible packing arrangements of the helices. Optional user-specified tertiary contacts are treated as springs, with the rigidity controlled by the user. Tertiary contacts can also be used to constrain entire regions where the geometry is known from a crystal structure.

More details about the implementation of NAST can be found in Jonikas et al. (2009a, b).

8.3.3 NAST's Role in Understanding RNA

8.3.3.1 Predicting Structures

The simplest use of NAST is to generate a 3D coarse-grained structure model that satisfies observed nucleotide-level RNA geometry from a primary sequence and secondary structure. If tertiary contacts are known or predicted, they can also be included in the modeling process. Given these inputs, NAST will start from an unfolded conformation of a coarse-grained representation of the target molecule (Fig. 8.5a). NAST will then attempt to satisfy the user-supplied information, while also satisfying the energy function derived from observed RNA crystal structure to generate a model (Fig. 8.5b).

Using only secondary structure and tertiary contact information, we have shown that NAST can be used to generate and identify representative structures of the yeast phenylalanine tRNA (a 76-residue structure) and the P4–P6 domain of the *Tetrahymena thermophila* group I intron (a 158-residue structure) that have strong topological similarity to their respective crystal structures (Jonikas et al. 2009a, b). Our studies used only information that was available before the crystal structures were solved. The highest ranking cluster of structures generated by NAST had an average RMSD of 8.0 ± 0.3 Å for the yeast tRNA and 16.3 ± 1.0 Å for the P4–P6 domain, a significant similarity especially given the resolution of the NAST coarse-graining. Furthermore, the NAST-generated model for the P4–P6 domain was achieved despite incorrect secondary structure information; 26% of the base pairs used as input to NAST were incorrect. This suggests that NAST can be useful even in cases where the secondary structure information is uncertain and could potentially be employed as a screening tool to determine what additional experimental data would enhance the 3D modeling of a given RNA.

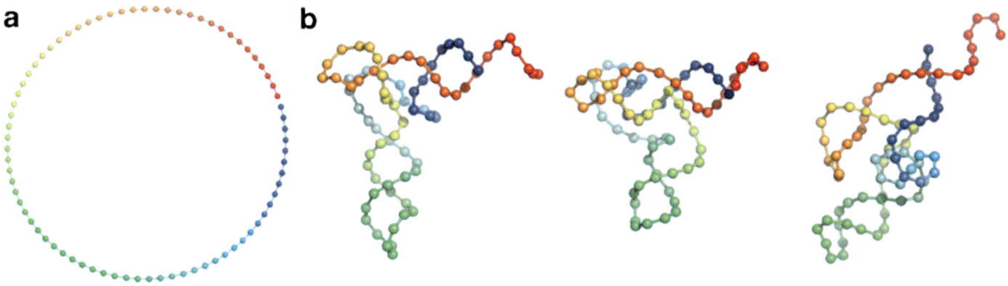

Fig. 8.5 (**a**) NAST starts the modeling process from an unfolded conformation of a coarse-grained representation of the target molecule. (**b**) NAST will seek to satisfy the energy function derived from observed RNA crystal structure, as well as the secondary structure and tertiary contacts provided by the user. Because of the stochastic nature of NAST, repeating the process numerous times will yield an ensemble of low energy models, based on NAST's energy function. Residues are colored by index

8.3.3.2 Combining Information from Several Models

NAST is also useful for combining information from several different models, for example, for *T. thermophila*, where the crystal structure is missing several helices. In Jonikas et al. (2009a, b), we combined geometric information from a full atomic model generated by experts (Michel and Westhof 1990), with the truncated crystal structure to build a full-length 3D structure that satisfied both the crystal structure data and the modeling data for the missing helices and their positions relative to the rest of the molecule.

8.3.3.3 Generating Diverse Unfolded Conformations

Another use of NAST is to generate a diverse ensemble of unfolded 3D conformations of a given primary sequence and secondary structure. In Fig. 8.6, we show 26 unfolded models of the *Azoarcus* ribozyme generated by NAST, using only primary sequence and secondary structure information. Because of the stochastic nature of NAST and the limited 3D information provided to the system, repeating the simulation numerous times will result in a diverse ensemble of coarse-grained conformations of a molecule that satisfy the secondary structure provided by the user. These structures can then be filtered by additional information, such as radius of gyration and solvent accessibility, to determine a reasonable set of unfolded conformations. These unfolded conformations can then be used as starting points for modeling folding pathways by adding additional distance constraints with NAST.

8.3.4 Using NAST's Coarse-Grained Models for Further Simulation and Analysis

The 3D structures generated by NAST are coarse-grained with a single atom (C3′) representation per nucleotide. Although this simple representation allows NAST to

Fig. 8.6 Examples of 26 unfolded conformations generated by NAST for the *Azoarcus* ribozyme. Only secondary structure was constrained in the generation of these structures. All structures were aligned to minimize RMSD to help judge the diversity of these unfolded conformations. Residues are colored by index

build 3D models very quickly, it limits the usefulness of the model. The C2A tool (Jonikas et al. 2009a, b) is a KB tool for adding full-atomic detail into coarse-grained models. C2A can be used with any coarse-grained 3D structure model, including those representations with more than one atom per residue. In Fig. 8.7, we show nine different full-atomic models that were built using the C2A tool, based on a single NAST coarse-grained model.

Once full atomic detail has been added to a coarse-grained model, it can be energy-minimized using the OpenMM Zephyr tool described in the next section of this chapter. This minimization process will eliminate some of the chemically unrealistic features of the full atomic structure, in particular unusually long bonds and unusually small distances between atoms. The resulting energy-minimized structures can then be used for further simulation and analysis using full-atomic physics-based modeling tools.

8.3.5 Scaling

Unlike RNABuilder, the computational cost of modeling RNA structures using NAST is not a straightforward relationship, dependent just on molecule size. Instead, it depends on a number of factors, including the length of the primary sequence, the number of helices, the number of base pairs in each helix, and the number of tertiary contacts. Running a 50-ps NAST simulation (10,000 time steps) of tRNA, which consists of 76 residues, 4 helices, 22 base pairs and 4 tertiary contacts, takes 3 s of CPU

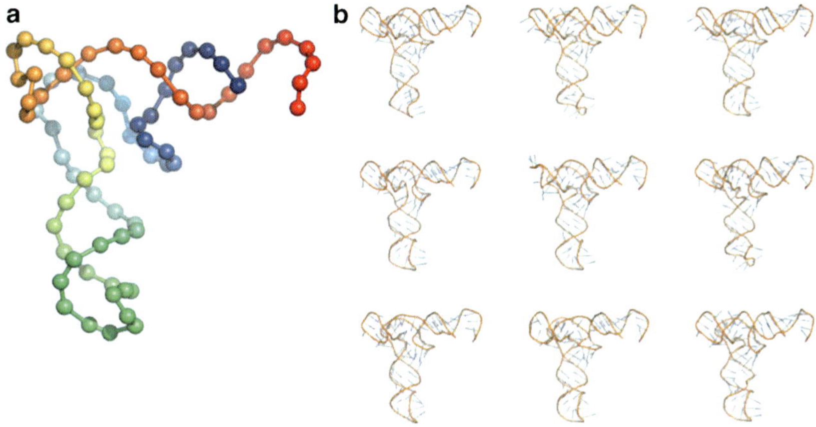

Fig. 8.7 Addition of full-atomic detail to a NAST tRNA coarse-grained model using C2A. Due to the KB stochastic nature of C2A, a single coarse-grained model (**a**) will result in an ensemble of full-atomic structures (**b**), all of which are within ~1 Å RMSD of the coarse-grained template

time, running on a single CPU. Meanwhile, the same amount of simulation for P4–P6, which consists of 7 helices, 58 base pairs, and 2 tertiary contacts, takes 11 s. In the published results for modeling tRNA and P4–P6 with NAST, we chose to run 300 CPU hours of simulation for each molecule to generate a larger diversity of structures.

8.3.6 *Advantages and Limitations of Modeling with NAST*

The simplified representation used by NAST makes it more appropriate for some applications than others. The speed gained by the simple representation makes it a useful tool for fast modeling of large molecules that require large conformational changes, such as from an unfolded state to a folded state. Additionally, full-atomic detail may not be as crucial in these cases, although it can be added either with C2A or by threading with RNABuilder.

There are clearly trade-offs in order to achieve this fast modeling. For instance, the NAST energy function does not have any physics-based contributions, so on-the-fly detection of possible base pairing is not feasible. Only those base pairs that are identified by the user will be constrained. There are no differentiating properties between different flavors of nucleotides, such as with RNABuilder, FARNA (Das and Baker 2007), and MC-Sym (Parisien and Major 2008). Every nucleotide is simply modeled as a sphere, and the entire molecule is essentially a "string of beads" where each bead behaves the same. Despite these limitations, the ability to generate large conformational changes in a short amount of time, combined with the ability to add full atomic detail to the final models, makes NAST a useful tool for speedy modeling of RNA molecules that are more complex than a single hairpin-loop structure. Even for structures with only two helical regions, NAST can help the user explore the conformational space available geometrically based on the input constraints.

8.3.7 Downloading NAST and C2A

Both NAST and C2A can be freely downloaded. The Web site for NAST is http://simtk.org/home/nast. The C2A Web site is located at http://simtk.org/home/c2a.

8.4 Molecular Simulation of RNA with OpenMM Zephyr

NAST and RNABuilder introduce novel paradigms for modeling 3D RNA structures, as described earlier. OpenMM Zephyr's focus is different. It provides an interface to full-atomic, molecular dynamics (MD) simulations, complementing 3D RNA modeling tools like NAST and RNABuilder. Although these MD simulations are typically difficult to set up, requiring a high level of expertise, OpenMM Zephyr simplifies the process significantly, enabling researchers, even those without any MD experience, to easily access the most state-of-the-art full-atomic simulation techniques. These techniques can augment the information and understanding gained from 3D modeling tools, refining the 3D RNA model or simulating the dynamics of the structure. Below, we describe the OpenMM Zephyr application, its design, and how it has been applied to 3D RNA structures.

8.4.1 Zephyr Leverages OpenMM, GROMACS, and VMD

Zephyr is a desktop molecular simulation application based on three tools: the OpenMM library for accelerating molecular dynamics on GPUs (Friedrichs et al. 2009), GROMACS for the molecular dynamics infrastructure (van der Spoel et al. 2005; Hess et al. 2008), and VMD for visualization (Humphrey et al. 1996). Zephyr ties these three tools together to provide a productive and educational workflow.

The GROMACS molecular dynamics package, upon which Zephyr is based, uses a physics-based approach to simulate the dynamics of a molecule over time. It evaluates the Newtonian equations of motion for systems with hundreds to millions of particles over tiny increments of time, typically femtoseconds for molecules, repeating the calculations over and over to simulate the motion of a molecule. Typically, these types of simulations require setting a large number of parameters and have a steep learning curve.

OpenMM Zephyr simplifies the process, leading the user through the steps required to set up and run a simulation, as well as to visualize the simulation results using VMD, a popular molecular visualization program. In addition, OpenMM Zephyr displays the specific GROMACS commands being used, enabling motivated

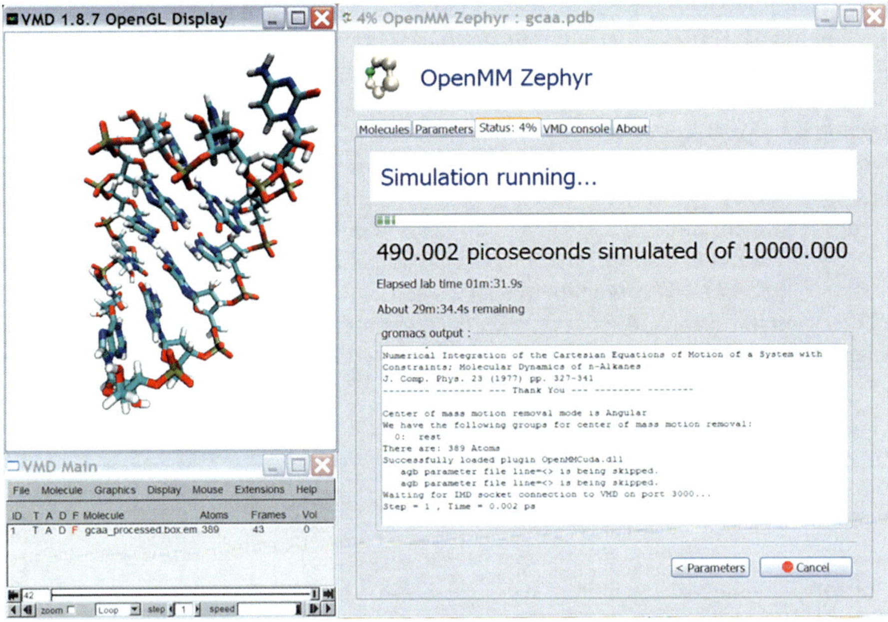

Fig. 8.8 OpenMM Zephyr screen during a simulation. Live animation of the simulation trajectory viewed through Zephyr using the program VMD is shown on the *left*. The Zephyr user interface is shown on the *right*

users to teach themselves how to run GROMACS from the command line so that they can eventually use the other more advanced options within GROMACS (Fig. 8.8).

Zephyr streamlines the simulation process by automatically chaining together sequential steps of a molecular simulation. The user specifies an input PDB molecular structure file and, if desired, can adjust a small number of parameters for the simulation. At that point, Zephyr manages all the subsequent steps: it automates the process of transforming the PDB file to match the conventions of the molecular force field; it invokes several tools to minimize the energy of the structure, followed by another series of tools to perform the dynamic simulation; and finally it saves all of the output to a working area previously specified by the user. In addition, it automatically launches the VMD application and sends the appropriate information to it, so that the simulation can be viewed live.

8.4.2 GPU Acceleration of Dynamic Simulation

Though physics based, full atomic simulations are computationally intensive, GROMACS has included many algorithmic optimizations to improve its performance. Zephyr is built upon a modified version of GROMACS. Currently, GROMACS only supports explicit solvent, where the water molecules and ions surrounding the RNA are modeled as individual particles. We have implemented a version of GROMACS that supports implicit solvent—a method of representing the solvent as a continuous medium instead of as individual "explicit" solvent molecules. Because of this difference, the performance of our modified version cannot be directly compared to that of GROMACS. In general, implicit solvent models are inherently faster. Moreover, our modified version of GROMACS incorporates the OpenMM library, which enables MD simulations to run on computers with suitable GPUs, yielding another significant speed-up. We can, however, compare OpenMM Zephyr's performance to AMBER, a package that uses the same implicit solvent model as implemented within Zephyr. The OpenMM Zephyr simulations are over 100 times faster on a GPU as compared to AMBER on a single CPU (Friedrichs et al. 2009), with even greater speed-ups for large molecules.

8.4.3 Zephyr and Usability

One of the goals of Zephyr is to make best-of-breed simulation tools, such as OpenMM and GROMACS, available to a wider audience by simplifying the install process and by significantly reducing the initial learning curve. A key component of this is the interface. While academic software engineering projects, particularly those geared more for experts in the field, typically provide an extensive array of choices and functionality, Zephyr purposely simplifies its interface, identifying and allowing users to choose from a minimal set of necessary parameters. In addition to a well-designed and tested interface and installer, Zephyr provides a user manual and uses a modern software engineering infrastructure provided through simtk.org for source code revision control, automated building and testing of the code, bug tracking, and user communication, e.g., mailings lists and discussion forums. These features result in a robust, user-friendly software with ongoing support.

8.4.4 Zephyr's Guiding Principles: Discoverability, Feedback, and Convention

Zephyr's emphasis is on learning. Learning is aided in Zephyr by three guiding principles: discoverability, feedback, and expert convention. Zephyr emphasizes ease-of-use, but it is not a black box. Discoverability means that the user is encouraged to learn more and more about the details of molecular simulation by investigating the provided interface, which explains the details of each step of the simulation workflow. Feedback means that the user remains aware of the current state of the simulation workflow, including error conditions. Both real-time and retrospective visualization of trajectories is provided through integration with the VMD program. Convention means the best practices of expert users are built in to the default workflow path. By creating a default simulation workflow that follows a conventional path from model building to energy minimization to dynamic simulation to trajectory analysis, the first-time user can be guided by expert knowledge to learn to manage a working simulation. The simplest task flow is the conventional task flow, but it is not the only task flow available within Zephyr.

8.4.5 Using Zephyr with RNA Structures

OpenMM Zephyr is useful for minimizing the energy of a full-atomic RNA model. The process of energy minimization can eliminate residual geometric oddities in a structure. This can be useful in situations where a model has been created computationally. For instance, the C2A fragment construction procedure can create unrealistic local geometry by connecting unrelated fragments. Zephyr can be used to relax these unrealistic structures through energy minimization. To do this, simply run Zephyr with only one step of dynamic simulation. This takes advantage of the fact that Zephyr automatically minimizes the energy of a structure before beginning a simulation.

OpenMM Zephyr's primary use, though, is as a tool for accelerated full-atomic physics-based simulation of RNA and other molecular structures. The GPU-accelerated molecular dynamics available through Zephyr is particularly useful for moderate size RNA structures.

8.4.6 Downloading Zephyr

Zephyr can be freely downloaded from https://simtk.org/home/zephyr for Windows, Linux, and Mac computers.

8.5 Conclusions

Molecular size, folding time, and charge are important challenges for structural modeling of RNA today, and a variety of approaches have been proposed to address them. Coarse-grained and MRM methods are both able to model large systems with long folding times by using lower-resolution versions of the molecule. Using KB force fields is another approach, which can implicitly treat electrostatic charge and offers speed advantages over a physics-based approach, especially for larger molecules. With these new methodologies, though, come questions about the level of accuracy that can be achieved and their applicability to a broad range of problems. To address some of these issues and highlight the trade-offs with using these techniques, we presented two recently developed tools for RNA modeling: RNABuilder and NAST/C2A.

RNABuilder is a KB approach that builds molecular structure by enforcing interactions between bases, which the user provides based on experiments, theory, or even conjecture. We showed that limited experimental data can be turned into atomic 3D structure with RMSDs ranging from 4.6 to 12 Å, depending on the molecule and the amount and quality of the experimental data. Because of its KB force field, RNABuilder is able to implicitly deal with charge. Importantly, RNABuilder uses the Simbody library, leading to order-N scaling with system size and thus the potential to treat large systems with long folding times. Being a single-model parallel MRM method, RNABuilder also allows converged regions to be rigidified, leading to considerable additional computational savings.

NAST is able to model 3D RNA structures in a short time ($<<1$ day) by using a single pseudo-atom to represent each residue and a KB force field. The KB force field is limited so that only those base pairs identified by the user will be constrained, and there are no differentiating properties between different nucleotides, as with several other modeling programs. However, the KB force field does allow NAST to capture the effects of charge implicitly. Further, despite these drawbacks, NAST is able to recapitulate the tRNA coarse-grained structure with an RMSD of 8.0 ± 0.3 Å and the P4–P6 domain with an RMSD of 16.3 ± 1.0 Å in a very short amount of time. While NAST only generates coarse-grained RNA structures, an associated program, C2A, has been developed that can add the fine-grained atoms back into the model by sampling from a database of structural fragments.

Lastly, we discussed OpenMM Zephyr, which provides a user-friendly graphical interface to an accelerated modified version of the popular GROMACS MD engine. The GROMACS MD engine uses a physics-based force field to simulate the motion of a full-atomic molecule. This type of simulation can be used to provide insights into the dynamics and function of a molecule, complementing the structural information obtained from programs like NAST and RNABuilder. It can also be used to refine the structures obtained from 3D RNA modeling programs. Basic principles of usability and learning guided the design of OpenMM Zephyr, bringing state-of-the-art full-atomic simulation techniques to researchers without experience with MD. It is especially worth noting Zephyr's incorporation of the freely available OpenMM library for GPU acceleration, which yields significant speed improvement over a single-CPU MD implementation. This provides yet another avenue to the challenge of simulating larger systems and longer time scales.

While there are many challenges to modeling 3D RNA structures, we are encouraged by the progress we have observed. New methodologies, such as those presented here, are being developed to address these challenges, each with their unique advantages and limitations, and we are excited to see what new knowledge and understanding they will undoubtedly bring to the field.

References

Atilgan AR et al (2001) Anisotropy of fluctuation dynamics of proteins with an elastic network model. Biophys J 80:505–515
Ayton GS, Noid WG, Voth GA (2007) Multiscale modeling of biomolecular systems: in serial and in parallel. Curr Opin Struct Biol 17:192–198
Bowman GR et al (2008) Structural insight into RNA hairpin folding intermediates. J Am Chem Soc 130:9676–9678
Cate JH et al (1996) Crystal structure of a group I ribozyme domain: principles of RNA packing. Science 273:1678–1685
Das R, Baker D (2007) Automated de novo prediction of native-like RNA tertiary structures. Proc Natl Acad Sci USA 104:14664–14669
Ferre-D'Amare AR, Zhou K, Doudna JA (1998) Crystal structure of a hepatitis delta virus ribozyme. Nature 395:567–574
Flores S, Altman R (2010) Turning limited experimental information into 3D models of RNA. RNA 16(9):1769–1778
Flores SC, Wan Y, Russell R, Altman RB (2010) Predicting RNA structure by multiple template homology modeling. Proceedings of the pacific symposium on biocomputing, pp 216–227
Friedrichs MS et al (2009) Accelerating molecular dynamic simulation on graphics processing units. J Comput Chem 30:864–872
Furtig B et al (2007) Time-resolved NMR studies of RNA folding. Biopolymers 86:360–383
Hess B, Kutzner C, van der Spoel D, Lindahl E (2008) GROMACS 4: algorithms for highly efficient, load balanced, and scalable molecular simulation. J Chem Theor Comput 4:435–447
Humphrey W, Dalke A, Schulten K (1996) VMD: visual molecular dynamics. J Mol Graph 14 (33–8):27–38
Jonikas MA et al (2009a) Coarse-grained modeling of large RNA molecules with knowledge-based potentials and structural filters. RNA 15:189–199
Jonikas MA, Radmer RJ, Altman RB (2009b) Knowledge-based instantiation of full atomic detail into coarse-grain RNA 3D structural models. Bioinformatics 25:3259–3266
Leontis NB, Stombaugh J, Westhof E (2002) The non-Watson-Crick base pairs and their associated isostericity matrices. Nucleic Acids Res 30:3497–3531
Levitt M (1969) Detailed molecular model for transfer ribonucleic acid. Nature 224:759–763
Macke T, Case DA (1998) Modeling unusual nucleic acid structures. In: Leontes NB, SantaLucia J Jr (eds) Molecular modeling of nucleic acids. American Chemical Society, Washington, DC, pp 379–393
Major F, Gautheret D, Cedergren R (1993) Reproducing the three-dimensional structure of a tRNA molecule from structural constraints. Proc Natl Acad Sci USA 90:9408–9412
Michel F, Westhof E (1990) Modelling of the three-dimensional architecture of group I catalytic introns based on comparative sequence analysis. J Mol Biol 216:585–610
Parisien M, Major F (2008) The MC-Fold and MC-Sym pipeline infers RNA structure from sequence data. Nature 452:51–55
Rangan P, Masquida B, Westhof E, Woodson SA (2003) Assembly of core helices and rapid tertiary folding of a small bacterial group I ribozyme. Proc Natl Acad Sci 100:1574–1579
Sippl M (1993) Boltzmann's principle, knowledge-based mean fields and protein folding. An approach to the computational determination of protein structures. J Comput Aided Mol Des 7:473–501
Stroud J (2010) The make-na server (http://structure.usc.edu/make-na/)
Taketomi H, Ueda Y, Go N (1975) Studies on protein folding, unfolding, and fluctuations by computer simulation. Int J Pept Protein Res 7:445–459

Tijerina P, Bhaskaran H, Russell R (2006) Nonspecific binding to structured RNA and preferential unwinding of an exposed helix by the CYT-19 protein, a DEAD-box RNA chaperone. Proc Natl Acad Sci USA 103:16698–16703

Tung CS, Joseph S, Sanbonmatsu KY (2002) All-atom homology model of the *Escherichia coli* 30 S ribosomal subunit. Nat Struct Biol 9:750–755

Van Der Spoel D et al (2005) GROMACS: fast, flexible, and free. J Comput Chem 26:1701–1718

derived structural information into our structure determination protocol. Long-range constraints for RNA modeling can be inferred from a variety of biochemical and bioinformatic techniques, ranging from chemical footprinting and cross linking to sequence covariation (Gutell et al. 1992; Juzumiene et al. 2001; Michel and Westhof 1990; Ziehler and Engelke 2001). Experimental constraints derived from these biochemical and bioinformatics techniques are generally of lower than atomic resolution, but can be readily incorporated into the coarse-grained RNA model for structure determination. The all-atom RNA model can then be reconstructed from the coarse-grained structural model.

First, we will describe our coarse-grained representation of RNA models for DMD simulations. Then, we will describe and evaluate the applications of the DMD–RNA procedure to ab initio folding of a set of small RNA models and structure determination using experimental constraints.

9.2 Coarse-Grained RNA Modeling Using Discrete Molecule Dynamics

We use DMD as the conformational sampling engine. A detailed description of the DMD algorithm can be found elsewhere (Dokholyan et al. 1998; Rapaport 2004; Zhou and Karplus 1997). The difference between discrete molecular dynamics and traditional molecular dynamics is in the interaction potential functions. Interatomic interactions in DMD are governed by stepwise potential functions (Fig. 9.1a). Neighboring interactions (such as bonds, bond angles, and dihedrals) are modeled by infinitely high square well potentials (Fig. 9.1b). By approximating the continuous potential functions with step functions of pairwise distances, DMD simulations are reduced to event-driven (collision) molecular dynamics simulation. In a DMD simulation, atoms move with constant velocity until they collide with another atom. As soon as the potential of interaction between the two atoms changes (i.e., the pairwise distance is at the step of the stepwise potential function), the velocities of the two interacting atoms change instantaneously (Fig. 9.1a). These velocity changes are required to conform to the conservation laws of energy, momentum, and angular momentum. Each such collision is termed an "event." The sampling efficiency of DMD over traditional MD is mainly due to rapid processing of collision events and localized updates of collisions (only colliding atoms are updated at each collision). In the limit of infinitesimally small steps, the discrete step function approaches the continuous potential function, and DMD simulations become equivalent to traditional molecular dynamics.

We approximate the single-stranded RNA molecule as a coarse-grained "beads-on-a-string" polymer with three beads representing each nucleotide, one for sugar (S), one for phosphate (P), and one for nucleotide base (B) (Fig. 9.2a). The P and S beads are positioned at the centers of mass of the corresponding phosphate group and the 5-atom ring sugar, respectively. For both purines (adenine and guanine) and

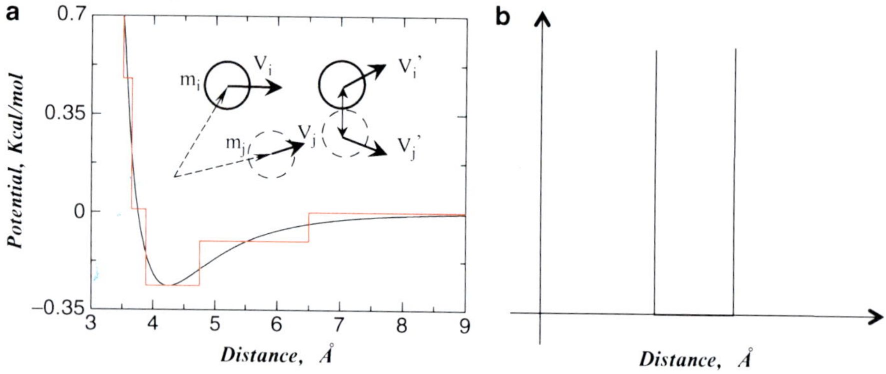

Fig. 9.1 Discrete molecular dynamics simulations. (**a**) Schematic of the DMD potential. The stepwise function used in DMD is the approximation of the continuous function in traditional molecular dynamics. The insert depicts the collision of two atoms with masses of m_i and m_j at the initial position of r_i and r_j, respectively. The two atoms move with constant velocities (v) until they meet at distance of R_{ij}. (**b**) Schematic of the potential energy of bonds in DMD. The atom pairs remain within the distance range during the simulation

pyrimidines (uracil and cytosine), we represent the base bead (B) as the center of the 6-atom ring. The neighboring beads along the sequence, which may represent moieties that belong to the same or a neighboring nucleotide, are constrained to mimic the chain connectivity and local chain geometry (Fig. 9.2a). Types of constraints include covalent bonds (solid lines), bond angles (dashed lines), and dihedral angles (dotted–dashed lines). The parameters for bonded interactions mimic the folded RNA structure and are derived from a high-resolution RNA structure database (Murray et al. 2003) (Table 9.1). Nonbonded interactions are crucial to model the folding dynamics of RNA molecules. In our model, we include base-pairing (Watson–Crick pairs of A–U and G–C and Wobble pair of U–G), base-stacking, short-range phosphate–phosphate repulsion, and hydrophobic interactions, which are described in the following section with the parameterization procedure.

Base Pairing. Two base-paired nucleotides have bases facing each other with the corresponding sugar and base beads aligned linearly. We use the "reaction" algorithm to model the orientation dependence of base-pairing interactions. The details of the algorithm can be found in (Ding et al. 2003). Briefly, to model the orientation dependence, we introduce auxiliary interactions in addition to the distance-dependent interactions between hydrogen bond donor and acceptor atoms (Fig. 9.2b). For example, when the two nucleotides (e.g., A–U, G–C, or U–G, represented as B_i and B_j in Fig. 9.2b) approach the interaction range, we evaluate the distances between S_iB_j and S_jB_i, which define the relative orientations of these two nucleotides. A hydrogen bond is allowed to form only when the distances fall within predetermined ranges. A schematic of the auxiliary interaction potential is shown in Fig. 9.2c, and the corresponding interaction parameters are listed in Table 9.2.

Hydrophobic Interactions and Overpacking. Buried inside the double-helix, the planar surface of bases are hydrophobic in nature. We include a weak attraction

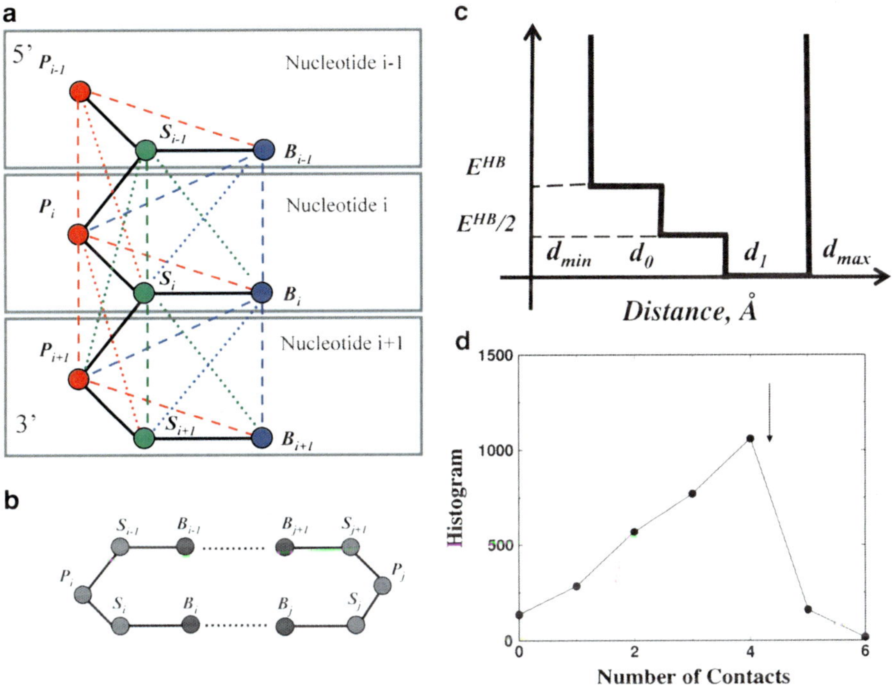

Fig. 9.2 Coarse-grained structural model of RNA employed in DMD simulations. (**a**) Three consecutive nucleotides, indexed $i-1$, i, $i+1$ are shown. The S, P, and B symbols correspond to loci of sugar, phosphate, and base beads in the RNA, respectively. Covalent interactions are shown as *thick lines*, angular constraints as *dashed lines*, and dihedral constraints as *dashed–dotted lines*. Additional steric constraints are used to model base stacking. (**b**) Hydrogen bonding in RNA base pairing. The base-pairing contacts between bases B_{i-1}:B_{j+1} and B_i:B_j are shown in *dashed lines*. A reaction algorithm is used (see Methods) for modeling the hydrogen-bonding interaction between specific nucleotide base pairs. (**c**) Schematic of the potential function for the auxiliary base-pairing interactions. (**d**) Histogram of the number of neighboring bases within a cutoff of 6.5 Å

between all the base beads. Due to the coarse-graining feature of our model, the assignment of attraction between bases results in overpacking (e.g., the symmetrically attractive interactions tend to form close packing). In order to avoid the artifact of overpacking, we first evaluate the packing observed in experimental 3D structures (http://ndbserver.rutgers.edu). We compute for each base the number of neighboring bases within a cutoff distance of 6.5 Å. The histogram of the number of neighbors is shown in Fig. 9.2d. Indeed, we find that the average number of neighbors is much smaller than that of close packing, 12. In order to avoid unrealistic close-packing due to the coarse-graining process, we introduce an effective energy term to penalize overpacking of bases:

$$E_{\text{overpack}} = \mathrm{d}E\Theta(n_c - n_{\max}), \tag{9.1}$$

where $\Theta(x)$ is a step function,

$$\Theta(x) = \begin{cases} x & x > 0 \\ 0 & x \leq 0 \end{cases}, \tag{9.2}$$

n_c is number of contacts, and n_{max} is the maximum number of contacts; dE is the repulsion coefficient. Based on the histogram of the number of base neighbors (Fig. 9.2d), we assign the value 4.2 for n_{max} and 0.6 kcal/mol for dE.

Base Stacking. To model stacking interactions, we assume that each base bead makes no more than two base–base stacking interactions and that three consecutively stacked base beads align approximately linearly. To determine the stacking interaction range between base beads, we compute center-to-center distances between base beads from known RNA structures. We find that distribution depends on base type (purine or pyrimidine) and identify stacking cutoff distances as 4.65 Å between purines, 4.60 between pyrimidines, and 3.80 Å between purine and pyrimidine. To approximately model the linearity of stacking interactions, two bases that form a stacking interaction to the same base are penalized for approaching closer than 6.5 Å. As a result, these three bases effectively define an obtuse angle. Next, we discuss the energy parameterization of base-stacking, base-pairing, and hydrophobic interactions.

Parameterization of Base-Pairing, Base-Stacking, and Hydrophobic Interactions. In order to determine the pairwise interaction parameters for stacking and hydrophobic interactions for all pairs of a base, we decomposed the sequence-dependent free energy parameters of the individual nearest-neighbor hydrogen bond model (INN-HB) (Mathews et al. 1999). We assume that the interaction of neighboring base pairs in INN-HB is the sum of all hydrogen-bond, base-stacking, and hydrophobic interactions. In a nearest neighboring base-pair configuration (Fig. 9.1), B_{i+1} and B_i (B_{j-1} and B_j) on one strand usually stack on top of each other. However, if both bases B_{i+1} and B_j are purines, we found that they tend to stack instead. The distance between bases B_i and B_{j-1} is usually greater than the cutoff distance of 6.5 Å for hydrophobic interactions. Therefore, we used the following equations to estimate the strength of pairwise interactions, where the first equation applies when B_{i+1}, B_j are both purines and the second equation applies otherwise:

$$E\begin{pmatrix} 5'B_iB_{i+1}3' \\ 3'B_jB_{j-1}5' \end{pmatrix} = \left(E^{HB}_{B_iB_j} + E^{HB}_{B_{i+1}B_{j-1}}\right) + E^{Stack}_{B_jB_{i+1}} + E^{hydrophobic}_{B_iB_{i+1}} + E^{hydrophobic}_{B_jB_{j-1}}, \tag{9.3}$$

$$E\begin{pmatrix} 5'B_iB_{i+1}3' \\ 3'B_jB_{j-1}5' \end{pmatrix} = \left(E^{HB}_{B_iB_j} + E^{HB}_{B_{i+1}B_{j-1}}\right) + E^{Stack}_{B_iB_{i+1}} + E^{stack}_{B_jB_{j-1}} + E^{hydrophobic}_{B_{i+1}B_j}. \tag{9.4}$$

Here, E^{stack}, E^{HB}, and $E^{hydrophobic}$ are the interaction strengths of base-stacking, base-pairing, and hydrophobic interactions, respectively. Given the experimentally tabulated energies between all possible neighboring base pairs (Mathews et al. 1999), we were able to determine values of E^{stack}, E^{HB}, and $E^{hydrophobic}$ that are consistent with experimental measurements using singular value decomposition (Khatun et al. 2004; Press et al. 2002). The interaction parameters are listed in Tables 9.2 and 9.3.

Table 9.1 The averages and standard deviations of the bonded atom pairs

Bonded atom pair	Distance range (Å)
$P_i\ S_i$	4.55 ± 0.09
$S_i\ P_{i+1}$	4.10 ± 0.07
$S_i\ A_i$	4.85 ± 0.15
$S_i\ U_i$	3.74 ± 0.08
$S_i\ G_i$	4.81 ± 0.14
$S_i\ C_i$	3.70 ± 0.13
$P_i\ P_{i+1}$	6.25 ± 0.95
$S_i\ S_{i+1}$	5.72 ± 0.45
$P_i\ A_i$	7.45 ± 0.45
$P_i\ U_i$	5.57 ± 0.37
$P_i\ Gi$	7.43 ± 0.43
$P_i\ C_i$	5.57 ± 0.37
$A_i\ P_{i+1}$	7.25 ± 0.42
$U_i\ P_{i+1}$	6.40 ± 0.20
$G_i\ P_{i+1}$	7.20 ± 0.43
$C_i\ P_{i+1}$	6.40 ± 0.20
$P_{i-1}\ S_i$	9.25 ± 0.95
$S_{i-1}\ P_{i+1}$	8.96 ± 0.44
$A_{i-1}\ S_i$	5.68 ± 0.68
$U_{i-1}\ S_i$	6.38 + 0.73
$G_{i-1}\ S_i$	5.68 ± 0.68
$C_{i-1}\ S_i$	6.38 ± 0.73
$S_{i-1}\ A_i$	7.25 ± 0.60
$S_{i-1}\ U_i$	5.66 ± 0.54
$S_{i-1}\ G_i$	7.25 ± 0.60
$S_{i-1}\ C_i$	5.66 ± 0.54

All the bonds, angles, and dihedrals are effectively modeled using a bonded interaction in the DMD simulations (Fig. 9.1b). A, U, G, and C corresponds to four types of bases (B)

Table 9.2 The parameters for base pairing, modeled by hydrogen bonds between A–U, G–C, and U–G

Atom pair	$d_{\min}$ (Å)	d_0, (Å)	d_1, (Å)	$d_{\max}$ (Å)
C*i*–G*j* base pair				
Si Gj	7.70	8.08	8.63	9.00
Ci Sj	9.74	10.10	10.53	10.82
A*i*–U*j* base pair				
Si Uj	9.76	9.94	10.50	10.76
Ai Sj	7.72	7.92	8.82	9.00
U*i*–G*j* base pair				
Si Gj	7.00	7.44	8.24	8.70
Ui Sj	9.50	10.25	10.80	11.35

The details of the DMD algorithm for the hydrogen bond can be found in Ding et al. (2003). The schematic interaction potential is shown in Fig. 9.2c. The hydrogen bond strengths, E^{HB}, for A–U, G–C, and U–G are 0.5, 1.2, and 0.5 Kcal/mol, respectively. The interaction potential between the donor and acceptor is $-E^{\mathrm{HB}}$

Table 9.3 The stacking and hydrophobic interaction strengths, expressed in kcal/mol units

E^{Stack}	A_U	U_A	G_C	C_G	G_U	U_G
A_U	−0.45	−0.50	−0.75	−0.95	−0.42	−0.70
U_A	−0.50	−0.40	−0.55	−0.60	−0.35	−0.35
G_C	−0.75	−0.55	−0.81	−0.95	−0.48	−0.92
C_G	−0.95	−0.60	−0.95	−1.10	−0.47	−0.51
G_U	−0.42	−0.35	−0.48	−0.47	−0.52	0.62
U_G	−0.70	−0.35	−0.51	−0.51	0.62	−0.44
$E^{Hydrophobic}$	A_U	U_A	G_C	C_G	G_U	U_G
A_U	−0.25	−0.40	−0.40	−0.50	−0.25	−0.35
U_A	−0.40	−0.30	−0.25	−0.25	−0.25	−0.25
G_C	−0.40	−0.25	−0.25	−0.45	−0.25	−0.41
C_G	−0.50	−0.25	−0.45	−0.50	−0.25	−0.41
G_U	−0.25	−0.25	−0.25	−0.25	−0.30	0.25
U_G	−0.35	−0.25	−0.41	−0.41	0.25	−0.25

The subscript indicates that the base bead is paired. For example, A_U is a base bead A that has been paired with a U bead. The cutoff distance for stacking interactions is 6.0 Å. The cutoff distance for hydrophobic interactions is 6.5 Å. The hardcore distance between all beads is set as 3.0 Å

Loop Entropy. Loop entropy plays a pivotal role in RNA folding kinetics and thermodynamics (Tinoco and Bustamante 1999). Hence, RNA folding prediction methods should take this entropic effect into account, either implicitly as in all-atom MD simulations (Sorin et al. 2004) or explicitly as in Monte Carlo or dynamic programming methods (Mathews 2006; Rivas and Eddy 1999). However, the reduction of degrees of freedom in our simplified RNA model causes entropy to be underestimated in DMD simulations. For example, we often observe formation of large loops that traps RNA molecules in nonnative conformations for significant simulation times. To overcome such artifacts arising from the coarse-graining process, we developed a simple modification of DMD simulation to model loop entropy explicitly. We use the free energy estimations for different types of loops, including hairpin, bulge, and internal loops (Mathews et al. 1999). Loop free energies were obtained from experimental fitting for small loops and extended to arbitrary lengths according to polymer theory. We compute the effective loop free energy in DMD simulations based on the set of base pairs formed in simulations. Upon the formation or breaking of each base pair, the total loop free energy changes according to the changes in either the number or size of loops. We estimate the changes in loop free energy, ΔG^{loop}, for each base pair formed during the simulation and determine the probability of forming such a base pair by coupling to a Monte Carlo procedure using a Metropolis algorithm with probability $p = \exp(-\beta\Delta G^{loop})$. If the base pair is allowed to form stochastically, the particular base pair will form only if the kinetic energy is sufficient to overcome the possible potential energy difference before and after the base-pair formation. Upon breaking of a base pair, the stochastic procedure is not invoked since base-pair breakage is always entropically favorable. The breaking of a base pair is only governed by the conservation of momentum, energy, and angular momentum before and after the base-pair breakage.

The total potential energy, E, is obtained by adding all interaction terms, as given in (9.5):

$$E = E_{\text{Bonded}} + E_{\text{Hbond}} + E_{\text{Stack}} + E_{\text{Hydrophobic}} + E_{\text{overpacking}} + G_{\text{loop}}, \tag{9.5}$$

and is used to perform DMD simulations of RNA molecules. The energy landscape of RNA molecules is very rugged with a vast number of local minima due to the high degeneracy of nucleotide types (only 4 compared to the 20 different amino acids found in proteins). In order to efficiently sample the conformational space of RNAs, we utilize the replica-exchange sampling scheme (Okamoto 2004; Zhou et al. 2001).

Replica Exchange DMD. In replica exchange computing, multiple simulations or replicas of the same system are performed in parallel at different temperatures. Individual simulations are coupled through Monte Carlo-based exchanges of simulation temperatures between replicas at periodic time intervals. For two replicas, i and j, maintained at temperatures T_i and T_j and with energies E_i and E_j, temperatures are exchanged according to the canonical Metropolis criterion with exchange probability p, where $p = 1$ if $\Delta = \left(1/k_{\text{B}}T_i - 1 - k_{\text{B}}T_j\right)\left(E_j - E_i\right) \leq 0$, and $p = \exp(-\Delta)$, if $\Delta > 0$. For simplicity, we use the same set of eight temperatures in all replica exchange simulations: 0.200, 0.208, 0.214, 0.220, 0.225, 0.230, 0.235, and 0.240. The temperature is in the abstract unit of kcal/(mol k_{B}). Note that we approximate the pairwise potential energy between coarse-grained beads with the experimentally determined free energy of nearest neighboring base pairs, instead of the actual enthalpy. As a result, the temperature does not directly correspond to physical temperatures. In DMD simulations, we maintain constant temperature using an Anderson thermostat (Andersen 1980).

Since the DMD code is highly optimized, we have found that the computational timescales linearly with respect to the system size. The folding simulation of a 50-nucleotide-long RNA sequence (median size of RNA chains in the sample) for 2×10^6 DMD simulation time units takes approximately 7 h of total wall-clock time, utilizing eight 2.33-GHz Intel Xeon compute nodes.

9.3 Ab Initio Folding of Small RNA Molecules

For each RNA molecule, we initially generated a linear conformation using the nucleotide sequence alone. Starting from this extended conformation, we performed replica exchange simulations at different temperatures as described above. From the simulation trajectories, we extracted sampled RNA conformational states, including the lowest energy state, the folding intermediate state, and the corresponding thermodynamic data. In Fig. 9.3, we illustrate the folding trajectory of one of the replicas for a turnip yellow mosaic virus (TYMV) pseudoknot (PDB ID: 1A60). An RNA pseudoknot structure has nonnested base pairing and minimally comprises base pairing between a loop region and a downstream RNA segment. Pseudoknots serve diverse biological functions, including

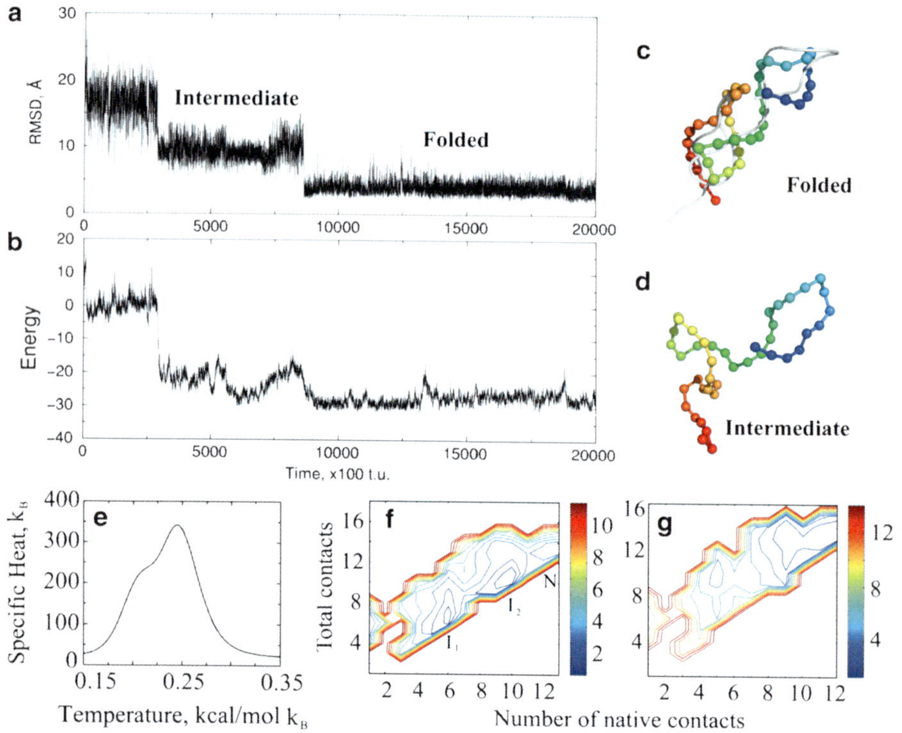

Fig. 9.3 Folding of a pseudoknot. For one replica, we present the RMSD (**a**) and energy (**b**) as the function of simulation time. Before folding into its native state (**c**), the molecule samples a folding intermediate state (**d**). (**e**) Specific heat is computed from the replica exchange trajectories using WHAM. (**f**) Two-dimensional potential of mean force 2D-PMF (potential mean force) for pseudoknot folding at $T^* = 0.245$ (corresponds to the major peak in the specific heat). The two intermediate states and the native state are indicated by I_1, I_2, and N, respectively. (**g**) The 2D-PMF plot at $T^* = 0.21$

formation of protein recognition sites that mediate replication and translational initiation, participation in self-cleaving ribozyme catalysis, and induction of frameshifts in translation of mRNA by ribosomes (Staple and Butcher 2005). For example, 1A60 is composed of a 5′-stem and a 3′-pseudoknot (Fig. 9.3c). From the simulation trajectory (Fig. 9.3), we observe folding of the RNA model within 5 Å root-mean-square deviation (RMSD) to the native state, and the lowest RMSD from the simulations is 2.03 Å. The lowest potential energy conformation, computed across all replicas using the effective free energy function in (9.5), has all native base pairs formed and an RMSD of 4.58 Å to the native state. Interestingly, we find that during the folding process the RNA molecule samples a stable folding intermediate state (Fig. 9.3a, b). The intermediate state forms a 5′-stem and a partially folded 3′-pseudoknot with one of the stems. Our identified folding intermediate state is consistent with the NMR studies of the solution structures of the TYMV pseudoknot and its 3′-stem (Kolk et al. 1998). Therefore, our DMD simulation not

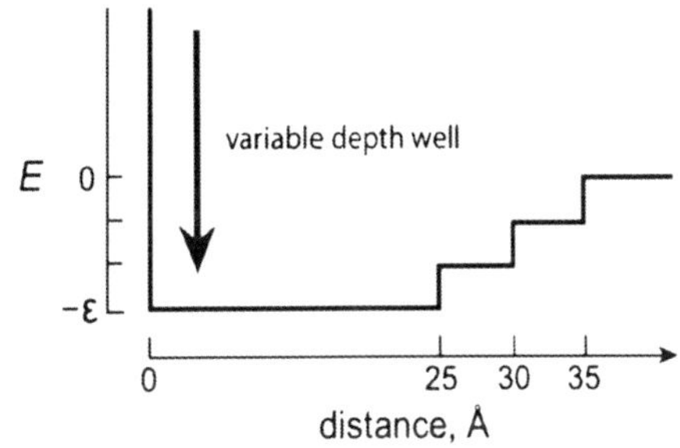

Fig. 9.5 Potential function used to convert experimental cleavage information into DMD potential energy constraints

(49 nts), a full-length hammerhead ribozyme from *S. mansoni* (HHR) (67 nts), *S. cerevisiae* tRNA^{Asp} (75 nts), and the P546 domain of the *T. thermophilia* group I intron (P546) (158 nts). Each of these RNAs has a complex three-dimensional fold, involving more than simple intrahelix interactions. Prior to publication of the high-resolution structures (Cate et al. 1996; Costantino et al. 2008; Martick and Scott 2006; Westhof et al. 1988), significant biochemical or bioinformatic data describing tertiary interactions were available for each RNA. The secondary structure was also known to high accuracy in each case. Only this prior information was used during DMD refinement. In all cases, we were able to generate a low-RMSD structure. The RMSD between the predicted structure and the native state for the CrPV, HHR, tRNA^{Asp}, and P546 RNAs are 3.6, 5.4, 6.4, and 11.3 Å, respectively (Lavender et al. 2010). Calculations were performed on a Linux workstation (Intel Pentium 4 processor, 3.2 GHz) and the CPU times ranged from 18 (CrPV, 49 nts) to 42 h (P546, 158 nts). Therefore, the combination of efficient DMD simulations and sufficient biochemical experiments can accurately determine RNA structure of arbitrary length.

9.5 Conclusions

We have developed a multiscale RNA modeling approach to model 3D structure and dynamics of RNAs having a wide range of lengths. We use a coarse-grained representation of the RNA to efficiently model the conformational space. For short RNA molecules (<50 nt), we are able to capture the folded state from the sequence alone. The availability of replica-exchange simulation trajectories at multiple temperatures allows for the characterization of folding thermodynamics as well as capture of the final folded state. To efficiently sample the exponentially increasing conformational space of large RNA molecules, we devised an automated modeling approach to determine large and complex RNA structures using experimentally derived structural information. A benchmark study (Lavender et al. 2010) highlights the application of combining DMD simulation and experimental structural information to yield native-like models for the diverse universe of functionally important RNAs whose structures cannot be characterized by conventional methods.

References

Andersen HC (1980) Molecular dynamics simulations at constant pressure and/or temperature. J Chem Phys 72:10

Baumgartner A (1987) Applications of the Monte-Carlo simulations in statistical physics. Springer, New York

Cannone J, Subramanian S, Schnare M, Collett J, D'Souza L, Du Y, Feng B, Lin N, Madabusi L, Müller K et al (2002) The comparative RNA web (CRW) site: an online database of comparative sequence and structure information for ribosomal, intron, and other RNAs. BMC Bioinformatics 3:e2

Cao S, Chen SJ (2005) Predicting RNA folding thermodynamics with a reduced chain representation model. RNA 11:1884–1897

Cao S, Chen SJ (2006) Predicting RNA pseudoknot folding thermodynamics. Nucleic Acids Res 34:2634–2652

Cate JH, Gooding AR, Podell E, Zhou K, Golden BL, Kundrot CE, Cech TR, Doudna JA (1996) Crystal structure of a group I ribozyme domain: principles of RNA packing. Science 273:1678–1685

Costantino DA, Pfingsten JS, Rambo RP, Kieft JS (2008) tRNA-mRNA mimicry drives translation initiation from a viral IRES. Nat Struct Mol Biol 15:57–64

Das R, Baker D (2007) Automated *de novo* prediction of native-like RNA tertiary structures. Proc Natl Acad Sci 104:14664–14669

De la Pena M, Gago S, Flores R (2003) Peripheral regions of natural hammerhead ribozymes greatly increase their self-cleavage activity. EMBO J 22:5561–5570

Deigan KE, Li TW, Mathews DH, Weeks KM (2009) Accurate SHAPE-directed RNA structure determination. Proc Natl Acad Sci USA 106:97–102

Ding F, Borreguero JM, Buldyrey SV, Stanley HE, Dokholyan NV (2003) Mechanism for the alpha-helix to beta-hairpin transition. Proteins 53:220–228

Ding F, Dokholyan NV (2005) Simple but predictive protein models. Trends Biotechnol 23:450–455

Ding F, Sharma S, Chalasani V, Demidov V, Broude NE, Dokholyan NV (2008) Ab initio RNA folding by discrete molecular dynamics: from structure prediction to folding mechanisms. RNA 14:1164–1173

Dokholyan NV, Buldyrev SV, Stanley HE, Shakhnovich EI (1998) Discrete molecular dynamics studies of the folding of a protein-like model. Fold Des 3:577–587

Eddy SR (2004) How do RNA folding algorithms work? Nat Biotechnol 22:1457–1458

Edwards TE, Klein DJ, Ferre-D'Amare AR (2007) Riboswitches: small-molecule recognition by gene regulatory RNAs. Curr Opin Chem Biol 17:273–279

Flor PJ, Flanegan JB, Cech TR (1989) A conserved base pair within helix P4 of the Tetrahymena ribozyme helps to form the tertiary structure required for self-splicing. EMBO J 8:3391–3399

Gherghe CM, Leonard CW, Ding F, Dokholyan NV, Weeks KM (2009) Native-like RNA tertiary structures using a sequence-encoded cleavage agent and refinement by discrete molecular dynamics. J Am Chem Soc 131:2541–2546

Gutell RR, Lee JC, Cannone JJ (2002) The accuracy of ribosomal RNA comparative structure models. Curr Opin Struct Biol 12:301–310

Gutell RR, Power A, Hertz GZ, Putz EJ, Stormo GD (1992) Identifying constraints on the higher-order structure of RNA: continued development and application of comparative sequence analysis methods. Nucleic Acids Res 20:5785–5795

Hajdin CE, Ding F, Dokholyan NV, Weeks KM (2010) On the significance of an RNA tertiary structure prediction. RNA 16:1340–1349

Hertzberg RP, Dervan PB (1982) Cleavage of double helical DNA by (Methidiumpropyl-EDTA) iron(II). J Am Chem Soc 104:313–315

Hofacker IL (2003) Vienna RNA secondary structure server. Nucleic Acids Res 31:3429–3431

Jan E, Sarnow P (2002) Factorless ribosome assembly on the internal ribosome entry site of cricket paralysis virus. J Mol Biol 324:889–902

Jossinet F, Westhof E (2005) Sequence to Structure (S2S): display, manipulate and interconnect RNA data from sequence to structure. Bioinformatics 21:3320–3321

Juzumiene D, Shapkina T, Kirillov S, Wollenzien P (2001) Short-range RNA-RNA crosslinking methods to determine rRNA structure and interactions. Methods 25:333–343

Kanamori Y, Nakashima N (2001) A tertiary structure model of the internal ribosome entry site (IRES) for methionine-independent initiation of translation. RNA 7:266–274

Khatun J, Khare SD, Dokholyan NV (2004) Can contact potentials reliably predict stability of proteins? J Mol Biol 336:1223–1238

Khvorova A, Lescoute A, Westhof E, Jayasena SD (2003) Sequence elements outside the hammerhead ribozyme catalytic core enable intracellular activity. Nat Struct Biol 10:708–712

Kolk MH, van der Graaf M, Fransen CT, Wijmenga SS, Pleij CW, Heus HA, Hilbers CW (1998) Structure of the 3′-hairpin of the TYMV pseudoknot: preformation in RNA folding. EMBO J 17:7498–7504

Kumar S, Bouzida D, Swendswn RH, Kollman PA, Rosenberg JM (1992) The weighted histogram analysis method for free-energy calculations on biomolecules .1. The method. J Computat Chem 13:11

Lavender CA, Ding F, Dokholyan NV, Weeks KM (2010) Robust and generic RNA modeling using inferred constraints: a structure for the hepatitis C virus IRES pseudoknot domain. Biochemistry 49:4931–4933

Major F, Gautheret D, Cedergren R (1993) Reproducing the three-dimensional structure of a tRNA molecule from structural constraints. Proc Natl Acad Sci U S A 90:9408–9412

Major F, Turcotte M, Gautheret D, Lapalme G, Fillion E, Cedergren R (1991) The combination of symbolic and numerical computation for three-dimensional modeling of RNA. Science 253:1255–1260

Martick M, Scott WG (2006) Tertiary contacts distant from the active site prime a ribozyme for catalysis. Cell 126:309–320

Massire C, Jaeger L, Westhof E (1998) Derivation of the three-dimensional architecture of bacterial ribonuclease P RNAs from comparative sequence analysis. J Mol Biol 279:773–793

Massire C, Westhof E (1998) MANIP: an interactive tool for modelling RNA. J Mol Graph Model 16(197–205):255–197

Mathews DH (2006) Revolutions in RNA secondary structure prediction. J Mol Biol 359:526–532

Mathews DH, Disney MD, Childs JL, Schroeder SJ, Zuker M, Turner DH (2004) Incorporating chemical modification constraints into a dynamic programming algorithm for prediction of RNA secondary structure. Proc Natl Acad Sci USA 101:7287–7292

Mathews DH, Sabina J, Zuker M, Turner DH (1999) Expanded sequence dependence of thermodynamic parameters improves prediction of RNA secondary structure. J Mol Biol 288:911–940

Merino EJ, Wilkinson KA, Coughlan JL, Weeks KM (2005) RNA structure analysis at single nucleotide resolution by selective 2′-hydroxyl acylation and primer extension (SHAPE). J Am Chem Soc 127:4223–4231

Michel F, Westhof E (1990) Modelling of the three-dimensional architecture of group I catalytic introns based on comparative sequence analysis. J Mol Biol 216:585–610

Mortimer SA, Weeks KM (2007) A fast-acting reagent for accurate analysis of RNA secondary and tertiary structure by SHAPE chemistry. J Am Chem Soc 129:4144–4145

Murphy FL, Cech TR (1994) GAAA tetraloop and conserved bulge stabilize tertiary structure of a group I intron domain. J Mol Biol 236:49–63

Murray LJ, Arendall WB 3rd, Richardson DC, Richardson JS (2003) RNA backbone is rotameric. Proc Natl Acad Sci U S A 100:13904–13909

Okamoto Y (2004) Generalized-ensemble algorithms: enhanced sampling techniques for Monte Carlo and molecular dynamics simulations. J Mol Graph Model 22:425–439

Parisien M, Cruz JA, Westhof E, Major F (2009) New metrics for comparing and assessing discrepancies between RNA 3D structures and models. RNA 15:1875–1885

Parisien M, Major F (2008) The MC-Fold and MC-Sym pipeline infers RNA structure from sequence data. Nature 452:51–55

Press WH, Teukolsky SA, Vetterling WT, Flannery BP (2002) Numerical Recipes in C, 2nd edn. Cambridge University Press, Cambridge

Rapaport DC (2004) The art of molecular dynamics simulation. Cambridge University Press, Cambridge

Rivas E, Eddy SR (1999) A dynamic programming algorithm for RNA structure prediction including pseudoknots. J Mol Biol 285:2053–2068

Shapiro BA, Yingling YG, Kasprzak W, Bindewald E (2007) Bridging the gap in RNA structure prediction. Curr Opin Struct Biol 17:157–165

Sorin EJ, Nakatani BJ, Rhee YM, Jayachandran G, Vishal V, Pande VS (2004) Does native state topology determine the RNA folding mechanism? J Mol Biol 337:789–797

Staple DW, Butcher SE (2005) Pseudoknots: RNA structures with diverse functions. PLoS Biol 3: e213

Tinoco I Jr, Bustamante C (1999) How RNA folds. J Mol Biol 293:271–281

Tsai HY, Masquida B, Biswas R, Westhof E, Gopalan V (2003) Molecular modeling of the three-dimensional structure of the bacterial RNase P holoenzyme. J Mol Biol 325:661–675

Vicens Q, Cech TR (2006) Atomic level architecture of group I introns revealed. Trends Biochem Sci 31:41–51

Wang B, Wilkinson KA, Weeks KM (2008) Complex ligand-induced conformational changes in $tRNA^{Asp}$ revealed by single nucleotide resolution SHAPE chemistry. Biochemistry 47:3454–3461

Wang C, Le SY, Ali N, Siddiqui A (1995) An RNA pseudoknot is an essential structural element of the internal ribosome entry site located within the hepatitis C virus 5′ noncoding region. RNA 1:526–537

Westhof E, Dumas P, Moras D (1988) Restrained refinement of 2 crystalline forms of yeast aspartic-acid and phenylalanine transfer-Rna crystals. Acta Crystallographica Sect A 44:112–123

White SA, Draper DE (1987) Single base bulges in small RNA hairpins enhance ethidium binding and promote an allosteric transition. Nucleic Acids Res 15:4049–4064

White SA, Draper DE (1989) Effects of single-base bulges on intercalator binding to small RNA and DNA hairpins and a ribosomal RNA fragment. Biochemistry 28:1892–1897

Wilkinson KA, Gorelick RJ, Vasa SM, Guex N, Rein A, Mathews DH, Giddings MC, Weeks KM (2008) High-throughput SHAPE analysis reveals structures in HIV-1 genomic RNA strongly conserved across distinct biological states. PLoS Biol 6:e96

Wilkinson KA, Merino EJ, Weeks KM (2006) Selective 2′-hydroxyl acylation analyzed by primer extension (SHAPE): quantitative RNA structure analysis at single nucleotide resolution. Nat Protocol 1:1610–1616

Zhou R, Berne BJ, Germain R (2001) The free energy landscape for beta hairpin folding in explicit water. Proc Natl Acad Sci U S A 98:14931–14936

Zhou Y, Karplus M (1997) Folding thermodynamics of a model three-helix-bundle protein. Proc Natl Acad Sci USA 94:14429–14432

Ziehler WA, and Engelke DR (2001). Probing RNA structure with chemical reagents and enzymes. Curr Protoc Nucleic Acid Chem *Chapter 6*, Unit 6 1

Zuker M (2003) Mfold web server for nucleic acid folding and hybridization prediction. Nucleic Acids Res 31:3406–3415

Table 10.2 Entropy parameters for loops L_1 and L_2

Stem size (bp)	Loop size (nt)											
	1	2	3	4	5	6	7	8	9	10	11	12
(S_2)	L_1 (across the major groove of S_2)											
2	–	–	–	6.2	6.4	6.4	6.6	6.8	6.9	7.1	7.2	
3	–	6.4*	6.4*	6.4	6.6	6.6	6.8	6.9	7.1	7.3	7.5	
4	4.5*	4.5*	4.5	5.4	5.6	6.0	6.3	6.6	6.9	7.1	7.3	
5	2.3	4.4	4.6	5.7	6.0	6.5	6.9	7.2	7.5	7.8	8.0	
6	2.3	4.4	4.8	5.8	6.0	6.5	6.8	7.1	7.4	7.6	7.8	
7	2.3	4.4	5.0	5.9	6.2	6.8	7.0	7.3	7.6	7.8	8.0	
8	–	4.4	5.2	5.7	6.4	6.7	7.1	7.3	7.5	7.7	7.9	
9	–	5.5*	5.5	6.4	6.7	7.2	7.5	7.9	8.1	8.3	8.5	
10	–	6.9*	6.9*	6.9	7.5	7.7	8.1	8.3	8.6	8.8	8.9	
11	–	–	–	–	8.7	8.8	8.9	9.1	9.2	9.3	9.3	
12	–	–	–	–	9.8	9.2	9.5	9.6	9.7	9.8	9.8	
(S_1)	L_2 (across the major groove of S_1)											
2	–	–	–	7.6	7.0	7.0	7.1	7.2	7.3	7.4	7.5	7.7
3	–	6.5*	6.5*	6.5	6.6	6.7	6.9	7.1	7.2	7.4	7.6	7.7
4	–	–	9.2*	9.2*	9.2	8.9	8.9	8.9	9.0	9.0	9.1	9.2
5	–	–	–	9.8*	9.8*	9.8	9.1	8.9	8.8	8.8	8.8	8.8
6	–	–	–	11.9*	11.9*	11.9*	11.9	11.0	10.4	10.1	9.9	9.8
7	–	–	–	–	12.4*	12.4*	12.4*	12.4	11.4	11.0	10.7	10.5
8	–	–	–	–	12.1*	12.1*	12.1*	12.1	11.6	11.4	11.2	11.1
9	–	–	–	–	–	13.7*	13.7*	13.7*	13.7	12.6	12.0	11.5
10	–	–	–	–	–	13.7*	13.7*	13.7	12.7	12.2	11.8	11.5
11	–	–	–	–	–	–	–	–	15.9	14.1	13.0	12.4
12	–	–	–	–	–	–	–	–	18.7	15.8	14.2	13.2

In the table, an entropy parameter ΔS_{loop} is given in the form of $-\Delta S_{\text{loop}}/k_B$, where k_B is the Boltzmann constant. The *asterisk* entries in the table indicate the loop conformations that cannot be realized in the diamond lattice but may be viable for a realistic pseudoknot. These loop conformations usually have a long stem and short loop. For the entropies of these restricted loops, we use the values of the minimal loop for the same helix length. The table is adapted from Cao and Chen (2006b)

While the above methods employ the same empirical thermodynamic parameters (the Turner rules) for secondary structural elements based on the nearest-neighbor base-pair interaction model, other models use knowledge-based scoring functions. For instance, the CONTRAfold (Do et al. 2006) model uses the energy parameters derived from a training set, and the MC-Fold (Parisien and Major 2008) model uses a scoring function that represents the probability of selecting a certain nucleotide cyclic motif (NCM) for the given sequence. The NCMs in the MC-Fold include the lone-pair loops and the double-stranded internal/bulge loops, which are extracted from known PDB structures. Benchmark tests show that CONTRA-fold and MC-Fold programs give better predictions than Mfold (Do et al. 2006; Parisien and Major 2008).

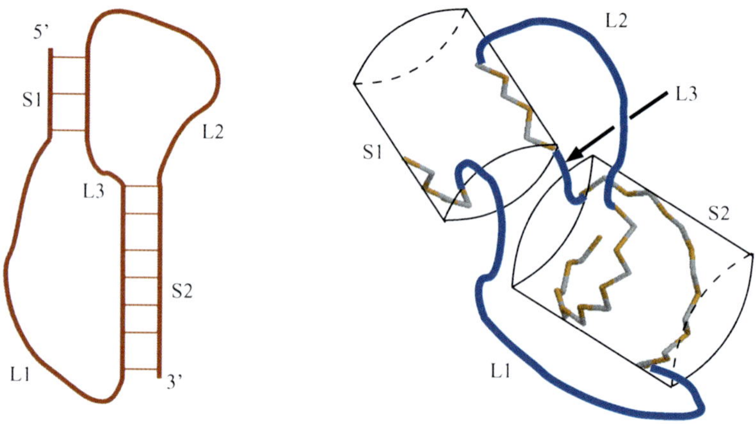

Fig. 10.2 A 2D structure and the 3D conformations for a pseudoknot with interhelix loop L_3

10.2.2 H-Type Pseudoknots: Free Energy Models

A general pseudoknot consists of two helical stems, S_1 and S_2, and three loops, L_1, L_2, and L_3 (see Fig. 10.2). Pseudoknots, which are neglected in most software for secondary structure prediction, play important structural and functional roles in many biochemical processes, such as viral replication (Brierley et al. 2007, 2008; Draper 1990; Gesteland and Atkins 1996; Giedroc et al. 2000; Giedroc and Cornish 2009; Staple and Butcher 2005), human telomerase RNA activity (Chen and Greider 2005; Comolli et al. 2002; Qiao and Cech 2008; Shefer et al. 2007; Theimer and Feigon 2006), and metabolite-sensing riboswitches (Kang et al. 2009; Klein et al. 2009; Spitale et al. 2009).

The free energy for a pseudoknot is equal to the sum of the free energies for the helical stems and loops: $\Delta G_{\text{pk}} = \Delta G_{S_1} + \Delta G_{S_2} + \Delta G_{\text{CS}} - T\Delta S_{\text{loops}}$, where ΔG_{S_1} and ΔG_{S_2} are the free energies of stems S_1 and S_2 and ΔG_{CS} is the coaxial stacking energy between stems S_1 and S_2. While the helix free energies can be evaluated from the nearest-neighbor model based on the empirical thermodynamic parameters for base stacks, to determine the loop entropy, ΔS_{loops} requires a physical model.

From statistical mechanics, $\Delta S_{\text{loops}} = -k_B \ln \Omega_{\text{Coil}}/\Omega$, where Ω is the number of the 3D conformations of the loops and Ω_{Coil} is the number of the corresponding coil conformations. The evaluation of the conformational entropy ΔS_{loops} is intrinsically a problem embedded in 3D space. The difficulty in the evaluation of the pseudoknot loop entropy comes from the conformational correlation between the loop and the helix. Specifically, the viability of a loop conformation is subject to the presence of the nearby helix due to the loop–helix excluded volume interaction and the end-to-end distance constraint of the loop set by the length of the helix. The presence of the nearby helix reduces the accessible space of the loop configuration and thus decreases the number of the viable loop conformations.

and then to construct atomistic 3D (native and alternative) structures from the low-resolution models. Given the huge conformational space available to an RNA molecule, such a multiscale approach has several distinctive advantages:

1. At the low-resolution level, many experimental questions are not concerned with the specific locations of the hydrogen or nitrogen atoms; instead, they are far more concerned with the global fold, backbone flexibility, and the potential for large structural rearrangements, as investigated by small-angle X-ray or neutron scattering and other low-resolution techniques (Chauhan et al. 2005; Deigan et al. 2009; Gherghe et al. 2009; Russell et al. 2002).
2. The key issue in prediction of the tertiary structural folding concerns the entropy of the global fold, which is largely determined by low-resolution properties such as the excluded volume and the chain connectivity effects. The reduced complexity of the low-resolution model allows us to maintain the rigor in physical principles when accounting for these properties. As a result, use of a low-resolution model enables first principles calculations for chain entropy, free energy, and the full free energy landscape for any given sequence.
3. The low-resolution structure provides a useful scaffold for the final all-atom folding model through structural refinements.
4. The predicted structure will provide highly needed guidance for experiments. For example, in NMR structural determinations of RNA, a severe limitation is that sequential resonance assignments rely heavily on Nuclear Overhauser Effect (NOE) data to establish connectivities ("NOE walks"), which often requires several months of data collection and analysis. The information on the nucleotide spatial proximity from the predicted (low-resolution) structure can provide useful constraints for enhancement in the efficiency and accuracy of resonance assignments.

Vfold is a recently developed low-resolution model based on the virtual bond representation of RNA conformation (Cao and Chen 2005; Cao et al. 2010; Chen 2008). As shown in Fig. 10.3a, the $P–O_5–C_5–C_4$ and the $C_4–C_3–O_3–P$ dihedrals tend to be planar and rigid because the torsional angles about the $C_5–O_5$ and $C_3–O_3$ bonds tend to remain in the relatively rigid *trans (t)* state. Therefore, the original six-torsion nucleotide backbone can be reduced to two "virtual bonds" spanning P to C_4 and C_4 to the next P in the chain (Olson and Flory 1972; Olson 1975, 1980). Calculations on the nucleotide atomic structures show that each virtual bond has a length of 3.9 Å. The third virtual bond ($C_4–N_1$ for pyrimidine or $C_4–N_9$ for purine) represents the orientation of the base. In addition, a survey of known RNA structures suggests that the distance between the N_1 (in pyrimidines) or N_9 (in purines) and C_4 atoms stays close to 3.9 Å and the torsion angle between plane $P_i–C_4–P_{i+1}$ and $P_i–C_4–N_1$ (N_9), close to the g^{-1} isomeric state. Thus, the $C_4–N_1$ (N_9) virtual bond is quite rigid. The three-vector virtual bond model leads to the following Vfold model for RNA folding:

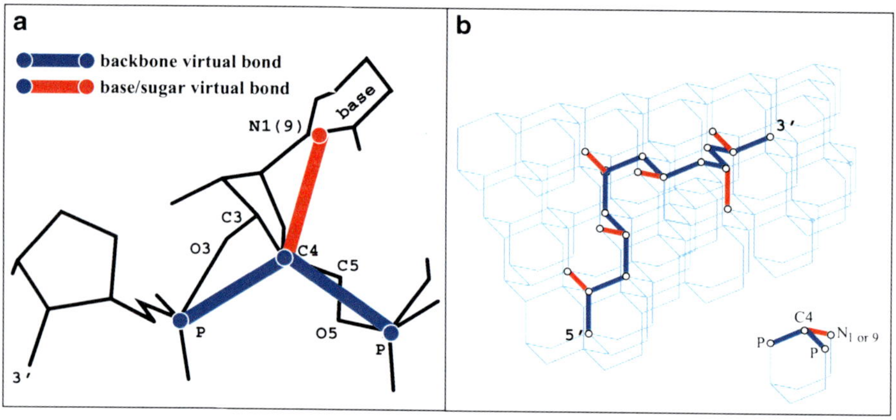

Fig. 10.3 (**a**) Each nucleotide has two *backbone* virtual bonds P–C_4–P (*blue*) and a *sugar–base* virtual bond (*red*) C_4–N_1 for pyrimidine or C_4–N_9 for purine. (**b**) An RNA conformation can be generated through random walks of the virtual bonds in a diamond lattice. In the diamond crystal, the four carbon atoms are located at the vertices and center of a tetrahedron. Four such tetrahedral connected at their vertices fit in a cube. Repetition of such cubes side by side generates a *diamond lattice*

1. The virtual bond structure for a helix is constructed from the atomic coordinates of an A-form RNA helix (Arnott and Hukins 1972).
2. The ensemble of virtual bond structures for loop conformations are generated by using the usual *gauche*$^+$ (g^+), *trans* (t), and *gauche*$^-$ (g^-) rotational isomeric states for a polymer. A survey on the existing known structures shows that such rotational isomeric states can well represent RNA loop conformations (Cao and Chen 2005; Duarte and Pyle 1998; Duarte et al. 2003; Richardson et al. 2008). The three isomeric states can be realized in the diamond lattice. Therefore, the virtual bonds of loop conformations are configured on the *diamond lattice*, where each lattice bond is a virtual bond (see Fig. 10.3b).
3. At the helix–loop junction, we fit the virtual bonds onto the diamond lattice with the minimum RMSD.

The Vfold model is fundamentally different from any of the simplified models such as the simple square or cubic lattice models used in other folding theories. Vfold is a realistic atomistic structural model because the virtual bonds are the realistic physical P–C_4 and C_4–P bonds in the structure, and the discretization (diamond lattice) of the virtual bond configurations for loop conformations is based on the principles of polymer physics as well as known RNA structures. So the Vfold model can directly predict experimentally measurable and biologically relevant structures and stabilities. The Vfold package is available for Windows and Unix users (to be released; URL: http://vfold.missouri.edu/chen-software02.html). For pseudoknotted folds, the CPU time (t seconds) for the Vfold-based structural prediction grows with the sequence length (l nucleotides) as $\ln(t) \approx -24.3 + 7.7 \ln(l)$ on a Intel(R) Xeon(R) CPU 5150 @ 2.66 GHz on Dell EM64T cluster system.

The most time-consuming part of the computation is the enumeration of the different stems and loops (2D structures).

10.3.2 Pseudoknot Structure and Stability

The predictive power of the Vfold model is shown by its ability to compute the entropy for complex folds such as a three-loop pseudoknot (Fig. 10.2). The computation involves three steps:

1. All the possible helix orientations are generated through the enumeration of (virtual bond) conformations of the loop L_3.
2. For each helix orientation, because the probability for a loop bumping into another loop is relatively small, the loops can be treated with the independent loop approximation (Cao and Chen 2009; Chen and Dill 2000; Poland and Scheraga 1970): The total conformational count (Ω) for the 3-loop system can be estimated as the product of the conformational count ($\Omega_{\text{loop } L_i}$) for each loop.

$$\Omega = \prod_i \Omega_{\text{loop } L_i}; \quad S_{\text{loop}} = k_B \ln \Omega. \tag{10.2}$$

   This approach is remarkable because it reduces the 3-loop conformational enumeration into conformational enumeration for one loop at a time, resulting in a dramatic reduction in the computer time from $T(\sum_n L_n)$ to $\sum_n T(L_n)$, where L_n is the length of the n-th loop and $T(L)$ is the computer time for counting conformations for a loop of length L.
3. The volume exclusion between a loop and the helices (grooves) is the key to the evaluation of the loop entropy. In the Vfold model, this can be explicitly taken into account by disallowing overlapping virtual bonds when the loop conformations are generated in the virtual bond diamond lattice.

The Vfold model leads to loop entropy parameter tables for canonical (Cao and Chen 2006a) 2-loop H-type pseudoknots and noncanonical 3-loop H-type pseudoknot (Cao and Chen 2009). Table 10.1 gives the entropy parameters for the canonical (2-loop) H-type pseudoknot. For the noncanonical (3-loop) H-type pseudoknot, the entropy tables are deposited at http://rnajournal.cshlp.org/content/15/4/696/suppl/DC1 (Cao and Chen 2009).

These loop entropy parameters allow for calculations of the folding free energy for a given pseudoknot (Cao and Chen 2006b, 2009; Chen 2008). For example, for the 3-loop pseudoknot shown in Fig. 10.4, the free energy is $\Delta G = \Delta G_1 + \Delta G_2 - T\Delta S(S_1, S_2, L_1, L_2, L_3) = (-7.1)$ kcal/mol $+ (-6.3)$ kcal/mol $+ k_B T\ (14.2) = -4.6$ kcal/mol, where S_1 and S_2 denote the length of stems 1 and 2 and L_1, L_2, and L_3 are the lengths of loops 1, 2, and 3, respectively. The entropy parameter $\Delta S(S_1, S_2, L_1, L_2, L_3)$ is transcribed from Table S1 of this reference: (Cao and Chen 2009).

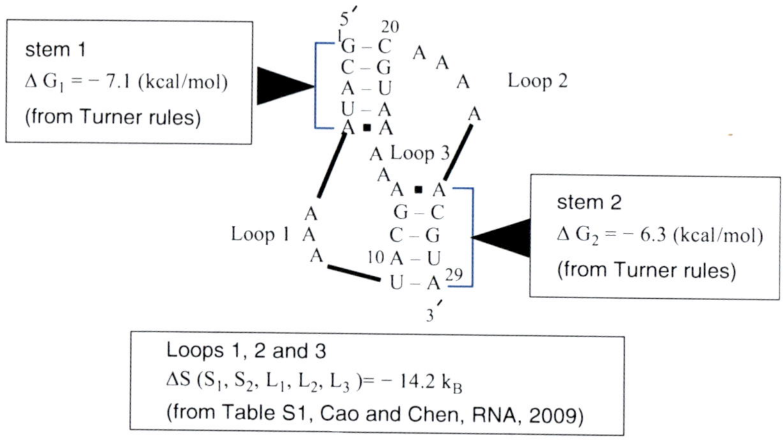

Fig. 10.4 The evaluation of the free energy for a 3-loop (noncanonical) H-type pseudoknot using the loop entropy parameters in Table S1 of reference (Cao and Chen 2009) and the Turner rule (Serra and Turner 1995)

Application of the Vfold model to the structure prediction of a pool of biologically significant RNA molecules ranging in length from 28 to 91 nucleotides (Ren et al. 2005) indicates that the Vfold model gives significant improvements in the accuracy of the predictions as compared to other existing RNA folding models (Cao and Chen 2009). Furthermore, the successful implementation of the Vfold-predicted entropy parameters in several other software packages indicates that their use can indeed lead to significantly improved accuracy in pseudoknot structural prediction (Andronescu et al. 2010; Sperschneider and Datta 2010; Liu et al. 2010). The calculation of entropies of more complex pseudoknotted structure is computationally demanding (Cao and Chen 2006a, 2009). Monte Carlo simulation may be a potentially useful method to generate virtual bond conformations for the evaluation of the entropy and free energy of more complex structures (Zhang et al. 2008, 2009).

10.3.3 Loop–Stem Base Triple Interactions

In RNA pseudoknots, the close spatial proximity between loops and stems facilitates formation of the tertiary contacts between loop and stem nucleotides. Specifically, a nucleotide in the loop surrounding the major or minor groove is prone to form base triple interactions with a base pair in the helix stem (see Fig. 10.5a, b). Analysis of the known X-ray structures reveals a large number of noncanonical tertiary interactions in RNA molecules (Xin et al. 2008). These tertiary interactions include A-minor motifs first identified in ribosomal RNA (Nissen et al. 2001), the base triples seen by X-ray in the pseudoknots (Su et al. 1999), and the ribose zipper interactions originally found in group I introns (Cate et al. 1996). Noncanonical tertiary interactions can be critical for the stabilization of the structures

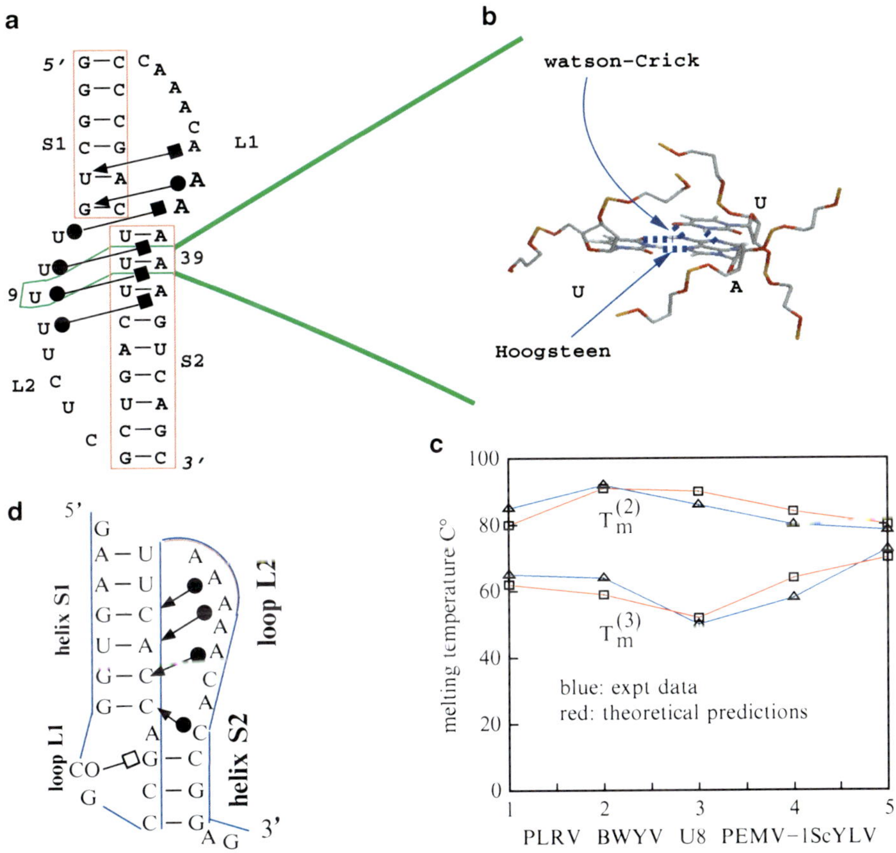

Fig. 10.5 (**a**) Human telomerase RNA (hTR) pseudoknot contains two base triples (*blue dashed lines*) between loop L2 and helix stem S1 and three base triples between loop L1 and helix stem S2 (Theimer et al. 2005). (**b**) The atomic configuration for the base triple between U9 and the U22-A39 base pair. (**c**) Theory–experiment test for the melting thermodynamics of five experimentally measured pseudoknot molecules: PLRV and PEMV-1 (Nixon et al. 2002), BWYV, and the U8 variant (Nixon and Giedroc 2000), ScYLV. The melting of these molecules shows two apparent transitions. The melting at the lower temperature T_{m}^{3} and the higher temperature T_{m}^{2} usually corresponds to the disruption of the loop–stem tertiary contacts and the secondary structure, respectively. In the theoretical predictions, the salt-dependent helix stability (in 1 M NaCl) is modified according to the experimental condition (0.5 KCl, pH 7.0) by using empirical formulas (SantaLucia 1998; Tan and Chen 2006). (**d**) The predicted (lowest free energy) 2D structure for the ScYLV pseudoknot agrees exactly with the NMR structure (Cornish et al. 2005, PDB: 1YG3). The *red lines* denote the loop–stem base triples

and RNA biological functions. For instance, mutations that disrupt the base triples can dramatically reduce the biological activity of telomerase (Theimer et al. 2005) and the efficiency of ribosomal frameshifting (Cornish et al. 2005; Kim et al. 1999; Su et al. 1999).

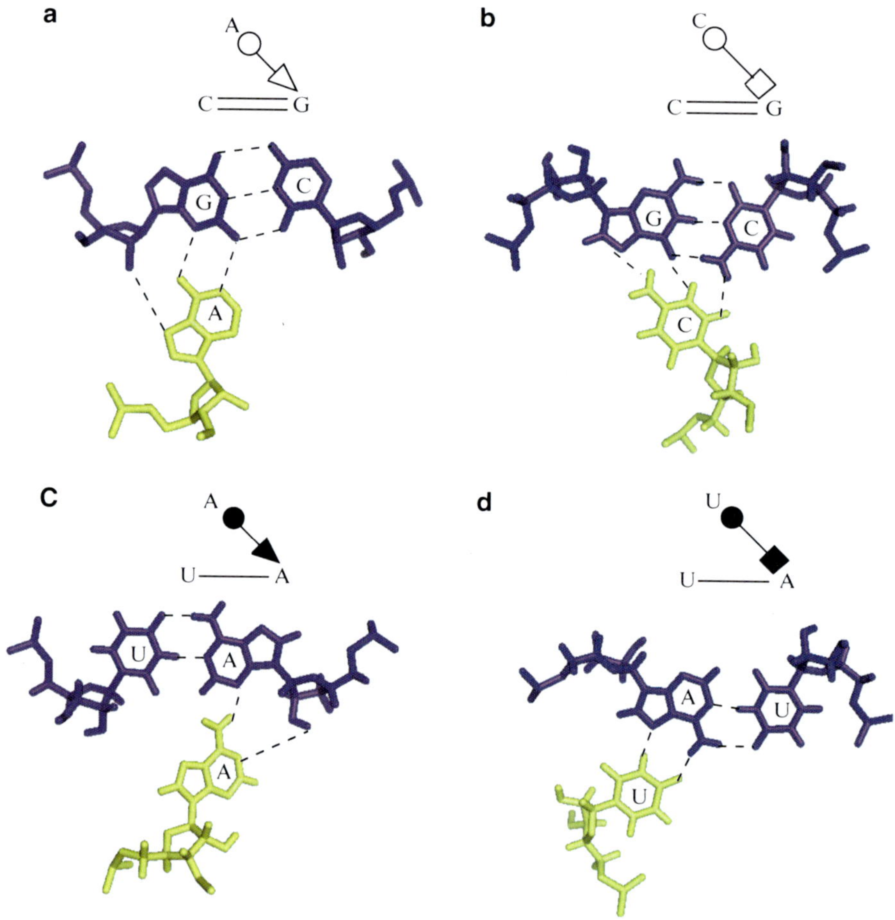

Fig. 10.6 Configurations of four base triples (a) A.(G–C), (b) C^+.(G–C), (c) A.(A–U), and (d) U.(A–U). They belong to *cis* W.W./*trans* S.W., *cis* W.W./*trans* H.W., *cis* W.W./*cis* S.W., and *cis* W.W./*cis* H.W. geometric families, respectively (Leontis and Westhof 2001; Almakarem et al. 2011)

There are eight different types of base triples in the known pseudoknot structures, (A.G–C), A.(C–G), C^+.(G–C), C^+.(C–G), U.(A–U), U.(U–A), A.(A–U), and A.(U–A). These base triples belong to different geometric families, defined by the families to which their component base pairs belong (Almakarem et al. 2011). Figure 10.6 shows four types of geometric families based on the nomenclature proposed by Leontis and Westhof (2001). For example, A.(G–C), C^+.(G–C), A.(A–U), and U.(A–U) belong to *cis* WW/*trans* SW, *cis* WW/*trans* HW, *cis* WW/*cis* SW, and *cis* WW/*cis* HW geometric families, respectively. According to the protonation properties, the eight base triples can be further classified into two types: the protonated [C^+.(G–C) and C^+.(C–G)] and the unprotonated base triples. The classification is based on the fact that the protonated base triples are more stable than the unprotonated base pairs (Cornish et al. 2005) due to strong electrostatic interactions.

Computational prediction of the noncanonical tertiary interactions is extremely difficult because of the lack of accurate energy parameters for tertiary interactions and the higher conformational complexity and the larger size of the conformational space (Das and Baker 2007; Ulyanov et al. 2007; Yingling and Shapiro 2006). Yingling and Shapiro predicted the base triples in the human telomerase RNA (hTR) pseudoknot using molecular dynamics simulations (Yingling and Shapiro 2006). Though the predicted structure is not exactly consistent with the NMR structure (Kim et al. 2008), the simulation was able to provide useful insights into the interplay between bulge formation and the base triple interactions in telomerase RNAs. Molecular dynamics simulation often relies on the initial input of the 3D structure. In contrast, the Vfold model, which uses the nucleotide sequence as the only input information, does not require any additional structural information.

The Vfold-based prediction of loop–stem tertiary interactions involves the following steps:

1. Through explicit enumeration of the virtual bond conformations, the Vfold model gives the loop entropy ΔS_{loop} as a function of loop length, helix length, and assignments of base triples (Fig. 10.5a, b). Based on the Vfold-predicted entropy parameter set, for a given structure that contains loop–stem tertiary contacts, the free energy ΔG can be evaluated as

$$\Delta G = \Delta G_{\text{helix}} - T\Delta S_{\text{loop}} + \sum_{i=1}^{n} (\Delta h^{(i)} - T\Delta s^{(i)}),$$

   where $\Delta h^{(i)}$ and $\Delta s^{(i)}$ are the enthalpy and entropy parameters for loop–stem contact i.
2. Theory–experiment comparisons for different systems with loop–stem contacts allow us to extract Δh and Δs parameters. As the lowest order approximation, we assume two sets of $(\Delta h, \Delta s)$ parameters for protonated and nonprotonated base triples, respectively. Fitting the melting curves for different pseudoknots that contain loop–stem base triples (Giedroc and Cornish 2009; Nixon et al. 2002; Nixon and Giedroc 2000) converged on the same set of $(\Delta h, \Delta s)$ parameters (Fig. 10.5c) below:

$$(\Delta h, \Delta s) = (-7\ \text{kcal/mol}, -19\ \text{cal/mol K})\ \text{for (A or U)}\\ \text{(A}-\text{U or U}-\text{A) and (A) (C}-\text{G or G}-\text{C)}, \quad (10.3)$$

$$\text{and} = (-14\ \text{kcal/mol},\ -38\ \text{cal/mol/K})(\text{pH } 7.0)\ \text{for C}^{+}(\text{C}-\text{G or G}-\text{C}). \quad (10.4)$$

The above Vfold approach has led to accurate predictions for the loop–stem tertiary contacts and thermodynamic stabilities for a series of experimentally determined pseudoknotted structures (Fig. 10.5c) (Cao et al. 2010).

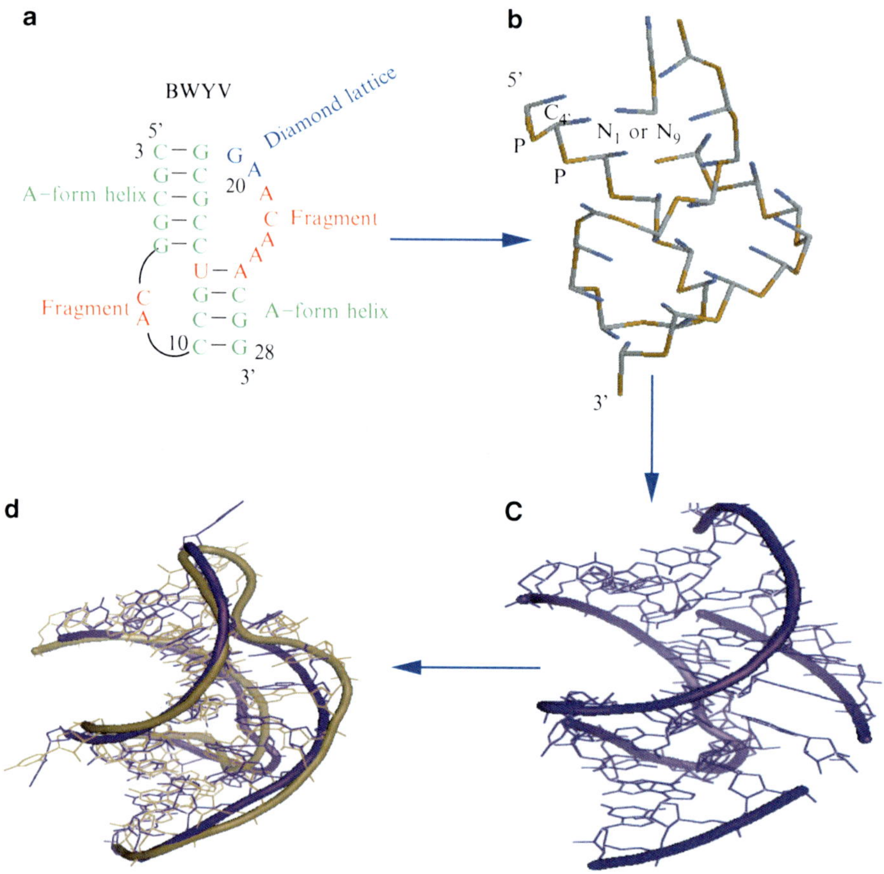

Fig. 10.7 (**a**) The Vfold-predicted Watson–Crick base pairs for the 2D structure of BWYV pseudoknot. (**b**) The predicted 3D low-resolution (virtual bond) structure built from the predicted 2D structure. (**c**) The all-atom 3D structure built from the virtual bond structure. (**d**) The all-atom structure refined by Amber 9. The figure is adopted from Cao et al. (2010). The root mean square deviation (RMSD) between the experimental structure (PDB code 473d, in *sand color*) and the predicted structure (in *purple-blue color*) is 2.7 Å over all heavy atoms

10.3.4 All-Atom 3D RNA Structures

Predicting RNA 3D structure is not a solved problem (Das and Baker 2007; Ding et al. 2008; Jonikas et al. 2009; Jossinet and Westhof 2005; Parisien and Major 2008; Shapiro et al. 2007; Tan et al. 2006). Currently, challenges include adequate treatment of the problem of conformational sampling (Das and Baker 2007) and the evaluation of the energetic parameters for tertiary contacts (Ding et al. 2008; Parisien and Major 2008). The Vfold model (Cao et al. 2010) can successfully predict the 2D structures of pseudoknots, including loop–stem tertiary interactions.

The Vfold model-predicted virtual bond structure provides a scaffold for the construction of all-atom models of the 3D structure. The prediction of the all-atom 3D structure from the Vfold-predicted 2D structures involves the following three steps (see Fig. 10.7):

1. Adding all atoms to the virtual bond structure. For nucleotides in each predicted helix, atoms are added according to the A-form helix atomic structure. The 3D conformations of the loop are generated from a combined fragment-based and diamond lattice-based method: The coordinates of the red nucleotides are adopted from the PEMV-1 fragment (PDB ID: 1KPX), and the remaining two nucleotides (blue) are generated by self-avoiding random walks on the diamond lattice. The method can effectively reduce the numbers of loop conformations to a few low-energy viable conformations. For nucleotides in the predicted loop conformations, atoms can be added using helix nucleotides as templates, by aligning the P, C_4, and $N_{1,9}$ atoms with those of a nucleotide in a helix. This step results in an "atomistic version" of the Vfold structure.
   The product of this initial refinement step is a prerefined atomic structure. The prerefined structure may contain some atoms/groups that clash sterically with each other. Such steric clashes can be readily resolved by the subsequent molecular dynamics simulation in the next step.
2. Energy minimization of the whole atomistic structure using AMBER molecular dynamics simulations. With the above prerefined structure as the initial state, molecular dynamics energy minimization with the AMBER molecular dynamics package (Case et al. 2005, 2006; Cornell et al. 1995; Pearlman et al. 1995) yield reliable predictions for all-atom 3D structures. In the energy minimization, the negative charges on phosphates are neutralized by Na^+ cations added to the solution. The nonbonded interactions are truncated at 12 Å. Water molecules are treated by the standard TIP3P model included in AMBER software.

As shown in Fig. 10.7, the above strategy gives reliable predictions for the all-atom 3D structures for simple tertiary folds such as pseudoknots.

10.4 Quantitative Prediction for RNA Function −1 Programmed Ribosomal Frameshifting

A −1 programmed ribosomal frameshift occurs when the reading frame is shifted by one nucleotide. The change of the reading frame causes the production of multiple proteins.

Ribosomal frameshifting is a mechanism used by many RNA viruses to regulate the relative number of copies of viral proteins (gag and pol). Maintaining a normal ratio of gag and gag–pol proteins is critical for viral replication. Alteration in frameshifting efficiency could lead to an abnormal ratio of gag and

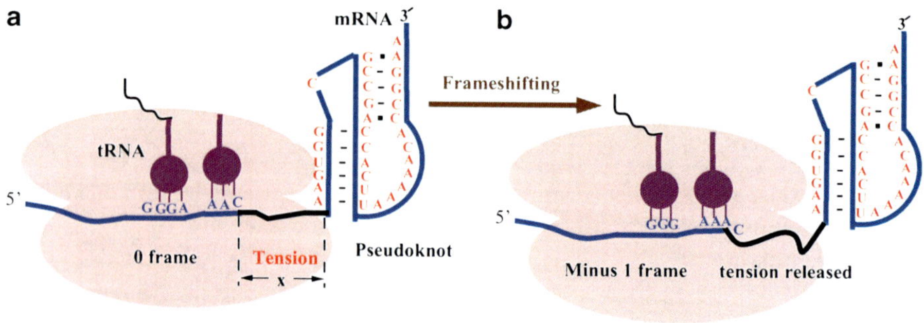

Fig. 10.8 A schematic diagram for the −1 ribosomal frameshifting induced by an RNA pseudoknot. The slippery sites of mRNA (*blue nucleotides*) is shifted in the $5' \to 3'$ direction by one nucleotide, which results in the reading frame to be shifted from the original (**a**) 0 frame to the new (**b**) −1 frame. The translated codons are (XXY,YYZ) in the 0 frame (**a**) and (XXX,YYY) (**b**) in the shifted −1 frame

gag–pol proteins, resulting in reduction or elimination of viral replication (Dinman et al. 1998).

The ribosomal frameshifting machinery consists of three coupled components (see Fig. 10.8a, b): (a) the slippery sites (nucleotides in blue color) with the sequence XXXYYYZ, where X can be any nucleotide, Y is A or U, and Z is A, U, or C; (b) a spacer (black thick line) that connects the $5'$ slippery sites and the $3'$ mRNA structure; and (c) the $3'$ mRNA structure which can fold into a pseudoknot structure (nucleotides in red color). According to the mechanical model for frameshifting (Hansen et al. 2007; Namy et al. 2006; Plant et al. 2003), the pseudoknot blocks the entrance of the mRNA into the ribosome and causes ribosomal pausing. During ribosomal pausing, the $3' \to 5'$ movement of the tRNA in the A/T to A/A aa-tRNA accommodation process can generate a tension force in the spacer (Plant et al. 2003; Yusupova et al. 2001). The tension, if sufficiently strong, could break the codon–anticodon pair of the 0 frame, causing a $5' \to 3'$ shift by one nucleotide of the mRNA chain and the subsequent shift of the codon–anticodon pair by one nucleotide.

The three components of the ribosomal frameshifting machinery are coupled through the spacer length m: The tension force in the spacer is a function of spacer length, which varies with the folding–unfolding of the downstream mRNA. The tension force in the spacer is also dependent on the end-to-end distance X of the spacer, which increases from 3.3 to 4.3 nm in the $3' \to 5'$ movement of the tRNA. If the downstream mRNA structure is robust against the large tension force, the spacer would more likely be subject to a large tension to induce frameshifting before unfolding of the downstream mRNA structure occurs. Otherwise, unfolding of the downstream mRNA structure would cause the relaxation of the tension, reducing the likelihood of frameshifting.

Quantitative prediction of ribosomal frameshifting requires modeling of the folding stability and the structure of the frameshifting machinery. The partition

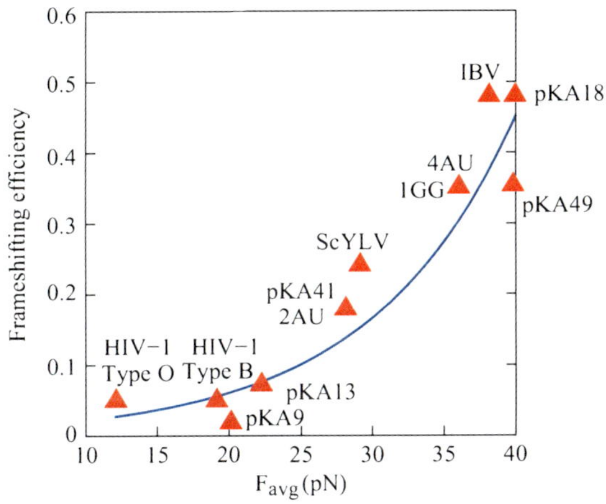

Fig. 10.9 The correlation between the frameshifting efficiency (η) and the mean force F_{avg} from $X = 3.3$–4.3 nm. We fitted a analytical expression (the blue thick line) for the correlation: $\eta \approx 0.0083e^{0.1F_{avg}}$. The experimental data of the frameshifting efficiency is adopted from references (Brierley et al. 1991; Napthine et al. 1999) for IBV and its mutant (pKA18, 1GG, 2UU, pKA41, 3 AU, 4 AU, pKA49, pKA13, and pKA9), reference (Dulude et al. 2002) for HIV-1 type B, reference (Baril et al. 2003) for HIV-1 type O, and reference (Cornish et al. 2005) for ScYLV

function for the 3-component frameshifting machinery is $Q(m,X) = Q_{ds}\,Q_{ss}\,Q_{codon}$. Here Q_{ds}, Q_{ss}, and Q_{codon} are the partition functions of the 3′ downstream mRNA structure ("structured" region, excluding the 5′ and 3′ tails), the (single-stranded) spacer of length m and end-to-end distance X, and the codon–anticodon base-pairing duplex (in the slippery region), respectively. While Q_{ss} and Q_{codon} can be evaluated using the extensible freely jointed chain model (EFJC) (Gerland et al. 2004; Hyeon and Thirumalai 2005; Liphardt et al. 2001; Smith et al. 1996; Strick et al. 2000) and the nearest-neighbor model (Serra and Turner 1995), respectively, the computation of Q_{ds} requires a statistical mechanical model such as the Vfold model. The tension force predicted from the partition function is given by $F(X) = d\Delta G(X)/\mathrm{d}X = -k_BTd\ \ln \sum_m Q(\mathrm{m},\ \mathrm{X})/\mathrm{dX}$.

The Vfold modeling for the system leads to an analytical relationship between the frameshifting efficiency and the mean tension force; see Fig. 10.9. It should be noted that recent single-molecule experimental data yields a highly similar relationship (Chen et al. 2009). Furthermore, because the experimental measurement (Chen et al. 2009) for human telomerase RNA and the Vfold-based theoretical predictions (Cao and Chen 2008) involve different frameshifting systems, the fact that similar analytical relationships are derived from independent theoretical and experimental studies suggests that the quantitative results for the frameshifting efficiency may be valid for a variety of systems.

10.5 Conclusions

For a long time, the bottleneck for RNA tertiary structural folding has been the inability to treat the free energy, especially the entropy, of structures with nonnested, long-range (tertiary) contacts between nucleotides distant in the 2D structure. Recent advances in the construction of low-resolution conformational models allow us to predict the entropy and the full free energy landscape for RNA tertiary global folds, as well as the 2D structures for the local and the global free energy minima. These 2D structures can further provide scaffolds for the construction of all-atom 3D models of the stable and metastable RNA folds through molecular dynamics calculations. Comparisons between theoretical predictions and experimental data for the 2D and 3D structures and the folding thermodynamics suggest that the statistical mechanical approach is reliable. One of the key factors that contribute to the predictive power of the statistical mechanical models is the rigorous conformational sampling/entropy.

With the rapidly growing size of the database of the experimentally measured RNA structures, fragment-based methods show promise, especially when the homologous conformations of modular components of the RNA of interest can be identified in the PDB database (Cao and Chen 2011; Das and Baker 2007; Parisien and Major 2008). However, compared to the number of the deposited protein structures in the PDB database, the number of known RNA structures remains relatively small. Therefore, fragment-based methods for RNA 3D prediction may fail if no known homologous conformations can be found in the PDB database. In that case, a de novo construction of the (low-resolution) conformations (especially for the junctions/loops regions) is the only viable approach. The Vfold model introduced in this chapter is one such model that can build the conformations de novo.

The current form of the Vfold theory can successfully treat pseudoknotted folds. However, the problem of predicting more complex pseudoknotted folds, such as the internal ribosome entry site (IRES) of the cricket paralysis-like viruses (Filbin and Kieft 2009) and other larger tertiary folds, remains. Further development of the model should go beyond the simple loop–stem base triples by including more complex tertiary interactions, such as tetraloop–receptor interactions and kissing loop interactions (Chitsaz et al. 2009; Huang et al. 2009).

Acknowledgment The authors thank Liang Liu for many useful discussions. This work was supported by the NSF through grants MCB0920411 and MCB0920067 and the NIH through grant R01-GM063732.

References

Aalberts DP, Hodas NO (2005) Asymmetry in RNA pseudoknots: observation and theory. Nucleic Acids Res 33:2210–2214

Alan C, Karakoc E, Nadeau JH, Sahinalp SC, Zhang KH (2006) RNA-RNA interaction prediction and antisense RNA target search. J Comput Biol 13:267–282

Almakarem AS, Petrov AI, Stombaugh J, Zirbel CL, Leontis NB (2011) Comprehensive survey and geometric classification of base triples in RNA structures. Nucleic Acids Res 40:1407–1423
Andronescu M, Zhang Z, Condon A (2005) Secondary structure prediction of interacting RNA molecules. J Mol Biol 345:987–1001
Andronescu MS, Pop C, Condon A (2010) Improved free energy parameters for RNA pseudoknotted secondary structure prediction. RNA 16:26–42
Arnott S, Hukins DWL (1972) Optimised parameters for RNA double-helices. Biochem Biophys Res Commun 48:1392–1399
Ban N, Nissen P, Hansen J, Moore PB, Steitz TA (2000) The complete atomic structure of the large ribosomal subunit at 2.4 Å resolution. Science 289:905–920
Baril M, Dulude D, Steinberg SV, Brakier-Gingras L (2003) The frameshift stimulatory signal of human immunodeficiency virus type 1 group O is a pseudoknot. J Mol Biol 331:571–583
Bartel DP (2009) MicroRNAs: target recognition and regulatory functions. Cell 2:215–233
Brierley I, Rolley NJ, Jenner AJ, Inglis SC (1991) Mutational analysis of the RNA pseudoknot component of a coronavirus ribosomal frameshifting signal. J Mol Biol 220:889–902
Brierley I, Pennell S, Gilbert RJC (2007) Viral RNA pseudoknots: versatile motifs in gene expression and replication. Nat Rev Micro 5:598–610
Brierley I, Gilbert RJC, Pennell S (2008) RNA pseudoknots and the regulation of protein synthesis. Biochem Soc Trans 36:684–689
Cao S, Chen S-J (2005) Predicting RNA folding thermodynamics with a reduced chain representation model. RNA 11:1884–1897
Cao S, Chen S-J (2006a) Free energy landscapes of RNA/RNA complexes: with applications to snRNA complexes in spliceosomes. J Mol Biol 357:292–312
Cao S, Chen S-J (2006b) Predicting RNA pseudoknot folding thermodynamics. Nucleic Acids Res 34:2634–2652
Cao S, Chen S-J (2008) Predicting ribosomal frameshifting efficiency. Phys Biol 5:016002
Cao S, Chen S-J (2009) Predicting structures and stabilities for H-type pseudoknots with interhelix loops. RNA 15:696–706
Cao S, Chen S-J (2011) Physics-based de Novo prediction of RNA 3D structures. J Phys Chem B 115:4216–4226
Cao S, Giedroc DP, Chen S-J (2010) Predicting loop-helix tertiary structural contacts in RNA pseudoknots. RNA 16:538–552
Case DA, Cheatham TE, Darden T, Gohlke H, Luo R, Merz KM, Onufriev A, Simmerling C, Wang B, Woods RJ (2005) The Amber biomolecular simulation programs. J Comput Chem 26:1668–1688
Case DA, Darden TA, Cheatham TE III, Simmerling J, Wang RE, Duke R, Luo KM (2006) AMBER 9. University of California, San Francisco
Cate JH, Gooding AR, Podell E, Zhou K, Golden BL, Kundrot CE, Cech TR, Doudna JA (1996) Crystal structure of a group I ribozyme domain: principles of RNA packing. Science 273:1678–1685
Chastain M, Tinoco I Jr (1991) Structural elements in RNA. Prog Nucleic Acid Res Mol Biol 41:131–177
Chauhan S, Caliskan G, Briber RM, Perez-Salas U, Rangan P, Thirumalai D, Woodson SA (2005) RNA tertiary interactions mediate native collapse of a bacterial group I ribozyme. J Mol Biol 353:1199–1209
Chen S-J (2008) RNA folding: Conformational statistics, folding kinetics, and ion electrostatics. Annu Rev Biophys 37:197–214
Chen S-J, Dill KA (2000) RNA folding energy landscapes. Proc Natl Acad Sci USA 97:646–651
Chen JL, Greider CW (2005) Functional analysis of the pseudoknot structure in human telomerase RNA. Proc Natl Acad Sci USA 102:8080–8085

Chen G, Chang K-Y, Chou M-Y, Bustamante C, Tinoco I Jr (2009) Triplex structures in an RNA pseudoknot enhance mechanical stability and increase efficiency of -1 ribosomal frameshifting. Proc Natl Acad Sci USA 106:12706–12711

Chitsaz H, Salari R, Sahinalp SC, Backofen R (2009) A partition function algorithm for interacting nucleic acid strands. Bioinformatics 25:i365–i373

Comolli LR, Smirnov I, Xu L, Blackburn EH, James TL (2002) A molecular switch underlies a human telomerase disease. Proc Natl Acad Sci USA 99:16998–17003

Cornell WD, Cieplak P, Bayly CI, Gould IR, Merz KM, Ferguson DM, Spellmeyer DC, Fox T, Caldwell JW, Kollman PA (1995) A 2nd generation force-field for the simulation of proteins, nucleic-acids, and organic-molecules. J Am Chem Soc 117:5179–5197

Cornish PV, Hennig M, Giedroc DP (2005) A loop 2 cytidine-stem 1 minor groove interaction as a positive determinant for pseudoknot-stimulated -1 ribosomal frameshifting. Proc Natl Acad Sci USA 102:12694–12699

Das R, Baker D (2007) Automated de novo prediction of native-like RNA tertiary structures. Proc Acad Natl Sci USA 104:14664–14669

Das R, Kudaravalli M, Jonikas M, Laederach A, Fong R, Schwans JP, Baker D, Piccirilli JA, Altman RB, Herschlag D (2007) Structural inference of native and partially folded RNA by high-throughput contact mapping. Proc Natl Acad Sci USA 105:4144–4149

Deigan KE, Li TW, Mathews DH, Weeks KM (2009) Accurate SHAPE-directed RNA structure determination. Proc Natl Acad Sci USA 106:97–102

Dimitrov RA, Zuker M (2004) Prediction of hybridization and melting for double-stranded nucleic acids. Biophys J 87:215–226

Ding Y (2006) Statistical and Bayesian approaches to RNA secondary structure prediction. RNA 12:323–331

Ding Y, Lawrence CE (2003) A statistical sampling algorithm for RNA secondary structure prediction. Nucleic Acids Res 31:7280–7301

Ding F, Sharma S, Chalasani P, Demidov VV, Broude NE, Dokholyan NV (2008) Ab initio RNA folding by discrete molecular dynamics: From structure prediction to folding mechanisms. RNA 14:1164–1173

Dinman JD, Ruiz-Echevarria MJ, Peltz SW (1998) Translating old drugs into new treatments: ribosomal frameshifting as a target for antiviral agents. Trends Biotechnol 16:190–196

Dirks RM, Pierce NA (2003) A partition function algorithm for nucleic acid secondary structure including pseudoknots. J Comput Chem 24:1664–1677

Dirks RM, Bois JS, Schaeffer JM, Winfree E, Pierce NA (2007) Thermodynamic analysis of interacting nucleic acid strands. SIAM Rev 49:65–88

Do CB, Woods DA, Batzoglou S (2006) CONTRAfold: RNA secondary structure prediction without physics-based models. Bioinformatics 22:e90–e98

Draper DE (1990) Pseudoknots and the control of protein synthesis. Curr Opin Cell Biol 2:1099–1103

Duarte CM, Pyle AM (1998) Stepping through an RNA structure: a novel approach to conformational analysis. J Mol Biol 284:1465–1478

Duarte CM, Wadley LM, Pyle AM (2003) RNA structure comparison, motif search and discovery using a reduced representation of RNA conformational space. Nucleic Acids Res 31:4755–4761

Dulude D, Baril M, Brakier-Gingras L (2002) Characterization of the frameshift stimulatory signal controlling a programmed -1 ribosomal frameshift in the human immunodeficiency virus type 1. Nucleic Acids Res 30:5094–5102

Ehresmann C, Baudin F, Mougel M, Romby P, Ebel JP, Ehresmann B (1987) Probing the structure of RNAs in solution. Nucleic Acids Res 15:9109–9128

Filbin ME, Kieft JS (2009) Toward a structural understanding of IRES RNA function. Curr Opin Struct Biol 19:267–276

Fisher ME (1966) Effect of excluded volume on phase transitions in biopolymers. J Chem Phys 45:1469–1473

mechanics, normal-mode analysis, molecular-dynamics and free-energy calculations to stimulate the structural and energetic properties of molecules. Comp Phys Commun 91:1–41
Petri V, Brenowitz M (1997) Quantitative nucleic acids footprinting: thermodynamic and kinetic approaches. Curr Opin Biotechnol 8:36–44
Peyret N, Seneviratne PA, Allawi HT, SantaLucia J Jr (1999) Nearest-neighbor thermodynamics and NMR of DNA sequences with internal AA, CC, GG, and TT mismatches. Biochemistry 38:3468–3477
Plant EP, Jacobs KL, Harger JW, Meskauskas A, Jacobs JL, Baxter JL, Petrov AN, Dinman JD (2003) The 9 Å solution: how mRNA pseudoknots promote efficient programmed -1 ribosomal frameshifting. RNA 9:168–174
Poland DC, Scheraga HA (1966) Occurrence of a phase transitions in nucleic acid models. J Chem Phys 45:1464–1469
Poland DC, Scheraga HA (1970) Theory of the helix-coil transition. Academic, New York
Qiao F, Cech TR (2008) Triple-helix structure in telomerase RNA contributes to catalysis. Nat Struct Mol Biol 15:634–640
Reeder J, Giegerich R (2004) Design, implementation and evaluation of a practical pseudoknot folding algorithm based on thermodynamics. BMC Bioinformatics 5:104
Rehmsmeier M, Steffen P, Höchsmann M, Giegerich R (2004) Fast and effective prediction of microRNA/target duplexes. RNA 10:1507–1517
Ren J, Rastegari B, Condon A, Hoos HH (2005) HotKnots: heuristic prediction of RNA secondary structure including pseudoknots. RNA 11:1494–1504
Richardson JS, Schneider B, Murray LW, Kapral GJ, Immormino RM, Headd JJ, Richardson DC, Ham D, Hershkovits E, Williams LD, Keating KS, Pyle AM, Micallef D, Westbrook J, Berman HM (2008) RNA backbone: Consensus all-angle conformers and modular string nomenclature (an RNA Ontology Consortium contribution). RNA 14:465–481
Rivas E, Eddy SR (1999) A dynamic programming algorithm for RNA structure prediction including pseudoknots. J Mol Biol 285:2053–2068
Ruan J, Stormo GD, Zhang W (2004) An iterated loop matching approach to the prediction of RNA secondary structures with pseudoknots. Bioinformatics 20:58–66
Russell R, Zhuang XW, Babcock HP, Millett IS, Doniach S, Chu S, Herschlag D (2002) Exploring the folding landscape of a structured RNA. Proc Natl Acad Sci USA 99:155–160
SantaLucia J Jr (1998) A unified view of polymer, dumbbell, and oligonucleotide DNA nearest-neighbor thermodynamics. Proc Natl Acad Sci USA 95:1460–1465
SantaLucia J, Turner DH (1997) Measuring the thermodynamics of RNA secondary structure formation. Biopolymers 44:309–319
SantaLucia J Jr, Saro P, Aduri R, Matta V (2004) Progress toward accurate 3D structure prediction of RNA. Abstr Paper Am Chem Soc Natl Meet 227:U912–U912
Serra MJ, Turner DH (1995) Predicting thermodynamic properties of RNA. Methods Enzymol 259:242–261
Shapiro BA, Wu JC (1997) Predicting RNA H-type pseudoknots with the massively parallel genetic algorithm. Comput Appl Biosci 13:459–471
Shapiro BA, Yingling YG, Kasprzak W, Bindewald E (2007) Bridging the gap in RNA structure prediction. Curr Opin Struct Biol 17:157–165
Shefer K, Brown Y, Gorkovoy V, Nussbaum T, Ulyanov NB, Tzfati Y (2007) A triple helix within a pseudoknot is a conserved and essential element of telomerase RNA
Smith SB, Cui Y, Bustamante C (1996) Overstretching B-DNA: the elastic response of individual double-stranded and single-stranded DNA molecules. Science 271:795–799
Sperschneider J, Datta A (2010) DotKnot: pseudoknot prediction using the probability dot plot under a refined energy model. Nucleic Acids Res 38:e103
Spitale RC, Torelli AT, Krucinska J, Bandarian V, Wedekind JE (2009) The structural basis for recognition of the preQ0 metabolite by an unusually small riboswitch aptamer domain. J Biol Chem 284:11012–11016

Staley JP, Guthrie C (1998) Mechanical devices of the spliceosome: motors, clocks, springs, and things. Cell 92:315–326

Staple DW, Butcher SE (2005) Pseudoknots: RNA structures with diverse functions. PLoS Biol 3:e213

Strick T, Allemand JF, Croquette V, Bensimon D (2000) Twisting and stretching single DNA molecules. Prog Biophys Mol Biol 74:115–140

Su L, Chen L, Egli M, Berger JM, Rich A (1999) Minor groove RNA triplex in the crystal structure of a ribosomal frameshifting viral pseudoknot. Nat Struct Biol 6:285–292

Tan ZJ, Chen S-J (2006) Nucleic acid helix stability: effects of salt concentration, cation valence and size, and chain length. Biophys J 90:1175–1190

Tan RKZ, Petrov AS, Harvey SC (2006) YUP: a molecular simulation program for coarse-grained and multiscaled models. J Chem Theor Comput 2:529–540

Theimer CA, Feigon J (2006) Structure and function of telomerase RNA. Curr Opin Struct Biol 16:307–318

Theimer CA, Blois CA, Feigon J (2005) Structure of the human telomerase RNA pseudoknot reveals conserved tertiary interactions essential for function. Mol Cell 17:671–682

Tyagi R, Mathews DH (2007) Predicting helical coaxial stacking in RNA multibranch loops. RNA 13:939–951

Ulyanov NB, Shefer K, James TL, Tzfati Y (2007) Pseudoknot structures with conserved base triples in telomerase RNAs of ciliates. Nucleic Acids Res 18:6150–6160

Waldsich C, Pyle AM (2008) A kinetic intermediate that regulates proper folding of a group II Intron RNA. J Mol Biol 375:572–580

Watts JM, Dang KK, Gorelick RJ, Leonard CW, Bess JW Jr, Swanstrom R, Burch CL, Weeks KM (2009) Architecture and secondary structure of an entire HIV-1 RNA genome. Nature 460:711–716

Williams AL Jr, Tinoco I Jr (1986) A dynamic programming algorithm for finding alternative RNA secondary structures. Nucleic Acids Res 14:299–315

Wimberly BT, Brodersen DE, Clemons WM Jr, Morgan-Warren RJ, Carter AP, Vonrheln C, Hartsch T, Ramakrishnan V (2000) Structure of the 30 S ribosomal subunit. Nature 407:327–339

Woodson SA (2000) Recent insights on RNA folding mechanisms from catalytic RNA. Cell Mol Life Sci 57:796–808

Xin Y, Laing C, Leontis NB, Schlick T (2008) Annotation of tertiary interactions in RNA structures reveals variations and correlations. RNA 14:2465–2477

Yingling YG, Shapiro BA (2006) The prediction of the wild-type telomerase RNA pseudoknot structure and the pivotal role of the bulge in its formation. J Mol Graph Mod 25:261–274

Yusupov MM, Yusupova GZ, Baucom A, Lieberman K, Earnest TN, Cate JHD, Noller HF (2001) Crystal structure of the ribosome at 5.5 Å resolution. Science 292:883–896

Yusupova GZ, Yusupov MM, Cate JH, Noller HF (2001) The path of messenger RNA through the ribosome. Cell 106:233–241

Zarrinkar PP, Williamson JR (1994) Kinetic intermediates in RNA folding. Science 265:918–924

Zhang J, Lin M, Chen R, Wang W, Liang J (2008) Discrete state model and accurate estimation of loop entropy of RNA secondary structures. J Chem Phys 128:125107

Zhang J, Dundas J, Lin M, Chen R, Wang W, Liang J (2009) Prediction of geometrically feasible 3D structures of pseudo-knotted RNA through free energy estimation. RNA 15:2248–2263

Zuker M (1989) On findings all suboptimal foldings of an RNA molecule. Science 244:48–52

Zuker M (2003) Mfold web server for nucleic acid folding and hybridization prediction. Nucleic Acids Res 31:3406–3415 including the role of modified bases, non-Watson–Crick base pairs, and the extreme electronegativity of RNA and its relation to interactions with water, ions, and proteins. The general discussion about preparation of large RNA–protein systems with divalent cations and modified bases is followed by a critical summary of methods for analyzing MD trajectories of RNA–protein complexes. We conclude with a short overview of coarse-grained simulation methods under development to simulate larger systems over longer times.

11.2 Evolutionary Analysis of Sequence and Structure

Highlighting the evolutionarily conserved features of RNA–protein systems focuses analysis on specific residues and structural elements that are important for folding, binding, catalytic activity, and intramolecular signaling. It also allows one to generalize results to homologous systems. Although it is widely known that sequence and structure conservation marks those RNA and protein regions that are functionally important, it remains a challenge to proceed from evolutionary information such as sequence and structure homology to an understanding of biomolecular function. Data from evolutionary analysis provide a steady stream of biological questions that are predominantly addressed through experiment, but this endeavor would be accelerated if computational techniques could provide more accurate predictions of function and interaction.

Evolutionary analysis is especially relevant for RNA–protein complexes as many appear to pre-date the last universal common ancestor and are widespread across the phylogenetic tree of life. Complexes in information processing are universal, and their phylogenetic variation follows for the most part the universal phylogenetic tree established by Woese based on 16 S rRNA (Woese 1987, 2000; Woese et al. 1990) (see Fig. 11.1a). Agreement with the universal phylogenetic tree for a particular molecule or group of molecules implies a canonical distribution across the organisms, and noncanonical distributions indicate possible lateral gene transfer in response to pressure to acquire additional functions. Canonical across all three domains of life are the rRNAs, elongation factors, and about half of the ribosomal proteins (r-proteins). The rest of the r-proteins are domain specific, but canonical within the particular domain. Since structure is more conserved than sequence, structure-based sequence alignments have been used extensively in the study of the translation machinery. For example, sequences of different specificities of aminoacyl-tRNA Synthetases (aaRS) have such low conservation that they cannot readily be aligned (Woese et al. 2000), but structure based alignments have been used to compare different catalytic domains of aaRSs and determine their phylogenetic divergence before the last universal common ancestor (Landes et al. 1995; O'Donoghue and Luthey-Schulten 2003). The evolutionary significance of RNA–protein interactions is especially clear when looking at "molecular signatures" of RNA and protein components of the ribosome ("rRNA" and "r-proteins").

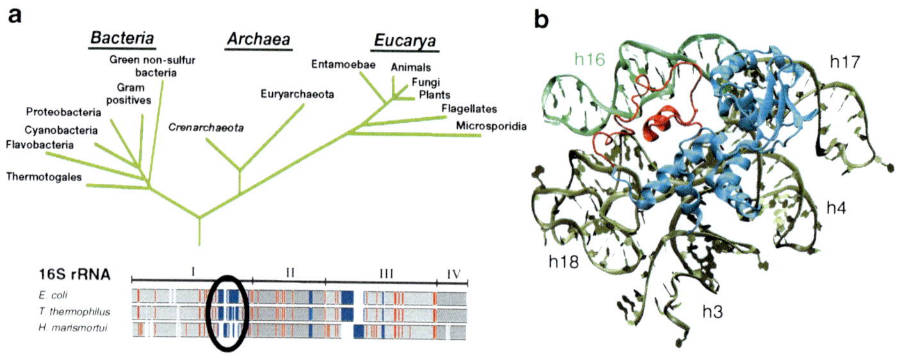

Fig. 11.1 Signatures in the ribosomal small subunit (**a**) The universal phylogenetic tree based on 16 S rRNA is shown above a plot of 16 S rRNA signatures. Sequence signatures are colored *red* and structure signatures are colored *blue*. The S4 binding site is *circled*. (**b**) Ribosomal protein S4 is shown with its binding site to the five-way junction in the 16 S rRNA (PDB ID 2I2P). The structural signatures are colored *red* in S4 and *green* in h16 (Chen et al. 2009)

11.2.1 Molecular Signatures in rRNA and r-Proteins

Ribosomal molecular signatures are idiosyncrasies in the ribosomal RNA (rRNA) and/or r-proteins characteristic of the individual domains of life. As such, insight into the early evolution of the domains can be gained from a comparative analysis of their respective signatures in the translational apparatus. Signatures in both the sequence and structure of the rRNAs contribute roughly 50% of the differences present in the universal phylogenetic tree providing a "bar code of life" that determines to which domain a given rRNA sequence belongs (Winker and Woese 1991; Roberts et al. 2008). It has been proposed that the observed ribosomal signatures are remnants of an evolutionary phase transition that occurred as the cell lineages began to coalesce, implying that they should also be reflected in corresponding signatures throughout the fabric of the cell and its genome. The presence of domain-specific r-proteins can be considered signatures in their own right, and correlations between the signatures of rRNAs and r-proteins show that the rRNA signatures coevolved with both domain-specific r-proteins and inserts in universal r-proteins. The question remains: what roles do these highly conserved elements play in the assembly and function of the ribosome?

The largest rRNA structural signature interacting with a protein partner appears in the binding site of the universal r-protein S4 on the bacterial ribosome. Both r-protein S4 and h16 in the 16 S rRNA contain bacterial signatures (see Fig. 11.1b), and these signatures interact with one another suggesting that they coevolved within early bacteria. S4 is known to be critical to the early assembly of the SSU (Held et al. 1974; Mulder et al. 2010), and recent work has been done to characterize its interactions with rRNA and to explain the role S4·h16 plays in the initial steps of rRNA folding (Bellur and Woodson 2009; Chen et al. 2010). Further experiments and simulations are required to determine what functional roles the other ribosomal signatures play in protein synthesis.

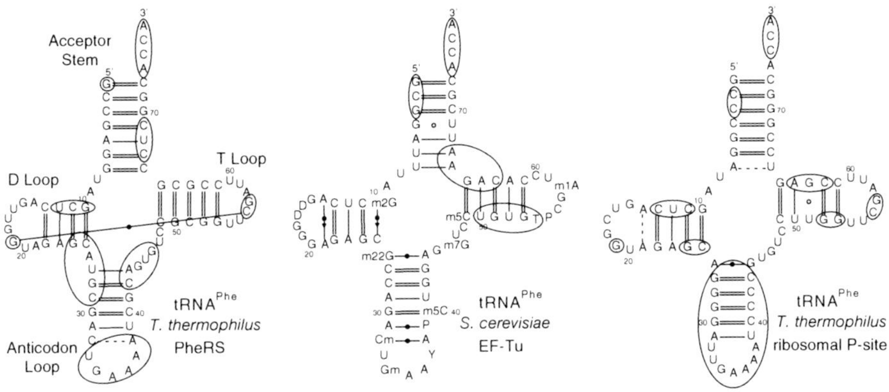

Fig. 11.2 Patterns of contacts in tRNA: nucleotides in bacterial tRNA^{Phe} that are within 4.5 A of PheRS $(\alpha\beta)_2$, EF-Tu, or r-proteins or rRNA at the ribosomal P-site. The contacts are determined from the PDB files 2IY5, 1OB5, and 2WRN, respectively. PheRS has an unusual tetrameric structure, and its contacts to the tRNA D and T loops are not representative of the class II aaRSs. The Leontis–Westhof classification of base pairs (Leontis and Westhof 2001) is used for the base pairs shown which were identified using RNAVIEW (Yang et al. 2003)

11.2.2 Binding Patterns for tRNA

As tRNA migrates from one complex to the next its binding partners use different modes of interaction. aaRSs and mRNA discriminate between tRNAs based on specificity or isoacceptor while EF-Tu and the ribosome must interact with all tRNAs. To bind these different molecules, each tRNA has evolved elements associated with particular specificities (Eigen and Winkler-Oswatitsch 1981) as well as features that are universal to all tRNAs. Based on crystal structures containing tRNA^{Phe}, Fig. 11.2 shows the molecular interactions that tRNA^{Phe} makes with PheRS, EF-Tu, and the ribosomal P-site. The dynamic variations in tRNA structure observed both experimentally and computationally are similar to the different tRNA conformations seen in various crystal structures (Alexander et al. 2010).

Using sequence and structure data available in online data repositories, it is possible to construct evolutionary profiles for the proteins and RNAs. An evolutionary profile is an alignment built from a nonredundant set of sequences in order to represent sequence diversity found throughout the tree of life while minimizing bias present in the databases (O'Donoghue and Luthey-Schulten 2005; Sethi et al. 2005). Bias in sequence sets occurs because certain groups of organisms are overrepresented; for example, many human pathogens come from the class *γ-proteobacteria* so their genomes are more likely to be sequenced. Panels a and b from Fig. 11.3 show the sequence identity of EF-Tu across evolutionary profiles for all three domains of life and then for bacteria (Eargle et al. 2008). Coloring the EF-Tu structure by conservation makes the tRNA-binding interface and GTP-binding pocket readily apparent. Profiles can also be taken from databases such as Rfam (Griffiths-Jones et al. 2003) and Pfam (Bateman et al. 2002), but the sequence sets may not be statistically well balanced (Sethi et al. 2005).

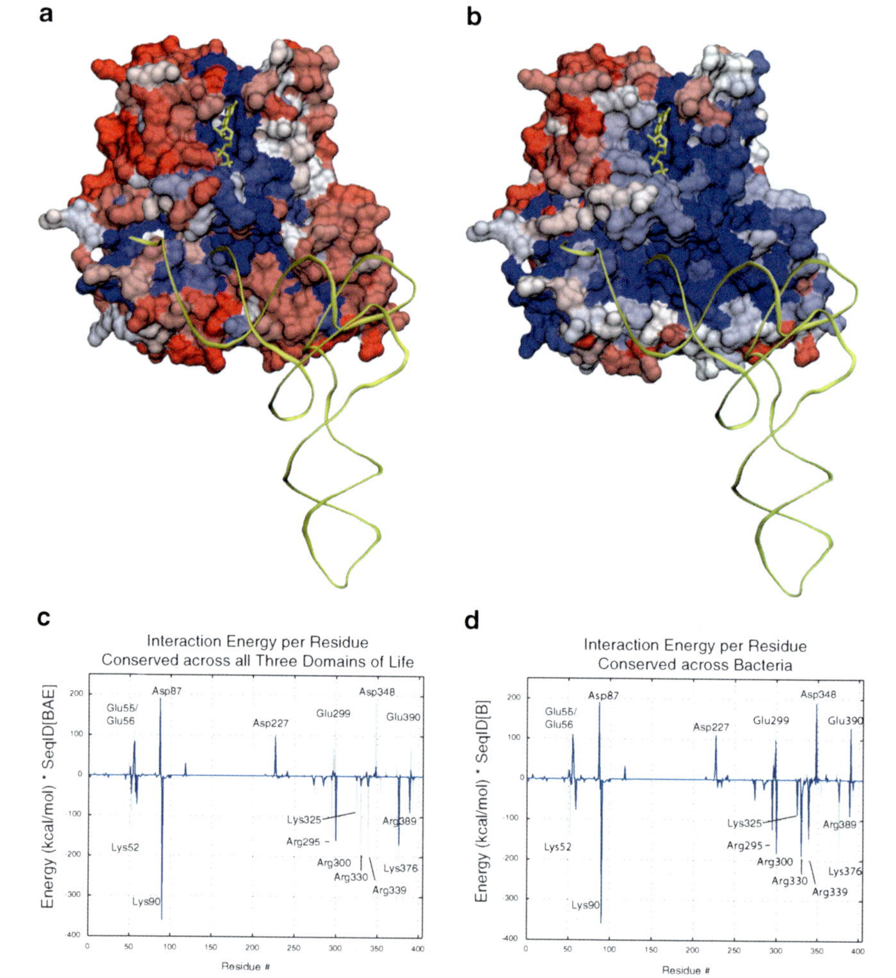

Fig. 11.3 Combining MD simulations and conservation: EF-Tu (PDB ID 1B23) colored by sequence identity across (**a**) all three domains of life and (**b**) bacteria only. *Red* is least conserved, *white* is 50% sequence identity, and *blue* is completely conserved. A GTP analogue and the tRNA backbone are colored *yellow*. EF-Tu nonbonded interaction energy per residue with $tRNA^{Phe}$ is shown in *light gray* in panels (**c**) and (**d**). In *blue*, the energies have been scaled by sequence identity across (**c**) all three domains of life and (**d**) bacteria (Eargle et al. 2008)

11.2.3 Practical Challenges in RNA Comparative Analysis

Current work with multiple sequence alignments is hampered by the difficulty of combining heterologous data types. Annotating sequence data with information about structure, folding, interaction interfaces (Fig. 11.3c, d), and other experimental results would ease the creation, maintenance, and interpretation of alignments. As suggested by the RNA Ontology Consortium[1] (Hoehndorf et al. 2011), these sorts of correspondence relations between RNA sequences and their attributes should be incorporated into new file formats and software applications (Brown et al. 2009). To aid the description and analysis of RNA–protein complexes, these same ideas should be extended to protein alignments as well. In addition, more freely available tutorials (Roberts et al. 2006; Li et al. 2009; Chen et al. 2011) are needed to disseminate bioinformatic data-handling techniques and analysis methods.

11.3 Molecular Dynamics Simulations of RNA–Protein Complexes

Even though the computational power of existing machines as well as improvements in methodology now allow all-atom MD simulations with explicit solvent to reach tens of microseconds for the folding of small proteins, probing of characteristics of RNA–protein complexes at atomic resolution is still limited to timescales of hundreds of nanoseconds. Bacterial protein synthesis proceeds at approximately 20 amino acids per second which makes it out of the reach of current all-atom simulations, but lower level processes that occur in hundreds of nanoseconds in the translation pathway include molecular recognition and binding (Eargle et al. 2008), the onset of tRNA migration (Black Pyrkosz et al. 2010), the formation of interaction networks in RNA–protein complexes (Sethi et al. 2009), and tRNA accommodation into the ribosomal binding sites (Sanbonmatsu et al. 2005; Trabuco et al. 2010). The ultimate goal is to understand how thermal fluctuations and motion at the molecular scale are rectified to produce directed motion of tRNAs and mRNA during protein synthesis, which, in turn, is influenced by hydrolysis of GTP, the interactions with the ions, and the presence or absence of modified bases.

Long simulations of protein folding have helped to probe existing force fields (Freddolino et al. 2009; Klepeis et al. 2009), revealing their weaknesses so that further improvements can be made. Advances in hardware and software now allow simulations extending to hundreds of nanoseconds for RNA systems (Garcia and Paschek 2008; Beššeová et al. 2010; Kirmizialtin and Elber 2010). Long-time simulations of isolated tRNA have shown that it experiences structural configurations similar to those found in tRNA on the ribosome (Li and Frank 2007). Due to their computationally intensive nature, it is challenging to set up and run MD simulations of large systems. Every step in the preparation takes proportionally greater effort, and the resulting data trajectories become unwieldy, requiring analysis using large computing clusters.

Many RNA–protein structures have been crystallized in buffers containing concentrations of salt and/or polyamines that are far above physiological levels. Also, RNA transcripts are frequently used because it is more difficult to obtain

[1] http://code.google.com/p/rnao/.

RNA with its full complement of modified nucleosides. Many RNA molecules are posttranscriptionally modified, and these modifications can modulate their resulting structure and dynamics.

11.3.1 Influence of Modified Nucleosides on RNA Dynamics

More than 100 different nucleoside modifications have been identified[2] (Limbach et al. 1994; Dunin-Horkawicz et al. 2006) with the majority appearing in various tRNAs and ribosomes. Modified nucleosides are frequently ignored because RNA samples prepared through in vitro transcription or expression in a different organism from the original host will not contain the wild-type modifications. However, many of these modifications are evolutionarily conserved and are interesting in their own right. Another issue related to the plethora of natural modified bases concerns their representation. Attempts to standardize the nomenclature and description of interaction patterns are being made through the RNA Ontology Consortium (Hoehndorf et al. 2011).

Modified bases can affect the structure, dynamics, and possible interaction partners of RNA molecules. For example, tRNA modifications are used by aaRSs to discriminate tRNAs of different specificities (Giege et al. 1998). In the tRNA anticodon, base modifications are frequently used in the wobble position to ensure correct interaction with cognate mRNA codons. Reviews of tRNA structure and dynamics (Helm 2006; Alexander et al. 2010) have covered how modified bases can change the secondary structure of tRNA by blocking the Watson–Crick edge and directing base pairing to form nonWatson–Crick base pairs involving the Hoogsteen or Sugar edges, as in the N1-methylation of the conserved adenosine in the T–Ψ–C loop. Nucleoside modifications in tRNA are shown in the secondary structure representation for the EF-Tu bound $tRNA^{Phe}$ in Fig. 11.2 and on the three-dimensional structure for $tRNA^{Cys}$ in Fig. 11.4.

Modified nucleosides affect RNA structural stability in various ways, depending on the surrounding nucleic acid sequence and structure, and no simple generalizations can be made about their roles in RNA function. They can both stabilize and destabilize local structures through interaction with nearby bases, water, or ions as exemplified by dihydrouridine (D) and pseudouridine (Ψ), two common modifications found in tRNA and rRNA. Dihydrouridine, which occurs in the D loop of tRNAs, is generated by fully saturating the C5–C6 double bond of uridine, by addition of two hydrogens. This nonplanar base is more flexible than uracil and does not stack well on other bases. Therefore it prefers to remain single stranded and unpaired. Pseudouridine is a C-glycoside isomer of uridine created by removing the uracil base and reattaching it at C4. It can stabilize local RNA structure by interacting with backbone phosphates through a high residency

[2] http://rna-mdb.cas.albany.edu/RNAmods.

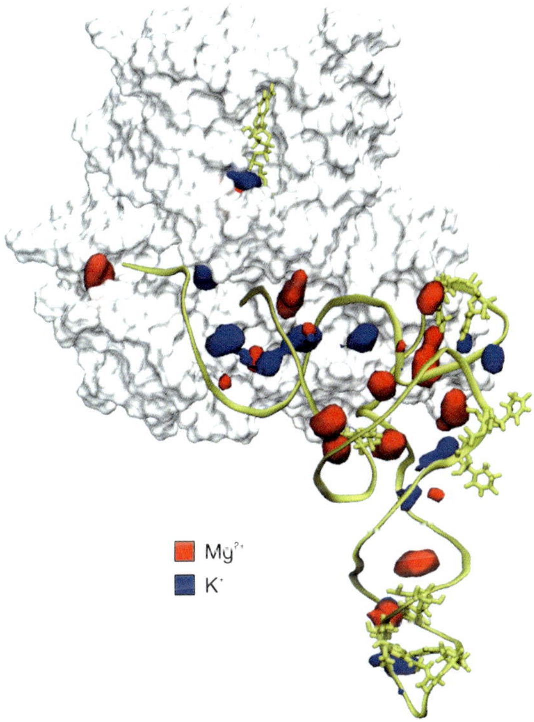

Fig. 11.4 Cation binding within tRNA: 5% occupancy across a 16-ns trajectory is shown for Mg^{2+} and K^+ which were present at 38 mM and 117 mM, respectively. Trajectory frames were aligned by the tRNA backbone atoms. Modified bases are shown in licorice representation (Eargle et al. 2008)

bridging water molecule (Auffinger and Westhof 1997; Charette and Gray 2000). Hypermodified bases have been shown by both NMR experiments and computation to be important for maintaining the anticodon loop conformation in $tRNA^{Phe}$ and $tRNA^{Cys}$ by disrupting base pairs that could occur between unmodified A37 and A38 with U32 and U33 (Cabello-Villegas et al. 2002; Eargle et al. 2008).

Incorporating modified nucleosides into MD simulations requires force field parameters for the specific modifications. While most force fields have parameters for the standard DNA and RNA nucleotides (Pranata et al. 1991; Foloppe and MacKerrell 2000; Oostenbrink et al. 2004), only **AMBER** has parameters for most of the known modified nucleotides (Wang et al. 2004). Otherwise, a literature search can reveal nucleoside parameters developed by various laboratories, or one can parametrize them through quantum chemistry calculations or by analogy with molecules already present in the force field.

11.3.2 RNA Interaction with Water and Ions

Energy landscapes of RNA folding are extremely sensitive to solvent and ions (Russell et al. 2002; Lipfert et al. 2010). Water and cations such as K^+ and Mg^{2+} provide the electrostatic screening necessary for the negatively charged phosphate backbone to condense and allow secondary and tertiary interactions to form (Draper 2008; MacKerell and Nilsson 2008). Higher water densities around RNA relative to protein also contribute to RNA flexibility (Roh et al. 2009). The classical force fields underestimate the interactions between hexahydrated Mg^{2+} and RNA (Ditzler et al. 2010), but the fact remains that physiological Mg^{2+} plays an enormous role in RNA structure and dynamics.

In mammalian cells, cytosolic concentrations of K^{+}, Na^{+}, and free Mg^{2+} are around 140, 10, and 1 mM, respectively. However, if all Mg^{2+} associated with nucleic acids is considered, the total concentration of Mg^{2+} within cells is closer to 30 mM (Cowan 1995). Magnesium ions bound to nucleoside triphosphates account for much of the difference, but local Mg^{2+} concentration is higher around DNA and large RNA molecules as well. The high charge density of Mg^{2+} results in very stable solvation shells. The exchange rate for water molecules in the first solvation shell of Mg^{2+} occurs at greater than μs timescales (Ohtaki 2001) so Mg^{2+} ions tend to remain with the water or RNA molecules they are initially bound to in MD simulations (Auffinger and Vaiana 2005). Due to its 12 potential hydrogen bond donors and relatively large radius, hexahydrated Mg^{2+} diffuses more slowly and has higher residency times around RNA than either K^{+} or Na^{+}. Since Mg^{2+} is important for RNA structure and dynamics in vivo, it is frequently included in RNA simulations, but it must be handled with care, especially during the system preparation.

There are many open questions about the nature of Mg^{2+} bound to nucleic acids. For a given molecule, how many Mg^{2+} ions are directly bound, and where are their binding sites? How many are diffusely bound, hexahydrated but still associated with the nucleic acid through the second or third solvation shell? The small radius of Mg^{2+} makes its determination difficult for X-ray crystallography, but sometimes the presence of the distinct octahedral arrangement of atoms in the first solvation layer lends support to assignment of Mg^{2+} to the central peak of the solvent electron density. Recent work has been done to elucidate the thermodynamics of Mg^{2+} binding (Grilley et al. 2006; Leipply and Draper 2010) and describe the ionic clouds around DNA in various buffers (Bai et al. 2007).

11.3.3 Challenges in the Set-Up of All-Atom Simulations

As there have been recent detailed protocols written for the set-up of RNA MD simulations (Hashem and Auffinger 2009), here we focus on issues particularly relevant to simulations of large, RNA–protein complexes. With RNA being such a highly charged molecule, both hydration and ion placement must be addressed in the system set-up to avoid a lengthy phase of constrained equilibration. Here we focus on the software that we use in our own studies, but alternative applications exist for many of the tasks required for system preparation. Further challenges arise due to the limited availability of computational tools and scripts for analysis of long MD simulations of large nucleic acids complexes, especially those containing modified bases. For example, while VMD (Humphrey et al. 1996) has been developed to visualize large assemblies containing millions of atoms, analysis of changes in RNA secondary structure and motion of RNA helices requires user-supplied scripts.

11.3.3.1 Preparing the Structure

For simulations based on existing structures, system preparation begins with a PDB file containing the 3D atomic coordinates for the structures in the system of interest. Typically, well-resolved water molecules and physiologically relevant ions should be kept, but unphysiological species such as sulfate or ammonium that are only present for purposes of crystallization can be removed. Hydration and placement of ions is described in more detail below. Frequently, mobile tails or loops are missing from crystal structures and must be modeled, perhaps by making use of existing homologous structures. To trap enzymes in intermediate states, sometimes structures contain substrate analogs that should be converted to the true substrate or removed. For ribosomal systems with available cryo-electron microscopy maps, models for further simulation have been constructed using MD flexible fitting (MDFF) whereby atomic structures are fitted to EM density maps while constraining secondary structure elements (Trabuco et al. 2008; Becker et al. 2009; Armache et al. 2010; Trabuco et al. 2010).

Assignment of the protonation states to titratable amino acids like histidine and aspartate is sensitive to neighboring residues and solvent accessibility. Local pK_a and protonation states for titratable residues and ligands are predicted by **PROPKA** 2.0 based on solvent exposure, potential hydrogen-bonding partners, and proximity to charged groups (Bas et al. 2008). The standard amino acids are included, but extra work is required to incorporate effects of nucleic acids or ligands. The latest version allows incorporation of generic chemical groups and their associated pK_a values as well as user knowledge for specific groups. If the bulk pK_a is known for a ligand, the user can provide this information before **PROPKA** is run. For example, the phosphate on AMP bound in the active site of GluRS can initially be assigned a bulk solution pK_a of 6.9 which is then modified through the iterative local pK_a calculation (Black Pyrkosz et al. 2010). Adenosine and cytidine may be protonated within RNA structures, but currently there appears to be no simple way to calculate protonation states of nucleic acids. A thorough but costly approach is to calculate local pK_a values using Poisson–Boltzmann solvers (Tang et al. 2007). It is good practice to visually inspect protonation assignments. Once protonation states have been determined, hydrogens and partial charges are added to the molecules by programs such as **psfgen**.

11.3.3.2 Adding Ions

One of the main problems that nucleic acids pose to MD set-up is the requirement for associated cations.

Without compensating positive charges, especially within the deep groove and other compact regions where phosphates are near each other, electrostatic repulsion will quickly distort and unfold the RNA. RNA simply placed in a water box with randomly distributed, neutralizing ions will denature without a costly constrained equilibration allowing ions to diffuse into the RNA structure. Cations added to the solvent, typically Mg^{2+} and either K^+ or Na^+, shield the negative charges of the sugar–phosphate backbone to allow the RNA to maintain its double-helical form during simulation. It has been shown that ions around DNA can take on the order of tens of nanoseconds to equilibrate (Ponomarev et al. 2004). Therefore, it is important to place ions as close to equilibrium states as possible when preparing the simulation.

Ions can be placed with tools like the program **ionize**.[3] Using the previously assigned atomic partial charges, **ionize** creates a three-dimensional lattice around the system and calculates Coulombic interaction energies for the placement of a charge at each lattice site. An ion is placed at the minimum energy site, and then lattice energies are regenerated for the next ion. This process is repeated until all ions have been added to the system. To make this problem tractable for large RNA–protein systems such as the ribosome, a parallelized version of this ion placement algorithm has been developed (Stone et al. 2007).

More rigorous but more time-consuming methods for ion placement include the use of Brownian dynamics (Hermann and Westhof 1998; Serra et al. 2002) or MD simulations (Kirmizialtin and Elber 2010) in which the RNA is fixed, and the ions are allowed to diffuse about the system. High-resolution X-ray structures were used to validate placement of Mg^{2+} in the coulomb lattice (Eargle et al. 2008), Brownian dynamics, and MD simulation methods. After placement of neutralizing ions, extra salt buffer can be added to achieve higher K^+ concentrations. As crystallization conditions often use higher concentrations for many cations than is seen in vivo, one should only select a subset of crystal ions corresponding to the lowest energy sites, when the goal is to simulate systems under physiologically relevant conditions.

Mg^{2+} ions diffuse slowly around RNA so their initial placement and hydration is especially important for the set-up of a stable system. Information from well-resolved crystal structures can be used to place Mg^{2+} ions with varying degrees of hydration in direct contact with the RNA. Closely placed Mg^{2+} ions, such as those present in the crystal structure, must be treated carefully. Since the first solvation shell of Mg^{2+} is so stable, it is common for initial contacts made with Mg^{2+} to remain for an entire MD simulation. If the six members of an Mg^{2+} atom's first solvation shell are not set during system set-up, the missing members will be pulled in from that atom's local environment and may result in spurious, long-lived contacts to the RNA. To prevent this, all Mg^{2+} ions should be fully solvated before production runs.

[3] http://www.ks.uiuc.edu/Development/MDTools/ionize/.

11.3.3.3 Hydrating the System

One needs to take into account the difference in water density of the solvent surrounding RNA and proteins (Roh et al. 2009) when solvating systems containing RNA–protein complexes. If there are cavities within the system, **DOWSER** (Zhang and Hermans 1996) can be used for the first round of solvation to ensure that these cavities are filled with water molecules. Additional atom dictionary files are needed for **DOWSER** to recognize the nucleotides in RNA, and these files are available through the **DOWSER** plugin to **VMD** (Gumbart et al. 2009). The first few layers of external solvent can be added by **Solvate 1.0**,[4] which uses a PDB file containing partial charge information to place and orient water molecules next to RNA and protein. These waters may then be verified for accuracy with **SwS** (Auffinger and Hashem 2007). Finally, equilibrated waters are added to fill the three-dimensional box used to carry out MD simulations with periodic boundary conditions.

11.3.4 Visualization and Analysis of Motions and Energetics

Once MD trajectories have been generated, many different analysis methods are available to characterize the dynamics. The majority of available analysis programs were written specifically for proteins, fewer have been written for nucleic acids, and even fewer were built with RNA–protein complexes in mind. While standalone applications exist to perform specific calculations on trajectory data, more flexible environments are useful for exploratory analyses and the creation of new methods. **VMD** (Humphrey et al. 1996), **Chimera** (Pettersen et al. 2004), and **PyMOL** (Schrödinger) are all equipped with programming language interpreters that allow access to structure and trajectory data so that new analysis scripts can be quickly written and tested.

11.3.4.1 Structural Dynamics

Standard analyses for RNA and protein molecules include root mean square deviation (RMSD) and root mean square fluctuation (RMSF) of atoms in the backbone or in residue side chains. The RMSD per residue for r-protein S4 is shown in Fig. 11.5. High RMSD in the signature region indicates that it is relatively disordered when S4 is not bound to rRNA. For RNA in particular, base pairing type and geometry can be calculated from single trajectory frames by programs like **RNAView** (Yang et al. 2003) and **3DNA** (Lu and Olson 2003). Because of relative motions between large substructures in RNA, it can be difficult to obtain good superpositions, and backbone RMSD tends to be less meaningful than for proteins (Alexander et al. 2010; Beššeová

[4] http://www.mpibpc.mpg.de/home/grubmueller/downloads/solvate/index.html.

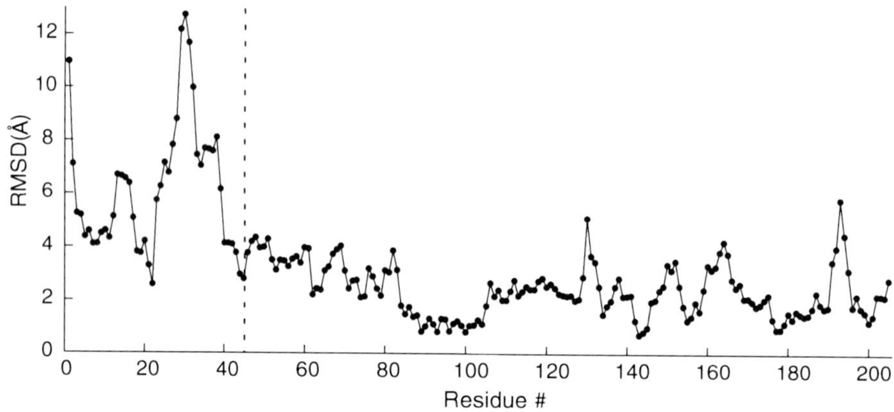

Fig. 11.5 Intrinsic disorder in bacterial structural signature of ribosomal protein S4: the first 45 amino acids correspond to the structural signature in S4, which is intrinsically disordered when not bound to h16 (Chen et al. 2010)

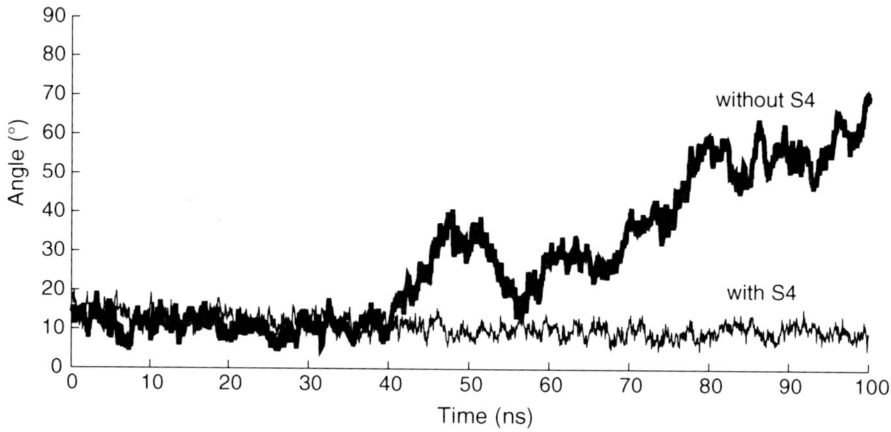

Fig. 11.6 rRNA interhelix angle: the angle between h16 and h18 isolated (*thick line*) and bound to S4 (*thin line*). The angles were obtained from the smallest principal axis of inertia calculated separately for each helix

et al. 2010). More useful analysis can be done at the level of individual helices. Description of the motions of larger RNA features, such as interarm angles, are routinely obtained through the use of programs like **CURVES+** (Lavery et al. 2009), by tracking the smallest principal axis of inertia for RNA helices (Trabuco et al. 2010), or by following specific atoms participating in the relevant motions (Beššeová et al. 2010). Figure 11.6 compares the interarm angle between rRNA helices h16 and h18 calculated for two 100-ns trajectories: one with the rRNA five-way junction free in solution and the other with S4 bound to the rRNA. Similar results can be obtained through scripts with calls to **VMD**'s "measure inertia" procedure.

11.3.4.2 Interaction Energies and Free Energy Landscapes

At RNA–protein-binding interfaces, the nonbonded interaction energy per residue is a good measure of a residue's importance for binding. Figure 11.3c, d show plots for this type of data averaged across a short (16 ns) MD trajectory. The gray lines represent the underlying energy data while the blue show that data scaled by percent sequence identity across all three domains of life and then across only bacteria. It is clear that most of the significant interactions at the EF-Tu·tRNA interface are conserved across bacteria, and about half are conserved across all of life. Dynamics are important in this type of analysis because individual contacts such as hydrogen bonds and salt bridges may regularly break and reform so that any single structure, such as the original crystal structure or one from a trajectory, does not have the necessary statistics to describe the average interaction energies.

Binding free energies can be generated from MD trajectories through methods such as molecular mechanics-Poisson Boltzmann surface area (MM-PBSA) (Froloff et al. 1997; Kollman et al. 2000; Rocchia et al. 2001; Pogorelov et al. 2007), which has already been used to calculate free energies for RNA–protein systems (Reyes and Kollman 2000; Yamasaki et al. 2007; Eargle et al. 2008; Black Pyrkosz et al. 2010) and systems containing Mg^{2+} ions with high residency times (Gohlke et al. 2003). At the moment, MM-PBSA requires pooling information gleaned from many different applications such as Poisson–Boltzmann solvers (Baker et al. 2001; Rocchia et al. 2002), MD programs, and entropy estimations, but long-time simulations of large systems can push some of these programs beyond their limits and resolution capabilities. Another drawback is that MM-PBSA is used primarily to predict trends in binding free energy across several different simulations, but the actual ΔG values obtained are not accurate. Other methods for obtaining free energy landscapes or reaction kinetics such as metadynamics, transition path sampling, forward flux simulation, Markov state modeling, and milestoning are more accurate but are limited to very small changes, require suitable reaction coordinates to be known beforehand, or take significantly more time to compute a usable trajectory. For a review of these methods, see (Schlick 2009).

11.3.4.3 Analysis of Correlated Motions

The dynamics of RNA–protein systems can be analyzed by evaluating the cross-correlation matrix, using the trajectory file from the simulation as input, to identify correlated motions of the specific amino acids and nucleotides. These analyses can be generated quickly by applications like **CARMA** (Glykos 2006). The correlation, or normalized covariance, C_{ij}, between two residues i and j is defined:

$$C_{ij} = \frac{\langle \Delta\vec{r}_i(t) \cdot \Delta\vec{r}_j(t) \rangle}{\sqrt{\langle \Delta\vec{r}_i(t)^2 \rangle \langle \Delta\vec{r}_j(t)^2 \rangle}}, \quad \text{where } \Delta\vec{r}_i(t) = \vec{r}_i(t) - \langle \vec{r}_i(t) \rangle. \tag{11.1}$$

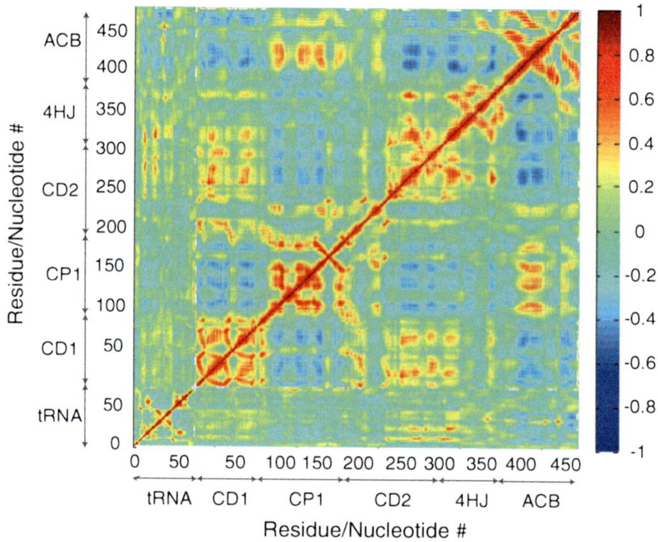

Fig. 11.7 Cross correlation map for GluRS·tRNA^{Glu}·Glu-AMP

In these expressions, $\vec{r}_i(t)$ is the position vector at time t of an atom chosen to represent residue i. Typically, this atom is the alpha-carbon for amino acids and the phosphorus atom for nucleotides, although any atom may be selected. The angle brackets indicate the time average of the quantity within the brackets, calculated over the entire trajectory or portions of it. Subsequent analyses of the correlations from different time segments of long time trajectories can give insight into the function of the composite systems.

Plotting the cross-correlation gives a high-level view of which parts of the biomolecular structures have coupled motion. The cross-correlation map for GluRS·tRNA^{Glu}·Glu-AMP is shown in Fig. 11.7. Regions of the tRNA are strongly correlated with the GluRS anticodon-binding domain as well as the catalytic domain. Through principal component analysis (PCA), unnormalized covariance matrices can be used to identify the dominant motions of the system that capture the majority of the fluctuations within a given time domain. As correlation data contains both the harmonic and anharmonic motions of the complexes in the presence of solvent and ions, it can also be used to study networks of interaction in RNA–protein complexes involved in molecular recognition and docking.

An alternative to the standard correlation is full correlation analysis (FCA) (Lange and Grubmüller 2008). Correlation analysis is based on only linear correlations, but through the use of mutual information, FCA is able to include nonlinear and higher-order correlations. Although the run time of FCA is significantly longer, it returns more complete information about the collective motions inherent in RNA and protein dynamics.

11.3.4.4 Dynamical Network Analysis

Allosteric signaling is ubiquitous throughout protein synthesis. For example, during aminoacylation of tRNA, the aaRSs must receive information about docking of the tRNA anticodon and transmit it to the active site. The proofreading step that occurs when EF-Tu·GTP·aa-tRNA docks to the ribosome requires the codon–anticodon base pairing information before EF-Tu can hydrolyze its bound GTP. Both of these reactions depend on binding interactions that occur nanometers away from the active site. Allosteric communication has previously been studied both experimentally (Goodey and Benkovic 2008) and through covariation analysis of multiple sequence alignments (Süel et al. 2002). More recently, signaling through RNA–protein complexes has been investigated using dynamical network analysis in which local correlated motions are used to identify pathways of communication (Sethi et al. 2009).

Biopolymers can be represented as basic contact networks by treating the monomers of the polymer (amino acids or nucleotides) as the nodes of the network, and the physical contacts between monomers as the edges (Aszói and Taylor 1993). Certain properties of these network models, such as shortest paths between pairs of nodes or the number of edges attached to a given node, have been shown to provide insight into biomolecular structure and function. Path lengths in these simple networks are defined as the number of edges traversed in a path from one node to another, and the shortest path is therefore the path with the shortest length, i.e., smallest number of edges. Analysis of the shortest paths through a network reveal the relative importance of different residues to molecular communication (del Sol et al. 2006). This type of analysis was performed on MetRS to find residues along the path from the anticodon binding site to the active site (Ghosh and Vishveshwara 2007). Similarly, the structure network for bacterial and archaeal rRNA was analyzed using various network metrics to identify nucleotides important for ribosomal function (David-Eden and Mandel-Gutfreund 2008).

MD simulation provides a way to refine the simple contact network to generate one that incorporates dynamical information so as to provide deeper insight into molecular communication. The transfer of information from one residue to a neighboring residue can be identified with the correlated motion between the two since knowledge of the movement of one provides knowledge about movement of the other (Sethi et al. 2009). The edges of a contact network can incorporate standard correlation or FCA data in the form of edge weights where smaller weights correspond to tighter coupling between two nodes. A network representation with correlation-weighted edges was used in the interpretation of RNA–protein contacts present in the ribosomal L1 stalk and showed that the L1 stalk is more closely associated with $tRNA^{Phe}$ than with $tRNA^{fMet}$ (Trabuco et al. 2010) (see Fig. 11.8). The $tRNA^{Phe}$ interface contacts are stronger and more localized to the G19·C56 base pair that stacks against the rRNA.

With edge weights between nodes i and j defined as $w_{ij} = -\log(|C_{ij}|)$, shortest path analysis can be carried out on this weighted network, where path length is the sum of the edge weights for edges along a path. These shortest paths will tend to travel through highly correlated residues (see Fig. 11.9a). Another useful metric

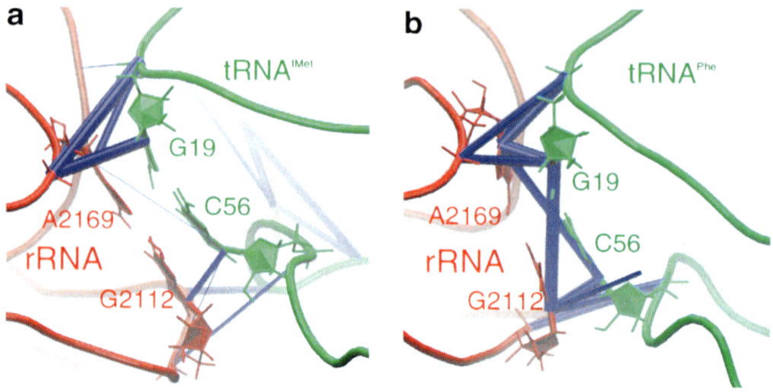

Fig. 11.8 Contacts (*blue*) between P/E hybrid tRNA and rRNA in the ribosomal L1 stalk (**a**) $tRNA^{fMet}$ has slightly more, but weaker, contacts to the rRNA than those between (**b**) $tRNA^{Phe}$ and the rRNA (Trabuco et al. 2010). Connections are shown in *blue* with thickness representing pairwise correlation

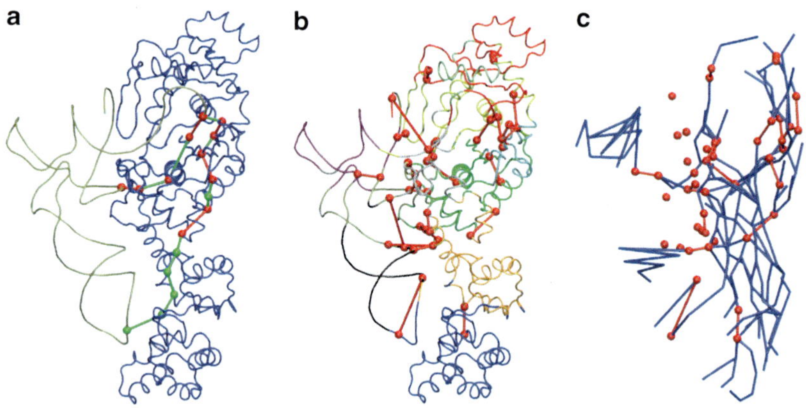

Fig. 11.9 Dynamic network analysis of GluRS·Glu-$tRNA^{Phe}$·AMP (PDB ID 1 N78) (**a**) Shortest, most highly correlated paths between A76 in the active site and the two identity elements U13 and U35. (**b**) The complex colored by community with critical intercommunity connections shown in *red*. (**c**) Top 10% of the edges with the highest betweenness in the total network. Each edge shown is present in at least 3,000 shortest paths between pairs of nodes (Sethi et al. 2009; Alexander et al. 2010)

derived from pairwise shortest paths is the "betweenness" of a network edge (Freeman 1979; Girvan and Newman 2002). Edge betweenness is defined as the fraction of shortest paths crossing an edge. This gives a measure of how central an edge is to the various communication pathways in a network.

Pairwise shortest paths are not the only interesting feature of these dynamical networks. A more global view of structural dynamics shows that groups of residues cluster together and move in concert with one another giving RNA–protein complexes a modular structure. In the language of network theory, these clusters of nodes are called communities, and the nodes in a community have more and stronger connections within the community than between communities. Network algorithms have been developed recently to partition networks into communities (Girvan and Newman 2002; Palla et al. 2005; Chennubhotla and Bahar 2006), and these algorithms can be used to elucidate the community structure of biomolecular complexes (see Fig. 11.9b) (Eargle et al. 2011).

One result of this community structure is that high betweenness edges responsible for connecting communities to each other are disproportionately important in allosteric signaling. High betweenness edges have been used previously to identify and visualize critical features of street maps, such as major roads and highways (Demšar et al. 2008). They act as communication bottlenecks within the network because shortest paths tend to flow through the highest correlation edges between communities. Residues participating in these critical edges have been shown to be highly conserved within the dynamical networks for GluRS · $tRNA^{Glu}$ (see Fig. 11.9a, b) and LeuRS · $tRNA^{Leu}$ (Sethi et al. 2009). There are similarities between the communicative role played by nodes involved in critical edges and those previously described as "active centers" in protein structure networks (Csermely 2008). A subnetwork consisting of edges with high betweenness is shown in Fig. 11.9c. Although this subnetwork has approximately one tenth the number of edges as the full network, it contains more than half of the critical edges connecting pairs of communities.

As these types of structure and dynamical network analyses are relatively new, they have not yet been fully evaluated. The main quantitative validation has come through the observed conservation of critical nodes and computational identification of residues, such as tRNA identity elements, which have been experimentally determined to be important for molecular communication (del Sol et al. 2006; Ghosh and Vishveshwara 2007; Sethi et al. 2009). Other technical details about network set-up and interpretation will need to be addressed in the future. How many nodes should represent RNA and protein molecules? Since nucleotides are larger than amino acids, should they have more nodes? How exactly should nodes be related to their underlying atomic substructure? Also, a statistical mechanical framework needs to be developed to connect network properties to experimental observables such as free energies of binding.

11.4 Reaching Longer Timescales Through Simplified Models

Coarse-grained MD simulations sacrifice atomic detail in order to study larger systems at longer timescales. Ribosomal assembly and tRNA translocation through the ribosome are processes that require seconds, and important dynamics occur at timescales unavailable with current all-atom MD methods. Two general methods of coarse graining easily take advantage of the flexibility in existing MD programs: particle-based coarse graining or force field coarse graining. For a recent review of coarse graining methods, see Freddolino et al. (2008).

Particle-based coarse graining comes in two flavors, both of which involve reducing the number of particles in the system. The first systematically replaces sets of atoms with single particles. This approach has been used recently to coarse grain RNA (Cui et al. 2006; Hyeon and Thirumalai 2007) as well as DNA (Sambriski et al. 2009). Amino acids and nucleotides are reduced to a few beads each; multiple water molecules or hexahydrated Mg^{2+} are merged into single beads. This can result in an order of magnitude decrease in the number of particles representing a system while maintaining the polymeric structure of the macromolecules as well as an explicit description of cations. Coarse-grained force fields can be derived from atomistic simulations through a force matching procedure (Izvekov and Voth 2005).

The second particle reduction scheme is structure-based coarse graining where the number of particles is specified up front, and the coarse-grained representation and associated force field parameters are generated automatically (Freddolino et al. 2008). For example, if a researcher wants to use 10–20 beads for each r-protein, depending on their relative sizes, the set of beads representing a given r-protein would be scattered throughout the protein's 3D structure using Voronoi tessellation where each bead is associated with one Voronoi cell. Then the collective mass and charge of atoms within that cell are projected onto the bead, and spring-like connections are strung between beads in adjacent cells. Solvent is approximated through a continuum dielectric.

Researchers can also coarse grain the energy landscape to reach timescales relevant for folding and assembly processes. Go-like potentials effectively smooth folding and binding energy landscapes by biasing the molecules in a given system toward their native structures. Go potentials have been used extensively to study protein folding dynamics and more recently to study tRNA movement within the ribosome (Whitford et al. 2009). Frequently, Go potentials are used in conjunction with particle-based coarse graining, but all-atom Go potentials have also been used (Clementi et al. 2003; Pogorelov and Luthey-Schulten 2004; Whitford et al. 2009). Go is applied through a nonbonded, pairwise contact potential. Beginning with a reference structure, native contacts within a set cutoff are given attractive, Lennard–Jones-like potentials while atoms or beads outside the cutoff receive hard sphere potentials. For RNA–protein systems, three sets of Go parameters are needed: RNA–RNA for contacts within the RNA structure, protein–protein for those within the protein, and RNA–protein for contacts at the interface between the two. Development of appropriate Go parameters for RNA is a continuing process. Although coarse-grained simulations for systems containing RNA are still in their infancy, the long time and length scales seen in RNA dynamics require computational scientists to improve RNA coarse-graining methods.

11.5 Conclusion

Molecular detail is required for characterizing and understanding the dynamics of RNA–protein complexes. Starting from atomic resolution biomolecular structures, molecular simulation can be used to verify experimental results, to provide molecular details that explain biomolecular function, and finally to predict outcomes for future experiments. The ongoing improvement of all-atom force fields and molecular simulation techniques is essential to better relate MD trajectories to questions of mechanism and to reach time and length scales relevant to processes like RNA folding, macromolecular assembly, and protein synthesis. For more rapid progress in the field of RNA–protein simulation, it is also essential to create and use standards for data representation and more effectively communicate and disseminate the techniques and tools used in data analysis.

Acknowledgments The authors are supported by NSF Grants MCB-0844670 and PHY-0822613 and NIH P41 RR005969. We thank Ke Chen for creating the figures related to S4·h16 and Eduard Schreiner for providing the VMD Tcl script for calculating the angle between RNA helices.

References

Alexander RW, Eargle J, Luthey-Schulten Z (2010) Experimental and computational determination of tRNA dynamics. FEBS Lett 584(2):376–386

Andresen K, Das R, Park H, Smith H, Kwok L, Lamb J, Kirkland E, Herschlag D, Finkelstein K, Pollack L (2004) Spatial distribution of competing ions around DNA in solution. Phys Rev Lett 93(24):248103

Armache JP, Jarasch A, Anger AM, Villa E, Becker T, Bhushan S, Jossinet F, Habeck M, Dindar G, Franckenberg S, Marquez V, Mielke T, Thomm M, Berninghausen O, Beatrix B, Söding J, Westhof E, Wilson DN, Beckmann R (2010) Localization of eukaryote-specific ribosomal proteins in a 5.5-Å cryo-EM map of the 80 S eukaryotic ribosome. Proc Natl Acad Sci USA 107(46):19754–19759

Aszói A, Taylor WR (1993) Connection topology of proteins. Comput Appl Biosci 9:523–529

Auffinger P, Hashem Y (2007) SwS: a solvation web service for nucleic acids. Bioinformatics 23(8):1035–1037

Auffinger P, Vaiana A (2005) Molecular dynamics simulations of RNA systems. Handbook of RNA biochemistry. Wiley-VCH Verlag, Weinheim, pp 560–576

Auffinger P, Westhof E (1997) RNA hydration: three nanoseconds of multiple molecular dynamics simulations of the solvated tRNA(Asp) anticodon hairpin. J Mol Biol 269(3):326–341

Auffinger P, Cheatham TE III, Vaiana AC (2007) Spontaneous formation of KCl aggregates in biomolecular simulations: a force field issue? J Chem Theor Comput 3(5):1851–1859

Bai Y, Greenfeld M, Travers KJ, Chu VB, Lipfert J, Doniach S, Herschlag D (2007) Quantitative and comprehensive decomposition of the ion atmosphere around nucleic acids. J Am Chem Soc 129(48):14981–14988

Baker NA, Sept D, Joseph S, Holst MJ, McCammon JA (2001) Electrostatics of nanosystems: application to microtubules and the ribosome. Proc Natl Acad Sci USA 98(18):10037–10041

Ban N, Nissen P, Hansen J, Moore PB, Steitz TA (2000) The complete atomic structure of the large ribosomal subunit at 2.4 A resolution. Science 289(5481):905–920

Bas DC, Rogers DM, Jensen JH (2008) Very fast prediction and rationalization of pKa values for protein-ligand complexes. Protein Struct Funct Genet 73:765–783

Bateman A, Birney E, Cerruti L, Durbin R, Etwiller L, Eddy SR, Griffiths-Jones S, Howe KL, Marshall M, Sonnhammer ELL (2002) The Pfam protein families database. Nucleic Acids Res 30(1):276–280

Becker T, Bhushan S, Jarasch A, Armache JP, Funes S, Jossinet F, Gumbart J, Mielke T, Berninghausen O, Schulten K, Westhof E, Gilmore R, Mandon EC, Beckmann R (2009) Structure of monomeric yeast and mammalian Sec61 complexes interacting with the translating ribosome. Science 326:1369–1373

Bellur DL, Woodson SA (2009) A minimized rRNA-binding site for ribosomal protein S4 and its implications for 30 S assembly. Nucleic Acids Res 37(6):1886–1896

Berk V, Zhang W, Pai RD, Cate JHD (2006) Structural basis for mRNA and tRNA positioning on the ribosome. Proc Natl Acad Sci USA 103(43):15830–15834

Beššeová I, Otyepka M, Réblová K, Šponer J (2009) Dependence of A-RNA simulations on the choice of the force field and salt strength. Phys Chem Chem Phys 11:10701–10711

Beššeová I, Réblová K, Leontis NB, Šponer J (2010) Molecular dynamics simulations suggest that RNA three-way junctions can act as flexible RNA structural elements in the ribosome. Nucleic Acids Res 38(18):6247–6264

Black Pyrkosz A, Eargle J, Sethi A, Luthey-Schulten Z (2010) Exit strategies for charged tRNA from GluRS. J Mol Biol 397:1350–1371

Brown JW, Birmingham A, Griffiths PE, Jossinet F, Kachouri-Lafond R, Knight R, Lang BF, Leontis N, Steger G, Stombaugh J, Westhof E (2009) The RNA structure alignment ontology. RNA 15(9):1623–1631

Cabello-Villegas J, Winkler ME, Nikonowicz EP (2002) Solution conformations of unmodified and A(37)N(6)-dimethylallyl modified anticodon stem-loops of Escherichia coli tRNA(Phe). J Mol Biol 319(5):1015–1034

Charette M, Gray MW (2000) Pseudouridine in RNA: what, where, how, and why. IUBMB Life 49(5):341–351

Chen Y, Varani G (2005) Protein families and RNA recognition. FEBS Lett 272(9):2088–2097

Chen K, Roberts E, Luthey-Schulten Z (2009) Horizontal gene transfer of zinc and non-zinc forms of bacterial ribosomal protein S4. BMC Evol Biol 9:179–195

Chen K, Eargle J, Sarkar K, Gruebele M, Luthey-Schulten Z (2010) The functional role of ribosomal signatures. Biophys J 99(12):3930–3940

Chen K, Magis A, Eargle J, Roberts E, Luthey-Schulten Z (2011) Evolution of translation: EF-Tu: tRNA. http://www.scs.illinois.edu/schulten/tutorials/ef-tu/

Chennubhotla C, Bahar I (2006) Markov propagation of allosteric effects in biomolecular systems: application to GroEL-GroES. Mol Syst Biol 2:36

Clementi C, Garcia AE, Onuchic JN (2003) Interplay among tertiary contacts, secondary structure formation and side-chain packing in the protein folding mechanism: all-atom representation study of protein L. J Mol Biol 326:933–954

Cowan JA (1995) The biological chemistry of magnesium. VCH, New York

Csermely P (2008) Creative elements: network-based predictions of active centres in proteins and cellular and social networks. Trends Biochem Sci 33(12):569–576

Cui Q, Tan RK-Z, Harvey SC, Case DA (2006) Low-resolution molecular dynamics simulations of the 30 S ribosomal subunit. Multiscale Model Simul 5(4):1248–1263

Das R, Mills T, Kwok L, Maskel G, Millett I, Doniach S, Finkelstein K, Herschlag D, Pollack L (2003) Counterion distribution around DNA probed by solution X-ray scattering. Phys Rev Lett 90(18):188103

David-Eden H, Mandel-Gutfreund Y (2008) Revealing unique properties of the ribosome using a network based analysis. Nucleic Acids Res 36(14):4641–4652

del Sol A, Fujihashi H, Amoros D, Nussinov R (2006) Residues crucial for maintaining short paths in network communication mediate signaling in proteins. Mol Syst Biol 2:19

Demšar U, Špatenková O, Virrantaus K (2008) Identifying critical locations in a spatial network with graph theory. Trans GIS 12(1):61–82

Ditzler MA, Otyepka M, Šponer J, Walter NG (2010) Molecular dynamics and quantum mechanics of RNA: conformational and chemical change we can believe in. Acc Chem Res 43(1):40–47

Draper DE (2008) RNA folding: thermodynamic and molecular descriptions of the roles of ions. Biophys J 95(12):5489–5495

Dunin-Horkawicz S, Czerwoniec A, Gajda MJ, Feder M, Grosjean H, Bujnicki JM (2006) MODOMICS: a database of RNA modification pathways. Nucleic Acids Res 34(Database issue):D145–D149

Eargle J, Black AA, Sethi A, Trabuco LG, Luthey-Schulten Z (2008) Dynamics of recognition between tRNA and elongation factor Tu. J Mol Biol 377(5):1382–1405

Eargle J, Li L, Luthey-Schulten Z (2011) Dynamical network analysis. http://www.scs.illinois.edu/schulten/tutorials/network/

Eigen M, Winkler-Oswatitsch R (1981) Transfer-RNA: the early adaptor. Naturwissenschaften 68(5):217–228

Foloppe N, MacKerrell AD Jr (2000) All-atom empirical force field for nucleic acids: I. Parameter optimization based on small molecule and condensed phase macromolecular target data. J Comput Chem 21:86–104

Freddolino PL, Arkhipov A, Shih AY, Yin Y, Chen Z, Schulten K (2008) Application of residue-based and shape-based coarse graining to biomolecular simulations, Coarse-graining of condensed phase and biomolecular systems. Chapman and Hall, London

Freddolino PL, Park S, Bt R, Schulten K (2009) Force field bias in protein folding simulations. Biophys J 96(9):3772–3780

Freeman LC (1979) Centrality in social networks conceptual clarification. Soc Network 1(3):215–239

Froloff N, Windemuth A, Honig B (1997) On the calculation of binding free energies using continuum methods: application to MHC class I protein-peptide interactions. Protein Sci 6(6):1293–1301

Garcia AE, Paschek D (2008) Simulation of the pressure and temperature folding/unfolding equilibrium of a small RNA hairpin. J Am Chem Soc 130(3):815–817

Ghosh A, Vishveshwara S (2007) A study of communication pathways in methionyl-tRNA synthetase by molecular dynamics simulations and structure network analysis. Proc Natl Acad Sci USA 104(40):15711–15716

Giege R, Sissler M, Florentz C (1998) Universal rules and idiosyncratic features in tRNA identity. NucleicAcids Res 26(22):5017–5035

Girvan M, Newman M (2002) Community structure in social and biological networks. Proc Natl Acad Sci USA 99:7821–7826

Glykos NM (2006) CARMA: a molecular dynamics analysis program. J Comput Chem 27(14):1765–1768

Gohlke H, Kiel C, Case DA (2003) Insights into protein-protein binding by binding free energy calculation and free energy decomposition for the Ras-Raf and Ras-RalGDS complexes. J Mol Biol 330(4):891–913

Goodey NM, Benkovic SJ (2008) Allosteric regulation and catalysis emerge via a common route. Nat Chem Biol 4(8):474–482

Griffiths-Jones S, Bateman A, Marshall M, Khanna A, Eddy SR (2003) Rfam: An RNA family database. Nucleic Acids Res 31(1):439–441

Grilley D, Soto AM, Draper DE (2006) Mg^{2+}-RNA interaction free energies and their relationship to the folding of RNA tertiary structures. Proc Natl Acad Sci USA 103(38):14003–14008

Gumbart JC, Trabuco LG, Schreiner E, Villa E, Schulten K (2009) Regulation of the protein-conducting channel by a bound ribosome. Structure 17(11):1453–1464

Hashem Y, Auffinger P (2009) A short guide for molecular dynamics simulations of RNA systems. Methods 47(3):187–197

Held WA, Ballou B, Mizushima S, Nomura M (1974) Assembly mapping of 30 S ribosomal proteins from *Escherichia coli*: further studies. J Biol Chem 249(10):3103–3111

Helm M (2006) Post-transcriptional nucleotide modification and alternative folding of RNA. Nucleic Acids Res 34(2):721–733

Hermann T, Westhof E (1998) Exploration of metal ion binding sites in RNA folds by Brownian-dynamics simulations. Structure 6(10):1303–1314

Hoehndorf R, Batchelor C, Bittner T, Dumontier M, Eilbecke K, Knight R, Mungall CJ, Richardson JS, Stombaugh J, Westhof E, Zirbel CL, Leontis NB (2011) The RNA ontology (RNAO): an ontology for integrating RNA sequence and structure data. Appl Ontol 6:53–89

Humphrey W, Dalke A, Schulten K (1996) VMD – visual molecular dynamics. J Mol Graph 14:33–38

Hyeon C, Thirumalai D (2007) Mechanical unfolding of RNA: from hairpins to structures with internal multiloops. Biophys J 92(3):731–743
Izvekov S, Voth G (2005) A multiscale coarse-graining method for biomolecular systems. J Phys Chem B 109(7):2469–2473
Jiang W, Hardy DJ, Phillips JC, MacKerell AD Jr, Schulten K, Roux B (2011) High-performance scalable molecular dynamics simulations of a polarizable force field based on classical Drude oscillators in NAMD. J Phys Chem Lett 2(2):87–92
Kirmizialtin S, Elber R (2010) Computational exploration of mobile ion distributions around RNA duplex. J Phys Chem B 114(24):8207–8220
Klepeis JL, Lindorff-Larsen K, Dror RO, Shaw DE (2009) Long-timescale molecular dynamics simulations of protein structure and function. Curr Opin Struct Biol 19(2):120–127
Kollman PA, Massova I, Reyes C, Kuhn B, Huo S, Chong L, Lee M, Lee T, Duan Y, Wang W, Donini O, Cieplak P, Srinivasan J, Case DA, Cheatham TE III (2000) Calculating structures and free energies of complex molecules: combining molecular mechanics and continuum models. Acc Chem Res 33(12):889–897
Landes C, Perona JJ, Brunie S, Rould MA, Zelwer C, Steitz TA, Risler JL (1995) A structure-based multiple sequence alignment of all class I aminoacyl-tRNA synthetases. Biochemie 77(3):194–203
Lange OF, Grubmüller H (2008) Full correlation analysis of conformational protein dynamics. Protein Struct Funct Genet 70(4):1294–1312
Lavery R, Moakher M, Maddocks J, Petkeviciute D, Zakrzewska K (2009) Conformational analysis of nucleic acids revisited: curves+. Nucleic Acids Res 37(17):5917–5929
Lee EH, Hsin J, Sotomayor M, Comellas G, Schulten K (2009) Discovery through the computational microscope. Structure 17(10):1295–1306
Leipply D, Draper DE (2010) Dependence of RNA tertiary structural stability on Mg^{2+} concentration: interpretation of the hill equation and coefficient. Biochemistry 49(9):1843–1853
Leontis NB, Westhof E (2001) Geometric nomenclature and classification of RNA base pairs. RNA 7(04):499–512
Li W, Frank J (2007) Transfer RNA in the hybrid P/E state: correlating molecular dynamics simulations with cryo-EM data. Proc Natl Acad Sci USA 104(42):16540–16545
Li L, Sethi A, Luthey-Schulten Z (2009) Evolution of translation: class I aminoacyl-tRNA synthetase:tRNA complexes. http://www.scs.illinois.edu/schulten/tutorials/evolution/
Limbach PA, Crain PF, McCloskey JA (1994) Summary: the modified nucleosides of RNA. Nucleic Acids Res 22(12):2183–2196
Lipfert J, Sim AYL, Herschlag D, Doniach S (2010) Dissecting electrostatic screening, specific ion binding, and ligand binding in an energetic model for glycine riboswitch folding. RNA 16(4):708–719
Lu XJ, Olson WK (2003) 3DNA: a software package for the analysis, rebuilding and visualization of three-dimensional nucleic acid structures. Nucleic Acids Res 31(17):5108–5121
MacKerell AD, Nilsson L (2008) Molecular dynamics simulations of nucleic acid-protein complexes. Curr Opin Struct Biol 18(2):194–199
Morozova N, Allers J, Myers J, Shamoo Y (2006) Protein-RNA interactions: exploring binding patterns with a three-dimensional superposition analysis of high resolution structures. Bioinformatics 22(22):2746–2752
Mulder AM, Yoshioka C, Beck AH, Bunner AE, Mulligan RA, Potter CS, Carragher B, Williamson JR (2010) Visualizing ribosome biogenesis: parallel assembly pathways for the 30 S subunit. Science 330(6004):673–677
O'Donoghue P, Luthey-Schulten Z (2003) On the evolution of structure in the aminoacyl-tRNA synthetases. Microbiol Mol Biol Rev 67:550–573
O'Donoghue P, Luthey-Schulten Z (2005) Evolutionary profiles derived from the QR factorization of multiple structural alignments gives an economy of information. J Mol Biol 346(3):875–894
Ohtaki H (2001) Ionic solvation in aqueous and nonaqueous solutions. Monatsh Chem 132:1237–1268

Oostenbrink C, Villa A, Mark AE, Van Gunsteren WF (2004) A biomolecular force field based on the free enthalpy of hydration and solvation: the GROMOS force-field parameter sets 53A5 and 53A6. J Comput Chem 25(13):1656–1676

Pabit SA, Qiu X, Lamb JS, Li L, Meisburger SP, Pollack L (2009) Both helix topology and counterion distribution contribute to the more effective charge screening in dsRNA compared with dsDNA. Nucleic Acids Res 37(12):3887–3896

Palla G, Derényi I, Farkas I, Vicsek T (2005) Uncovering the overlapping community structure of complex networks in nature and society. Nature 435(7043):814–818

Pettersen EF, Goddard TD, Huang CC, Couch GS, Greenblatt DM, Meng EC, Ferrin TE (2004) UCSF Chimera-a visualization system for exploratory research and analysis. J Comput Chem 25(13):1605–1612

Pogorelov TV, Luthey-Schulten Z (2004) Variations in the fast folding rates of the lambda-repressor: a hybrid molecular dynamics study. Biophys J 87:207–214

Pogorelov TV, Autenrieth F, Roberts E, Luthey-Schulten ZA (2007) Cytochrome c_2 exit strategy: dissociation studies and evolutionary implications. J Phys Chem B 111(3):618–634

Ponomarev SY, Thayer KM, Beveridge DL (2004) Ion motions in molecular dynamics simulations on DNA. Proc Natl Acad Sci USA 101(41):14771–14775

Pranata J, Wierschke SG, Jorgensen WL (1991) OPLS potential functions for nucleotide bases. Relative association constants of hydrogen-bonded base pairs in chloroform. J Am Chem Soc 113(8):2810–2819

Reyes CM, Kollman PA (2000) Structure and thermodynamics of RNA-protein binding: using molecular dynamics and free energy analyses to calculate the free energies of binding and conformational change. J Mol Biol 297(5):1145–1158

Roberts E, Eargle J, Wright D, Luthey-Schulten Z (2006) MultiSeq: unifying sequence and structure data for evolutionary analysis. BMC Bioinformatics 7:382

Roberts E, Sethi A, Montoya J, Woese CR, Luthey-Schulten Z (2008) Molecular signatures of ribosomal evolution. Proc Natl Acad Sci USA 105(37):13953–13958

Rocchia W, Alexov E, Honig B (2001) Extending the applicability of the nonlinear Poisson-Boltzmann equation: multiple dielectric constants and multivalent ions. J Phys Chem B 105:6507–6514

Rocchia W, Sridharan S, Nicholls A, Alexov E, Chiabrera A, Honig B (2002) Rapid grid-based construction of the molecular surface and the use of induced surface charge to calculate reaction field energies: applications to the molecular systems and geometric objects. J Comput Chem 23(1):128–137

Roh J, Briber R, Damjanovic A, Thirumalai D, Woodson S, Sokolov A (2009) Dynamics of tRNA at different levels of hydration. Biophys J 96(7):2755–2762

Russell R, Zhuang X, Babcock HP, Millett IS, Doniach S, Chu S, Herschlag D (2002) Exploring the folding landscape of a structured RNA. Proc Natl Acad Sci USA 99(1):155–160

Sakharov DV, Lim C (2008) Force fields including charge transfer and local polarization effects: application to proteins containing multi/heavy metal ions. J Comput Chem 30(2):191–202

Sambriski E, Schwartz D, de Pablo J (2009) A mesoscale model of DNA and its renaturation. Biophys J 96(5):1675–1690

Sanbonmatsu KY, Joseph S, Tung CS (2005) Simulating movement of tRNA into the ribosome during decoding. Proc Natl Acad Sci USA 102(44):15854–15859

Schlick T (2009) Molecular dynamics-based approaches for enhanced sampling of long-time, large-scale conformational changes in biomolecules. F100 Biol Rep 1:51

Schluenzen F, Tocilj A, Zarivach R, Harms J, Gluehmann M, Janell D, Bashan A, Bartels H, Agmon I, Franceschi F, Yonath A (2000) Structure of functionally activated small ribosomal subunit at 3.3 Å resolution. Cell 102(5):615–623

Schrödinger LLC The PyMOL molecular graphics system, version 1.3

Serra MJ, Baird JD, Dale T, Fey BL, Retatagos K, Westhof E (2002) Effects of magnesium ions on the stabilization of RNA oligomers of defined structures. RNA 8(3):307–323

Sethi A, O'Donoghue P, Luthey-Schulten Z (2005) Evolutionary profiles from the QR factorization of multiple sequence alignments. Proc Natl Acad Sci USA 102(11):4045–4050

Sethi A, Eargle J, Black AA, Luthey-Schulten Z (2009) Dynamical networks in tRNA: protein complexes. Proc Natl Acad Sci USA 106(16):6620–6625

Silvian LF, Wang J, Steitz TA (1999) Insights into editing from an Ile-tRNA synthetase structure with $tRNA^{Ile}$ and mupirocin. Science 285:1074–1077

Stone JE, Phillips JC, Freddolino PL, Hardy DJ, Trabuco LG, Schulten K (2007) Accelerating molecular modeling applications with graphics processors. J Comput Chem 28(16):2618–2640

Süel GM, Lockless SW, Wall MA, Ranganathan R (2002) Evolutionarily conserved networks of residues mediate allosteric communication in proteins. Nat Struct Mol Biol 10(1):59–69

Tang CL, Alexov E, Pyle AM, Honig B (2007) Calculation of pKas in RNA: on the structural origins and functional roles of protonated nucleotides. J Mol Biol 366(5):1475–1496

Tinoco I Jr, Bustamante C (1999) How RNA folds. J Mol Biol 293:271–281

Trabuco LG, Villa E, Mitra K, Frank J, Schulten K (2008) Flexible fitting of atomic structures into electron microscopy maps using molecular dynamics. Structure 16(5):673–683

Trabuco LG, Schreiner E, Eargle J, Cornish P, Ha T, Luthey-Schulten Z, Schulten K (2010) The role of L1 stalk: tRNA interaction in the ribosome elongation cycle. J Mol Biol 402:741–760

Wang J, Wolf RM, Caldwell JW, Kollman PA, Case DA (2004) Development and testing of a general AMBER force field. J Comput Chem 25(9):1157–1174

Whitford PC, Noel JK, Gosavi S, Schug A, Sanbonmatsu KY, Onuchic JN (2009) An all-atom structure-based potential for proteins: bridging minimal models with all-atom empirical forcefields. Protein Struct Funct Bioinf 75(2):430–441

Wimberly BT, Brodersen DE, Clemons WM, Morgan-Warren RJ, Carter AP, Vonrhein C, Hartsch T, Ramakrishnan V (2000) Structure of the 30 S ribosomal subunit. Nature 407(6802):327–339

Winker S, Woese CR (1991) A definition of the domains Archaea, Bacteria and Eucarya in terms of small subunit ribosomal RNA characteristics. Syst Appl Microbiol 14(4):305–310

Woese CR (1987) Bacterial evolution. Microbiol Rev 51(2):221–271

Woese CR (2000) Interpreting the universal phylogenetic tree. Proc Natl Acad Sci USA 97(15):8392–8396

Woese CR, Kandler O, Wheelis ML (1990) Towards a natural system of organisms: proposal for the domains Archaea, Bacteria, and Eucarya. Proc Natl Acad Sci USA 87(12):4576–4579

Woese CR, Olsen GJ, Ibba M, Söll D (2000) Aminoacyl-tRNA synthetases, the genetic code, and the evolutionary process. Microbiol Mol Biol Rev 64(1):202–236

Yamasaki S, Nakamura S, Terada T, Shimizu K (2007) Mechanism of the difference in the binding affinity of E. coli $tRNA^{Gln}$ to glutaminyl-tRNA synthetase caused by noninterface nucleotides in variable loop. Biophys J 92:192–200

Yang H, Jossinet F, Leontis N, Chen L, Westbrook J, Berman H, Westhof E (2003) Tools for the automatic identification and classification of RNA base pairs. Nucleic Acids Res 31(13):3450–3460

Zhang L, Hermans J (1996) Hydrophilicity of cavities in proteins. Protein Struct Funct Genet 24(4):433–438

Chapter 12
Quantum Chemical Studies of Recurrent Interactions in RNA 3D Motifs

Jiří Šponer, Judit E. Šponer, and Neocles B. Leontis

Abstract High-quality quantum mechanical (QM) calculations provide physically based descriptions of molecular systems that are free of empirical parameters. This contrasts with force-field computations based on simple and entirely nonphysical, analytical functions that must be completely parametrized for a given purpose. The costs of high-quality QM computations, however, can be enormous, limiting them to small model systems with ~50+ atoms. Thus, a major challenge of the QM approach is how to extrapolate data computed on model systems to intact biomolecules of biological interest. QM calculations have been used to study the basic molecular forces in nucleic acids. A notable accomplishment of these studies has been to clarify the nature of aromatic base stacking. Another important application of modern QM computations is to furnish reference data for parametrizing molecular modeling force fields. In this chapter, we provide a summary of the nature of QM calculations, their strengths, limitations, and relation to other methods. Then, we review the use of high-level ab initio (first principles) QM methods to calculate geometries and energies of fundamental nucleotide interactions in RNA 3D structures.

J. Šponer (✉) • J.E. Šponer
Institute of Biophysics, Academy of Sciences of the Czech Republic, Kralovopolska 135, 612 65 Brno, Czech Republic

CEITEC – Central European Institute of Technology, Brno, Czech Republic
e-mail: sponer@ncbr.muni.cz

N.B. Leontis
Chemistry Department, Bowling Green State University, Bowling Green, OH 43403, USA

N. Leontis and E. Westhof (eds.), *RNA 3D Structure Analysis and Prediction*, Nucleic Acids and Molecular Biology 27, DOI 10.1007/978-3-642-25740-7_12,

12.1 Introduction

In structural biology and bioinformatics, the most commonly applied computational modeling methods rely on the use of molecular mechanics force fields. These force fields are simple analytic functions, parametrized to approximate the potential energies of molecules as functions of their geometries. Computations that use these force fields are limited by the functional form of the potential energy and by the accuracy of the parametrization. Force fields are nonphysical models of molecular systems and always involve considerable approximation of the properties of real molecules. Moreover, force fields rely heavily on compensation of errors. Despite these limitations, properly parametrized force fields can generate reasonable results and provide valuable insight into complex biomolecular systems (Ditzler et al. 2010). However, the capabilities of force-field calculations should not be overrated.

First, some properties of real systems cannot be adequately described within the limits of the technique. For example, the standard force-field approach, describing divalent cations such as zinc or magnesium as van der Waals Lennard-Jones spheres with +2 point charges localized in their centers, is very unrealistic (Sponer et al. 2000; Petrov et al. 2011a, b). The total energy of polarization and charge-transfer nonadditivities in the first ligand shell of a divalent cation is about 70 kcal/mol, 12 times the gas-phase binding energy of a water dimer (Sponer et al. 2000). Common biomolecular force fields completely lack appropriate terms for these effects, which can propagate far beyond the first ligand shell and dramatically affect the neighborhood of the divalent cation (Katz et al. 1996; Sponer et al. 2000). In other words, the force-field approximation breaks down completely for divalent cations such as Mg(II), and so necessarily do the Mg(II) dynamics seen in MD simulation studies, with the exception of long-range screening effects. This is a fact rarely acknowledged in the contemporary modeling and simulation literature (Dudev and Lim 2003, 2008; Ditzler et al. 2010).

Second, modeling the nucleic acid backbone poses even bigger problems for MD simulation: While multivalent ions are dispensable in simulations, the nucleic acid backbone is not. As is generally true for biological anions, the electron density of the phosphate group extends far beyond the nuclei, making it a highly polarizable chemical entity that is very sensitive to its environment. This sensitivity, together with the complex anomeric properties of the ribose sugar ring, to which the phosphate is connected, results in a "deadly cocktail" of electronic structure effects that are quite beyond the capabilities of simple force fields utilizing conformation-independent point charges localized at atomic centers. One approach to gain control over these conformational dependencies is to use dihedral angle profiles, but this can never be done perfectly (Banas et al. 2010; Mladek et al. 2010; Zgarbová et al. 2011b).

Quantum chemistry provides an alternative and complementary approach to describe molecules. Modern quantum chemistry is a mainstream technique of physical chemistry with major impacts in many areas of chemistry. High-quality quantum chemical calculations can now be carried out that provide accurate assessments of structures and energies of small systems, currently in the range ~50+ atoms. Less reliable QM methods are applicable to describe larger systems, although the errors of such calculations can exceed those of force-field computations. In contrast to force-field methods, first-principle ab initio QM calculations can provide physically correct descriptions of model systems, with quantitative accuracy. The best QM approaches do not require any parametrizations. More detailed technical description of QM approaches, including explanations of the basic terminology and levels of calculations for nonspecialists working on nucleic acids, can be found in these references (Sponer et al. 2006, 2010; Banas et al. 2009).

When properly interpreted, QM calculations can contribute to our understanding of RNA structure and energetics by providing physicochemical insight into local molecular interactions and intrinsic geometrical preferences of molecular fragments that cannot be obtained by any other theoretical or experimental method (Hobza and Sponer 1999; Sponer et al. 1996a, b, c, 2001, 2008, 2010; Mladek et al. 2010). QM methods use fundamental principles, based on the time-independent Schrödinger equation, to rigorously calculate molecular wave functions. Expert applications of QM calculations provide rigorous assessments to replace ad hoc speculations concerning the fundamental nature of stabilizing molecular interactions or the origins of stereoelectronic effects (Bugg et al. 1971; Hunter 1993; Egli and Gessner 1995). For example, QM calculations ruled out the hypothesized presence of specific out-of-plane π–π effects in base stacking (Sponer et al. 1996a). In addition, QM methods showed that in sugar–base stacking, there is no specific molecular orbital interaction between the O4′ atom and nucleobase aromatic rings (Sponer et al. 1997). Also, QM methods invalidated the idea of specific attractive forces created by stacking of polar exocyclic groups of nucleobases with the aromatic rings of adjacent nucleobases (Sponer et al. 1996a). Rather, these calculations showed that all these interactions can be understood simply as ordinary van der Waals complexes, well described by properly parametrized molecular mechanics force fields, with no need for any additional, specific terms.

On the other hand, QM calculations allow for inclusion of effects that are often overlooked or ignored by structural biologists, owing to the fact that they lie beyond the applicability of molecular mechanics force fields. These effects include electronic structure polarization and charge transfer (Sponer et al. 2000; Petrov et al. 2011b), as well as pyramidalization of nucleobase exocyclic amino groups (Sponer and Hobza 1994a,b, 2003; Luisi et al. 1998; Vlieghe et al. 1999). Due to their completeness and accuracy, ab initio QM calculations play leading roles in the parametrization and refinement of modern force fields for nucleic acid MD simulations (Cieplak et al. 1995, 2009; Cornell et al. 1995), including the latest refinements of the AMBER Cornell et al. force field (Perez et al. 2007; Banas et al. 2010; Zgarbová et al. 2011b). Regarding force-field parametrization, QM methods are used to evaluate reference potential energy surfaces, thus linking molecular structures with energies. The molecular mechanics force field is then fitted to reproduce the reference QM data as accurately as possible.

As with other scientific methods, QM calculations have their advantages and limitations, and these will be discussed below. When the limitations are understood and respected, the methodology can provide valuable data. In this chapter, we discuss the applications of QM methods to recurrent nucleotide interactions, identified in RNA 3D structures by bioinformatic clustering analyses (Leontis and Westhof 2001; Leontis et al. 2002; Stombaugh et al. 2009; Zirbel et al. 2009). QM calculations provide further insight into the nature and energetics of these interactions, as summarized in a recent review (Sponer et al. 2010).

As discussed elsewhere in this book, structured RNA molecules play a wide range of roles in bacterial as well as eukaryal cells, involving all facets of gene expression. To carry out their functions, many RNA molecules form complex structures exhibiting unique, evolutionarily conserved architectures. An important feature of such structures is the occurrence of modular and recurrent 3D motifs. Generally, these motifs correspond one to one to the nominally single-stranded "loops" apparent in RNA secondary structures (hairpin, internal, and multihelix junction loops). These modular units play important structural and functional roles, by forming stabilizing tertiary interactions that organize the 3D architectures of RNA molecules or by directly interacting with other molecules, including other nucleic acids or proteins, or small molecule substrates and signaling molecules. Modular motifs are themselves highly structured entities, composed of recurrent and modular nucleotide interactions, including edge to edge, non-Watson–Crick base pairs, face-to-face base-stacking interactions, and specific base–backbone interactions. More complex interaction patterns, such as base triples and quadruples, can be decomposed into combinations of these interactions (Abu Almakarem et al. 2011). Thus, as we will show, QM calculations provide important baseline data concerning these interactions, including information about their geometries and energetics.

At the chemical level, the RNA and DNA backbones differ almost trivially: RNA possesses a hydroxyl group at the 2′-position of the sugar unit where DNA has a hydrogen atom. This seemingly minor difference has large implications for RNA architecture. First, it affects the preferred sugar conformation and therefore the type of helix formed (A-form in RNA vs. B-form in DNA). Second, it makes possible a range of interactions involving the "sugar edge" of RNA nucleotides. The sugar edge includes the atoms of each base that are exposed on the minor groove of canonical double helices: the O2 atoms of cytosine and uracil, C(O2) and U(O2) and the A(H2), A(N3), G(N2), and G(N3) atoms, in addition to the 2′-OH group of each nucleotide. The sugar edges of RNA nucleotides can interact with the Watson–Crick, Hoogsteen, or sugar edges of other bases to form unique sets of base pairs that play important roles in RNA structure, especially in stabilizing tertiary interactions. Thus, a prime focus of recent QM studies has been those RNA interactions that involve the sugar edge, which include six base-pair families (Sponer et al. 2005a, b, c, 2007, 2009; Vokacova et al. 2007; Mladek et al. 2009; Sharma et al. 2010a, b). As previous work on DNA already surveyed many of the non-Watson–Crick base pairs involving the Hoogsteen edge (Sponer et al. 2004), recent QM calculations on RNA have focused on the six "sugar edge" base-pair families.

Note that the differences in electronic structure between U and T are minor, and therefore, their base pairings at the Watson–Crick edges are similar (Swart et al. 2004; Perez et al. 2005). However, the presence of the 5-methyl group, which U lacks, serves to prevent Hoogsteen edge pairing in T, while it may enhance the thermodynamic stability of DNA double helices.

QM analysis of base-pairing interactions involving the Hoogsteen and WC edges (known as mismatches in DNA) can be found in a reference study (Sponer et al. 2004). QM calculations for other H-bonding interactions in RNA have been reported from different groups (Brandl et al. 2000, 2001; Oliva et al. 2006, 2007; Roy et al. 2008; Sharma et al. 2008, 2009, 2010a, b), including structure-energy analysis of base–phosphate (BPh) interactions (Zirbel et al. 2009; Zgarbova et al. 2011a).

12.2 Overview of the Ab Initio QM Methodology

12.2.1 Comparison to Other Computational Methods

This section aims to explain the basic features of QM methods for nonspecialists. Because ab initio QM calculations are based on first-principle theory and require no empirical parameters, they are fundamentally different from semiempirical QM approaches, as well as all force-field modeling methods (Sponer and Lankas 2006; Banas et al. 2009). Semiempirical QM approaches also rely extensively on specific parametrizations and therefore cannot be considered genuine electronic structure methods. Force-field methods employ effective, but nonphysical, analytic functions to model molecular forces and rely extensively on specific parametrizations. In contrast to parametrized computations, the accuracies of ab initio QM calculations can be systematically improved by (1) concerted and systematic improvement of the quality of atomic orbital basis sets and (2) by the inclusion of electron correlation effects. However, if the calculations are not properly balanced, they can fail miserably. It does not make sense to use small basis sets of atomic orbitals with high-quality electron correlation methods or vice versa. Both basis set size and level of electron correlation must be adequate to achieve quality computational results, as explained in detail elsewhere (Sponer and Lankas 2006; Banas et al. 2009). Finally, in contrast to force-field calculations, QM methods can be used to describe chemical reactions, which involve the breaking and creation of chemical bonds.

A unique feature of QM approaches is that above a certain level of theory, the calculations systematically converge to the hypothetical true values, that is, the values that in principle would be achieved by exact computations, which are understood to correspond to reality. The required level of theory depends on the nature of the system. Thus, the fundamental feature of modern ab initio QM computations is the absence of parameters and the theoretical guarantee of convergence, at least for systems with fully occupied electron shells and no unpaired electrons (closed-shell systems). Consequently, it is common practice in theoretical chemistry to treat high-quality QM computations on a par with reliable experimental data, although it is always important to specify exactly how the computations are carried out, especially the geometries utilized. The principle drawback of the approach is that the first-principle nature of the computations makes them very costly and they scale poorly with the number of atoms.

Traditionally, the term "ab initio QM calculation" has been applied to conventional, parameter-free, wave-function theory (WFT) computations (Sponer and Lankas 2006; Banas et al. 2009). The name reflects the fact that the wave functions, or molecular orbitals, are constructed as linear combinations of atomic orbitals. The basic level of WFT computations is the Hartree–Fock, self-consistent field approximation (HF/SCF). This approach is insufficient for most applications as it completely neglects electron correlation effects. Modest levels of electron correlation are included with the second-order Møller–Plesset method (MP2). For many applications, MP2 provides sufficient accuracy, especially when used with extrapolation to the complete (infinite) basis set (CBS) of atomic orbitals. CBS extrapolation is based on two MP2 calculations using large sets of atomic orbitals. The CCSD(T) variant of the coupled-cluster method includes a large portion of electron correlation and is the current gold standard for systems with dozens of atoms. The reader can find more details in our recent reviews (Sponer et al. 2008; Banas et al. 2009).

Conventional WFT QM calculations are now often replaced by cheaper density functional theory (DFT) methods. There has been a concerted and quite successful effort to improve DFT in the past few years (Zhao and Truhlar 2008; Banas et al. 2009; Rappoport et al. 2009; Grimme 2011). As a result of the efforts of many research groups, there are now more than 100 different DFT methods described in the literature, which consequently is difficult for nonspecialists to penetrate. One guiding comment is warranted regarding DFT methods: For a given application or chemical problem, the appropriate DFT method can be of comparable accuracy to the best WFT computations and at a tiny fraction of computer cost. However, none of the available DFT methods are sufficiently accurate for all types of applications simultaneously, as different DFT approaches were adjusted for different applications. Thus, the applicability of a given DFT to a new set of problems needs to be tested empirically. The practical relation of DFT and WFT computations is therefore the following: The highest-accuracy WFT calculations, such as CCSD(T) extrapolated to CBS, represent genuine benchmarks for calibrating DFT methods. Thus, benchmark databases of highly accurate WFT calculations are absolutely essential for tuning and further developing DFT methods (Jurecka et al. 2006).

12.2.2 Information Obtained from Ab Initio QM Calculations

QM calculations provide molecular wave functions for modeled systems, from which one can derive numerous physicochemical properties, including vibrational spectra, dipole and higher multipole moments, polarizabilities, proton affinities, and most fundamental NMR parameters, although QM methods do not yet exist to calculate all desired quantities with satisfactory accuracy. With regard to structural biology, the main achievement of QM calculations is the description of the nature of relevant molecular interactions and quantification of their energetics. QM results complement the purely structural data obtained by X-ray crystallography, the leading experimental approach of structural biology, from which the energetics of local molecular interactions can only be inferred indirectly. As noted above, interpretations of the observed contacts based on chemical intuition or subjective expectations can lead to inadequate accounts of the interaction patterns seen in structural data. By contrast, properly executed QM calculations of sufficiently high level can reliably describe molecular interactions in systems with closed-shell electronic structure and can be used safely to discuss the role of intrinsic molecular interactions in different contexts. While NMR studies of nucleic acids can provide dynamical data for macromolecular equilibria, from which, under suitable conditions, free energies can be determined, such solution experiments are inherently incapable of dissecting the intrinsic interactions (such as base-stacking and base-pairing energies) from the overall balance of forces (see below).

12.2.3 Nucleic Acid Systems Amenable to Study by QM Methods

In light of their high computational costs, QM calculations are generally applied to study small model systems, with the aim of accurately describing specific local forces that contribute to the structures, dynamics, and functions of macromolecular systems, including RNA molecules. Typical model systems include individual hydrogen-bonded base pairs, pairs of stacked bases ("base stacks"), covalently bonded dinucleotides, and bases interacting with backbone phosphate groups or with metal cations. Generally, the systems are studied in complete isolation, that is, in vacuo, the gas-phase condition. Gas-phase calculations reflect the intrinsic features of the electronic structure of the studied systems, with no perturbation from any other interactions. The payoff is the high accuracy and physicochemical completeness of the computational description, which cannot be achieved by any other technique. The reader should note that "electronic structure features" do not refer to special "quantum effects," irrelevant to biomolecular structure, but rather to the most fundamental descriptors, such as the basic ground state energy of base pairing and stacking.

12.2.4 Sources of RNA Geometries and Their Optimization for QM Calculations

QM calculations can provide meaningful energy data only when applied to appropriately selected and optimized geometries. Most commonly, one optimizes the geometry of the system before calculating the quantum mechanical interaction energy. The interaction energy is related to the binding energy, and its exact meaning will be explained below. For many systems, unconstrained gradient optimization leads to relevant structures. Modern QM software packages make it easy to carry out such calculations, in which all coordinates (or geometrical parameters) are optimized with respect to the electronic energy. Thus, the optimization algorithms locate the geometry on the potential energy surfaces that corresponds either to the local or the global energy minimum. This approach is suitable for systems where well-defined intrinsic energy minima of the isolated model systems correspond to biochemically relevant structures, as is the case for canonical Watson–Crick base pairs. On the other hand, the stacking patterns seen in experimental nucleic acid structures do not correspond to minima on the potential energy surfaces of isolated dimers of stacked nucleobases. In this case, point-by-point conformational scanning is preferred over geometry optimization (Sponer et al. 1996a, b, c).

An obvious option is to carry out QM energy calculations on experimentally determined structures. When they are available, atomic-resolution nucleic acid structures determined by X-ray crystallography are preferred over NMR structures. But even for X-ray structures, the atomic coordinates are generally not determined to sufficient accuracy to permit their direct use in QM energy computations (Sponer et al. 2008). First, the geometries of the individual monomers in PDB files are not sufficiently relaxed and will produce high electronic energies. Nonoptimal, intramolecular geometries can produce electronic distributions with incorrect (perturbed) electrostatic potentials, which can bias calculations of the intermolecular forces and binding energies. It is therefore necessary to replace (e.g., by overlay) the nucleobase monomers in the PDB files with QM-optimized monomer units.

Also, the internucleotide interaction geometries observed in X-ray structures may cause substantial errors in QM calculations. In particular, uncorrected steric clashes can produce large errors in the calculated energies. For example, errors arise because the base stacks in X-ray structures may be compressed or extended in the vertical direction due to inaccurate determination of the interbase dihedral angles. In fact, even small errors in the interbase distances, which may be acceptable for structural analysis, are not acceptable for QM analysis and can lead to considerable errors in calculated energies. This is true when the geometry falls within a range of interatomic separation distances where the short-range repulsions begin to dominate, as the calculated energy is a highly nonlinear function of the interatomic distance (r^{-6} and e^{-br} dependencies for dispersion attraction and short-range repulsion, respectively).

Similarly, the energies of H-bonded base pairs are sensitive to errors in the experimental geometry due to the very close approach of H-bonded atoms. Moreover, a bad geometry can result from the presence of two or more local substates, which are averaged in the experimentally derived geometry, leading to an unrealistic energy. Thus, models obtained by fiber diffraction cannot be recommended for direct calculations. Furthermore, caution is needed when using averaged geometries based on database studies, as they also can have unrealistic structures with poor energies. While this does not preclude studying the energies of experimental geometries, it does suggest that it is important to carefully check the geometries before carrying out energy calculations. In general, it is often necessary to adjust the starting geometries to eliminate unphysical contacts, while trying to stay as close as possible to the experimental geometries. The general rule is that all geometries should be assessed case by case before undertaking expensive QM calculations. If this is not done, the results are likely to be inaccurate irrespective of the quality of the QM method.

Furthermore, it is advisable to generate a range of structures around pairing as well as stacking geometries and to analyze the properties of the potential energy surfaces. With regard to base-stacking interactions, it remains an open question whether they can be characterized by a single geometry representing a unique energy minimum (Svozil et al. 2010). Most likely, stacking states correspond to a range of populated geometries, as evidenced by the significant coordinate fluctuations seen for stacked bases in explicit solvent molecular dynamics simulations, as well as in experimentally determined structures.

In many cases, we are interested in analyzing the local interactions which exist within a macromolecular complex of biological interest and which are substantially affected by the overall molecular context. In these cases, the best approach is to fix the local interaction geometry of interest accordingly and then to relax the monomers in situ (Vlieghe et al. 1999, Šponer et al. 2003). The intermolecular geometry can be frozen by fixing the coordinates defining the interaction, for example, as a set of six coordinates per dimer that includes the intermonomer distance, the two displacements orthogonal to the vector joining the centers of the two interacting units, and the three dihedral angles describing their relative orientations.

The sugar hydroxyl group at the 3′-position, which normally is covalently bonded to the 5′-carbon of the next residue, poses problems in QM computations of RNA base pairs. One option is methylation of the 3′-oxygen. In some cases, the phosphate groups participate in the interactions under study and also need to be included in the computations. This creates problems due to the strongly ionic nature of the associated interactions, which, in their biological context, are attenuated by solvent screening.

QM studies of RNA base or nucleotide interactions often require application of sophisticated geometrical constraints that need to be implemented case by case, to maintain the desirable features of the experimental structures. For electrically charged systems, optimization in the presence of continuum solvent may be a viable option, despite the limitations noted below. Continuum solvent models use a continuum dielectric to mimic the water environment (Tomasi et al. 2005). The continuum is polarized by the solute molecule and creates an electric field, which back-polarizes the solute electronic structure. The classical counterpart of QM continuum solvent models is based on Poisson–Boltzmann (PB) theory, or the simpler generalized born method (GB), which is used in molecular modeling. Classical force-field calculations obviously do not include polarization of the solute. While continuum solvent models accurately reflect the effect of solvent polarization on the solute electronic structure, they have one major weakness. The hydration energies are brutally sensitive to parameter settings, such as effective atomic radii. Inaccuracies in hydration energy estimates transfer to computations of binding energies, which is the fundamental reason, rarely admitted in the molecular modeling and simulation literature, that ligand binding energies calculated based on PB and GB approaches are so inaccurate (Špacková et al. 2003).

As noted above, when investigating complex interaction geometries, it is often necessary to apply suitable geometrical constraints to maintain a functionally relevant geometry. In previous work, QM methods were evaluated for their ability to reproduce experimentally observed structures of non-Watson–Crick base pairs (Sponer et al. 2005a, b, c, 2007, 2009; Vokacova et al. 2007; Mladek et al. 2009; Sharma et al. 2010a, b). For some base-pairing geometries, in spite of the use of geometrical constraints, full agreement with experimental structures was not obtained, even when taking into account the range of geometries observed experimentally. In these cases, caution is warranted when interpreting the interaction energies obtained. Thus, in some published work, energies are reported for structures that do not correspond to geometries seen in real RNA molecules, including changes in hydrogen-bonding patterns and deviations in intermonomer distances. Readers should always carefully examine the geometries used in QM calculations when evaluating the interaction energies obtained. The uncertainty in geometry concerns only some base-pairing patterns. Often, it is due to differences between the geometries of relaxed structures, calculated in isolation, and the geometries adopted by the same interaction in the context of intact, folded and solvated RNA molecules.

12.2.5 Advantages and Disadvantages of QM Methods

Although the gas-phase nature of QM computations provides the advantages of accuracy and completeness, it also poses certain disadvantages. The primary disadvantage is that the results of gas-phase calculations, although very accurate per se, are difficult to extrapolate to the solution state, to allow comparison with experimental data of nucleic acid stabilities in biologically relevant states. The experimental stabilities arise from the complex and context-dependent interplay of the intrinsic forces, which individually can be determined accurately by QM methods, with all the other effects that contribute to stability in solution environments, including solvation-related effects and entropy effects. In fact, the intrinsic contributions can be entirely masked by these other effects.

Nonetheless, the QM methodology has the advantage that it is the only tool that can reliably characterize the intrinsic conformational energetics of nucleic acid fragments and their interactions. Atomic-resolution structural studies reveal geometries but do not directly yield interaction energies. When two chemical functional groups are in contact in a structure, this does not necessarily mean there is a substantial attraction between them, attributable to the electronic structure. While solution thermodynamic (TD) measurements of nucleic acid stabilities provide key data for algorithms that predict secondary structures (Mathews et al. 1999; Mathews and Turner 2006), these measurements can only capture the overall stabilities associated with the studied systems and not the intrinsic base-stacking or base-pairing energies. It is often implicitly assumed in the literature that the trends in nucleic acid TD measurements can be interpreted in terms of the intrinsic interactions, even though the measurements cannot be decomposed into individual contributions. In fact, the relationship between the TD data and the strengths of intrinsic interactions is not known and has never been systematically investigated. When a meaningful correlation does not exist between TD data and the intrinsic forces, as is often the case, this should be clearly stated, to avoid explanations that are inconsistent with the basic physics of molecular interactions. The transferability of TD data to different systems, contexts or experimental conditions, where the balance of forces differs from the original measurements, is limited, because we do not know the balance between the intrinsic forces and extrinsic contributions. In fact, we remain far from achieving a true understanding of the factors that contribute to the measured TD data and parameters for RNA structure formation (Shankar et al. 2006; Siegfried et al. 2007; Hammond et al. 2010; Reblova et al. 2010). Interestingly, modern QM results obtained on these systems are generally not mentioned in the TD literature.

In summary, understanding the link between direct base-to-base interactions and the TD stabilities of nucleic acids remains a major challenge to which computational approaches can contribute (Yildirim and Turner 2005, Yildirim et al. 2009; Kopitz et al. 2008; Koller et al. 2010). The result could improve transferability of parameters, which is crucial for structure prediction (Yildirim and Turner 2005). Achieving reliable and quantitatively correct descriptions of the intrinsic interactions constitutes an important first step in this direction.

12.2.6 Comparison of QM Calculations to Gas-Phase Measurements of Nucleic Acid Interactions

QM calculations can only be compared directly with gas-phase association studies. The major advantages of QM calculations of biomolecular fragments over related gas-phase spectroscopic measurements include (1) that QM calculations allow one to simultaneously obtain structures and energies, which is not possible using any known experimental technique, and (2) that QM calculations allow one to directly investigate structures and geometries that are relevant to biology, by matching the geometries of the studied fragments to those observed in the intact biomolecules. By contrast, gas-phase experiments typically report on global equilibrium minima of the fragments of interest in the gas phase, which may be quite different from the geometries actually adopted in functionally relevant, macromolecular contexts. Field ionization mass spectrometry, first developed in the 1970s, is the only available experiment for measuring the enthalpies of nucleobase complexes in the gas phase, but it does not provide any information about their geometries (Yanson et al. 1979). Moreover, this experiment generally reflects a mixture of diverse structures and, thus, not only the expected Watson–Crick arrangement, as subsequently demonstrated by theoretical QM and force-field simulation studies (Kratochvil et al. 1998, 2000). The subsequent attempt to measure the energetics of base pairs in the 1990s using IR laser desorption into molecular beam expansions completely failed to achieve thermodynamic equilibrium, and the results were in striking disagreement with the earlier mass spectrometry data, as well as with QM data (Dey et al. 1994). These results were subsequently invalidated (Hobza et al. 1996). While the most advanced current experiments, based on laser desorption and IR–UV spectroscopy, do not report the energetics of base pairing, they do provide indirect information about the populated structures (Nir et al. 2000), although with some obvious limitations: The structures populated in the gas-phase experiments are not always the biologically relevant ones. For example, G and C nucleobases form rare tautomers in the gas phase that they do not form at all in polar solvents. The process by which the biomolecular building blocks are introduced into the gas phase (such as laser desorption) can also affect the structures observed in these experiments. Frequently, this generates excited state intermediates. Interactions that are important in the liquid phase, for example, base stacking, may not be significantly populated in the gas phase and, thus, are not detected spectroscopically. Given the flexible nucleic acids backbone, it is virtually impossible to execute gas-phase experiments in a manner that samples biochemically relevant backbone geometries.

In summary, QM is the only technique that can provide the intrinsic energetics of the nucleic acid molecular building blocks. The most accurate contemporary QM methods achieve an "expected accuracy" of ~0.5 kcal/mol for base-pairing or base-stacking interaction energies. Expected accuracy means that it is a qualified estimate with respect to the hypothetical (and intrinsically unknowable) values that can only be obtained by fully converged QM calculations. For comparison, optimal values of base-pair stacks are around −10 kcal/mol, while H-bonded base pairs are typically in the range from −10 to −30 kcal/mol. For comparison, the interaction energy of a water dimer in the gas phase, representing one hydrogen bond, is ~−5 kcal/mol (Feyereisen et al. 1996). For simple systems, such as the water dimer, where (relatively) unambiguous experimental energy data are available, the agreement between contemporary theory and experiments is, in fact, almost perfect.

12.2.7 Inclusion of Solvent Effects in QM Calculations

Inclusion of solvent effects in QM calculations is possible, but such calculations do not achieve the quality, accuracy, or reliability of gas-phase computations (Miller and Kollman 1996; Tomasi et al. 2005; Klamt et al. 2009). Typically, continuum solvation methods are used to mimic the solvent screening of electrostatic interactions. Alternatively, a small number of explicit water molecules are included to model key water molecules bridging nucleobases or hydrating cations. High-quality calculations show that polar solvent molecules generally do not substantially perturb the nature of intrinsic molecular interactions (i.e., the direct interactions between the biomolecular subsystems), although they do tend to electronically polarize the solute molecules. However, the added solvation free energy dramatically affects the overall free energy balance. For example, solvent screening erases large portions of the stabilization of base pairs arising from H-bonding, which is largely an electrostatic effect, and eliminates the effect of electrostatics on the dependence of base-stacking energy on the twist angle between the in-plane dipoles of two stacked bases. In other words, the coulombic parts of these interactions are counterbalanced by the solvent screening: The intrinsic base–base electrostatic interaction is still present but is mirrored by the energetically opposing term due to solvation energy. Base stacks with optimized intrinsic electrostatic terms have worse solvation energies than stacks with repulsive intrinsic electrostatics. The variations in the intrinsic and solvation electrostatic terms cancel each other (Florian et al. 1999). Water also suppresses the capabilities of guanine and cytosine to form rare tautomers, among other examples of solvent effects (Colominas et al. 1996). The recent prediction that guanine is surprisingly deprotonated in water at its N1 position (shifting the proton to N7) is example of incorrect solvent calculations (Hanus et al. 2003). It is very unlikely that tautomer species are substantially populated in biochemically relevant environments. Some nucleobases such as isoguanosine can readily form tautomers in water, but these nucleobases were rejected by evolution (Blas et al. 2004). These issues need to be considered when interpreting and extrapolating from QM calculations to experimental observations. Unfortunately, as noted above, the values of hydration energies obtained by classical as well as QM-based continuum solvent approaches are quite sensitive to the specific parametrization, including the choice of atomic radii. Consequently, we presently lack reliable methods for obtaining consistent and accurate estimates of hydration energies that are unambiguously applicable to base-pairing and base-stacking interactions inside nucleic acids. In the simplest QM approach to solvation, the model complex, for example, two paired or stacked bases, is fully immersed in a continuum solvent, parametrized to model bulk water. This leads to almost complete counterbalancing of the electrostatic interactions. However, within large nucleic acid complexes, the bases are less exposed to solvent, and we do not presently have any method that adequately accounts for local structural context to allow us to confidently study the effect of solvent screening within nucleic acids.

In summary, our current options for modeling solvent effects using continuum approaches are far from satisfactory. Classical continuum solvent calculations for modeling nucleic acids, like those known as MM–PBSA schemes ("molecular mechanics Poisson–Boltzmann surface area") (Kollman et al. 2000), are inherently inaccurate and should be used with caution (Ditzler et al. 2010). In light of these limitations, the best choice may be to avoid biasing accurate QM calculations by including inherently inaccurate solvent corrections and instead to interpret the QM results in light of genuine gas-phase data. However, there are situations where at least some inclusion of solvent effects is unavoidable, for example, in studying ionic systems, which contain a net electrical charge and which in the gas phase are obviously dominated by the electrostatic terms, an unrealistic situation for nucleic acids. Unfortunately, even the single negative charges of individual phosphate groups severely compromise our ability to carry out gas-phase calculations in a way that yields biologically relevant results. Many computational scientists carry out QM computations on larger and larger systems, in the belief that increasing the number of atoms increases the biological relevance of the calculations. However, this is not always the best choice, as increasing the size of the system may amplify rather than reduce the bias arising from the incompleteness of the studied system (Mladek et al. 2010). For example, more transferable QM results are obtained for base pairs by using model systems consisting of two bases rather than two complete nucleotides, including both phosphate groups.

12.2.8 Quality of QM Calculations

Quantum chemical calculations are based on solving the time-independent Schrödinger equation. The aim of the calculations is to achieve the highest possible accuracy and most physically complete description of the system in a given geometry or set of geometries. However, there is an unavoidable trade off in pursuing these goals: On the one hand, on smaller systems, accuracy superior to any other experimental or theoretical method can be achieved. The best QM methods achieve spectroscopic accuracy. On the other hand, this may come at the cost of omitting the larger molecular context.

Our ability to carry out high-quality QM calculations is the result of advances in computer technologies, both hardware and software. The first truly modern computations of base-pairing and base-stacking interactions between two nucleobases were published around 1995 (Hobza et al. 1995; Sponer et al. 1996a, b, c). Previously, such computations were simply not feasible.

The wider scientific community recognized the advances in ab initio and DFT QM techniques that made calculations such as these possible by awarding the Nobel Prize in Chemistry in 1998 to J.A. Pople and W. Kohn for establishing the basic foundations of quantum chemistry (http://nobelprize.org/nobel_prizes/chemistry/laureates/1998/). This award was made just a few years after the first reliable computations on chemically interesting systems became feasible. In fact, the overall impact of quantum chemistry in contemporary science is considerably larger than the impact of force-field modeling and molecular dynamics simulation. Moreover, the field has seen a steady improvement of available methods. Since the middle 1990s, the application of QM methods has become standard practice in many areas of science, as an indispensable complement to experimental approaches, often assisting experimentalists in identifying interesting systems for study. However, QM methods are less frequently applied to structural biology, largely because of the complexity of the systems of interest to this field. In addition, structural biologists are generally less aware of the capabilities of QM methods and tend to turn to force-field modeling and simulation.

To apply QM to questions relevant to structural biology, the modeled system must be carefully designed, an adequate level of QM treatment must be selected, and the resulting data must be properly interpreted, based on in-depth knowledge of the experimental system of interest. With the ready, off-the-shelf availability of powerful computers and state-of-the-art software codes, anyone can carry out QM computations of biomolecular interactions. However, lack of expertise may lead to errors, including use of inappropriate methods, design of incorrect or irrelevant systems, and invalid extrapolation of the results to the experimental situation. Such errors are not uncommon in the contemporary literature and have contributed to the perception that QM results are not particularly useful in structural biology. However, the fault lies not in the methodology itself but in errors in its application. We refer the reader to specialized reviews for technical details regarding how to appropriately execute QM calculations relevant to nucleic acids (Sponer and Lankas 2006; Banas et al. 2009). In summary, unlike the results of classical simulations, QM calculations of nucleic acids are not directly comparable to biologically relevant situations. On the other hand, simulations are not able to achieve the accuracy of QM calculations.

12.2.9 Physicochemical Interpretation and Meaning of Ab Initio QM Results

As noted above, high-level QM calculations are carried out on model systems of interest, for example, stacked or paired bases in complete isolation ("gas phase"), and thus, the calculations reflect the system's intrinsic properties. For H-bonded base pairs, we usually first determine the optimal geometry of the base pair using gradient geometry optimization of the electronic energy of the whole system (Sponer et al. 2004). In some cases, the optimization should be executed while imposing constraints to model geometries observed experimentally.

For base stacking, global gas-phase optimizations generally result in biochemically irrelevant geometries (Sponer et al. 1996a, b, c, 2001, 2008; Hobza and Sponer 1999). Thus, stacking calculations are carried out for geometries constrained to remain close to experimentally observed stacking interactions. Alternatively, sets of calculations are carried out, with systematic variation of geometric parameters, to map out a selected region of the potential energy surface of the interacting dimer.

After the desired geometry for the interacting bases is calculated or selected, the interaction energy is evaluated. The interaction energy, ΔE^{AB}, between two subsystems, A and B, in the given geometry, is the energy difference between the total electronic energy of the dimer in that geometry, E^{AB}, and the electronic energies E^{A} and E^{B} of isolated subsystems separated to infinity where they do not interact (12.1).

$$\Delta E^{AB} = E^{AB} - E^{A} - E^{B}. \tag{12.1}$$

The interaction energy reflects the electronic part of the molecular interaction, which is chemically the most interesting contribution to the stabilization. This number reports, for a given geometry, on the interaction between the electronic structures of the two monomers. The interaction energy calculated in this manner is equivalent to the energy calculated with force-field methods using the same equation (12.1). However, the QM method is more rigorous and complete than any force-field evaluation. Occasionally, the view is expressed that QM calculations are irrelevant for biological applications because they are primarily designed to capture specific or marginal quantum effects. This is a misunderstanding, since QM calculations derive the fundamental energetics, based on first principles, including all electronic contributions, and with an accuracy that considerably exceeds that of force-field calculations. In effect, QM calculations provide a tool for determining the total electronic structure energy, in a given geometry, of hypothetical gas-phase systems at absolute zero (0 K), from which electronic energy differences between different geometries can be determined.

12.3 Applications of Ab Initio QM Methods

12.3.1 Fundamental Understanding of the Nature of Base-Stacking Interactions

We illustrate the range of applicability of QM methods using several examples from published research, starting with base-stacking interactions, for which QM calculations were instrumental in obtaining a correct theory (Sponer et al. 1996a, c, 2001, 2008; Hobza et al. 1997; Hobza and Sponer 1999). At the beginning of the 1990s, several mutually contradictory theories of base stacking were circulating in the literature. One common view proposed that base stacking is a complex interaction in which the π-electron clouds of the bases play an integral role that cannot be properly captured by standard molecular mechanics force fields with in-plane charges (Hunter 1993). In 1996, high-level QM calculations demonstrated, quite unexpectedly, that base stacking can, in fact, be well described to a good approximation as a combination of the three most common contributions to molecular interactions: electrostatic attraction, dispersion attraction, and short-range exchange repulsion (Sponer et al. 1996a, c). The calculations ruled out the existence of any additional, specific "π–π" interactions that distinguish aromatic stacking from ordinary, nonaromatic van der Waals interactions. Extensive comparisons of rigorous QM calculations and simple force-field calculations lacking any "π–π" terms showed amazingly good agreement over the whole potential energy surface. The calculations demonstrated that base stacking can be described quite well using force fields that combine the van der Waals interaction, represented by a Lennard-Jones potential energy term, with the electrostatic interaction, represented by the Coulomb potential calculated using an appropriate set of atom-centered point charges. These results did not imply that the derived force fields are perfect, but they did provide the first theoretical verification supporting the qualitative correctness of AMBER type of force field, which now dominates contemporary molecular modeling (Cornell et al. 1995; Hobza et al. 1997), a significant theoretical accomplishment. These earlier, reference QM calculations were recently repeated with much improved methods (Jurecka et al. 2004; Sponer et al. 2006; Morgado et al. 2009). Although the calculated energies were quantitatively shifted to somewhat more stabilizing values, due to full inclusion of the dispersion energy, the basic physicochemical picture of base stacking remains unchanged in the light of the new calculations (see Fig. 12.1). This example nicely demonstrates the convergence of results obtained by modern quantum chemistry.

12.3.2 Elucidation of Role of Amino-Group Pyramidalization in Base Pairing

Some of the first QM studies of isolated nucleobases to include electron correlation revealed that the geometries of A, G, and C exocyclic amino groups are intrinsically nonplanar, adopting a partially sp^3 pyramidal geometry (Sponer et al. 1994). It took almost a decade to measure this theoretically predicted effect by an appropriate experiment, using IR spectroscopy and cooling to 0.37 K in liquid helium nanodroplets (Dong and Miller 2002). Amino-group pyramidalization effects are entirely neglected by developers of contemporary molecular mechanical force fields and are rarely considered by structural biologists, who assume that amino-group hydrogen atoms do not substantially deviate from the base planes, even in the presence of attractive out-of-plane interactions that can be readily inferred from structural data. The amino groups that are involved in primary in-plane H-bonds are, of course, planarized. This is the case for the amino groups in the canonical *cis* Watson–Crick (cWW) AU and GC Watson–Crick base pairs, where H-bonding shifts the electronic structure to essentially complete sp^2-hybridization of the amino nitrogen orbitals. However, the QM calculations showed that amino-group pyramidalization can stabilize certain less frequent interactions. An important example for RNA 3D structure is the intrinsically nonplanar cWW AG base pair, where, in a planar arrangement, the unpaired amino group of guanine would interact unfavorably with the C2–H group of adenosine. The base pair responds by large propeller twisting about an axis running along the Hoogsteen edges of the paired bases, shifting the guanine amino group away from the plane of the adenine, and

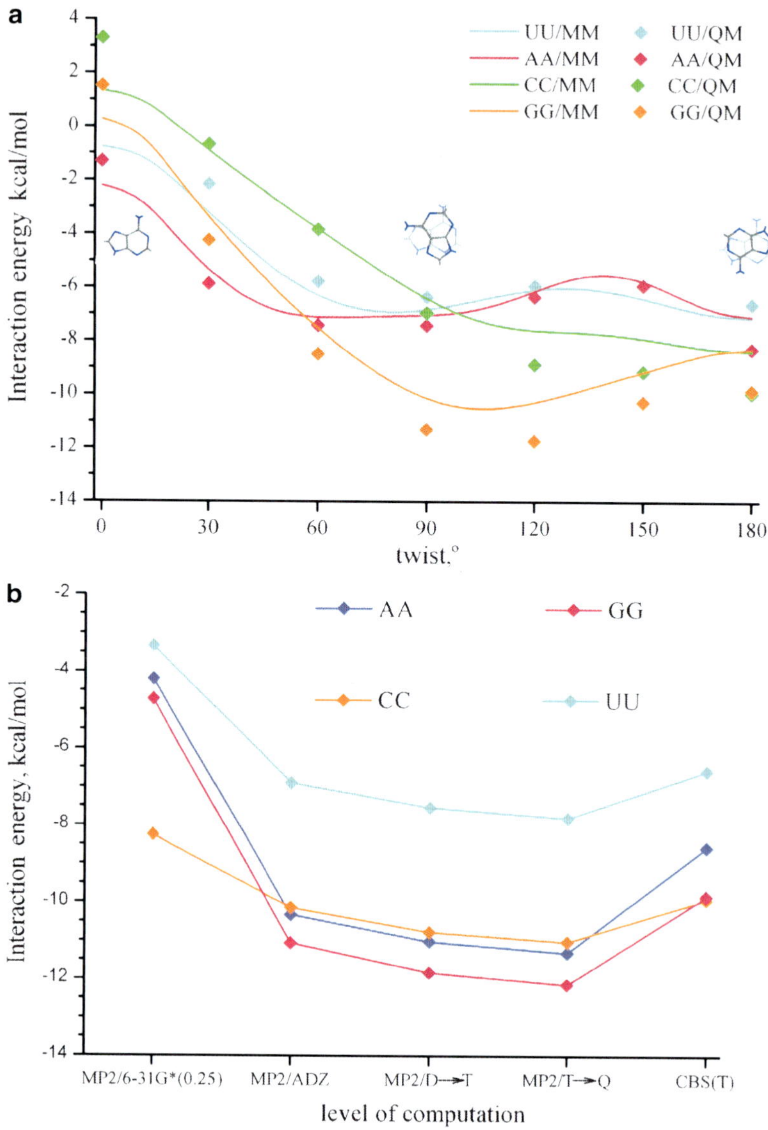

Fig. 12.1 (**a**) Variation of the base-stacking energy as a function of the twist angle between the nucleobases in A/A, U/U, C/C, and G/G base–base stacks. The actual twist angle is illustrated by the geometries of A/A stacks. The solid lines represent force-field calculations using the Cornell et al. (1995) (AMBER) MM force field. The quantum chemical data (represented by *filled diamond*) were obtained at the MP2 level of theory and were extrapolated to the complete basis set (CBS) of atomic orbitals. The energy data also include correction for higher-level electron correlation effects. (**b**) Dependence of the stacking energies on the quality of the theoretical approximation used. Stacking energies were computed for antiparallel undisplaced face-to-back arrangements of A/A, U/U, C/C, and G/G stacks. The following theoretical approaches were considered: MP2/6-31G*(0.25), MP2/aug-cc-pVDZ (ADZ) calculations, MP2/CBS calculations using aug-cc-pVDZ → aug-cc-pVTZ (D → T) and aug-cc-pVTZ → aug-cc-pVQZ (T → Q) extrapolations, and the final MP2/CBS T → Q calculations corrected for the CCSD(T) contribution with small basis set (CBS(T)). AXZ(X = D,T,Q) = aug-cc-pVXZ (Sponer et al. 2008)

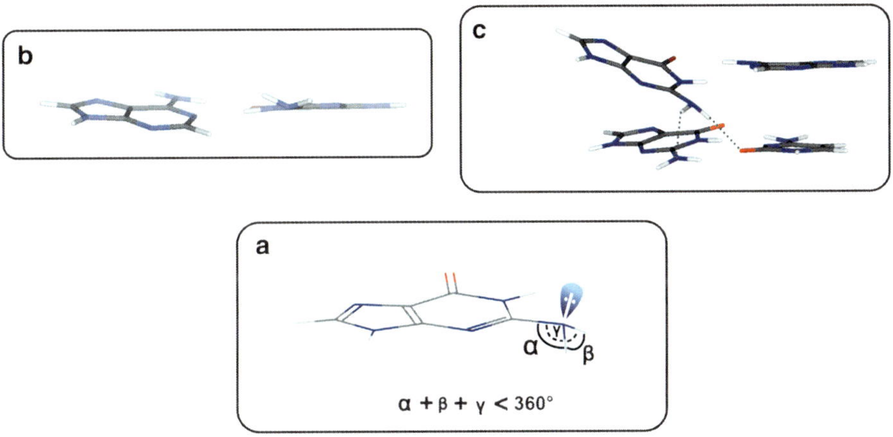

Fig. 12.2 The amino groups of nucleic acid bases are intrinsically nonplanar, that is, their nitrogens adopt a partial sp^3 hybridization in isolation (**a**) Pyramidalization means that the sum of the three valence angles around the amino-group nitrogen is less than 360° and a lone electron pair develops above the nitrogen. The amino groups of isolated nucleobases have two C_s-symmetry-related (inverted) minima, with the planar arrangement being the transition state between them. The balance between sp^2 and sp^3 hybridizations is very sensitive to molecular interactions with other molecules, that is, the amino groups are very flexible. The amino groups can form out-of-plane H-bonds as well as amino-acceptor interactions. Molecular mechanics force fields do not allow to describe the flexibility of the amino-group electronic structure and typically enforce the sp^2 planar arrangement. (**b**) The cWW AG base pair is intrinsically nonplanar, with propeller twisting roughly around the O6(G) and N6(A) axis. This, together with amino-group pyramidalization, alleviates repulsion between H2(A) and the amino group of G. The nonplanar amino-group hydrogens can then form out-of-plane H-bonds with O2 of GC or AU base pairs stacked below, as visualized in (**c**), where the base-pair geometries are taken from experimental structure while hydrogen positions are determined via QM (Sponer et al. 2003)

resulting in pyramidalization of this amino group, as shown in Fig. 12.2. This characteristic conformation has been seen many times in both RNA and DNA X-ray structures but, to the best of our knowledge, in all these studies, the nonplanarity was incorrectly attributed to the effects of base stacking, under the assumption that the base pair is intrinsically planar (Prive et al. 1987; Ennifar et al. 1999).

In reality, base stacking opposes the large, observed propeller twisting, because the intrinsic propeller twist of cWW A/G is so large that it reduces intrastrand stacking with the adjacent canonical base pairs, which have smaller propeller twist. By contrast, the canonical Watson–Crick base pairs are intrinsically planar, and their modest propeller twist is inherent to the A-RNA helix topology, as the intrastrand overlap of bases improves with helical twisting. Furthermore, interpretations of the structural studies overlooked the formation of stabilizing, out-of-plane, cross-strand H-bonds between the guanine 2-amino group and the O2 keto groups of pyrimidines of adjacent Watson–Crick base pairs. This example shows that incorrect theoretical premises concerning intrinsic interactions may easily lead to incorrect interpretations, even of local molecular interactions evident in atomic-resolution X-ray structures. In fact, out-of-plane

H-bonding interactions involving pyramidalization of 2-amino groups have been shown to affect the degree of AG to GA covariation in RNA sequences in motifs that form cWW base pairs and to contribute to the context dependence of the geometry of the resulting base pair (Sponer et al. 2003), as AG base juxtapositions adjacent to helices can result in other kinds of base pairs, for example, *trans* Hoogsteen/sugar edge (tHS), depending on context (Yildirim et al. 2009). Other interactions are also known to profit from activation of the partial sp^3 hybridization of the exocyclic amino nitrogen atoms of bases, which may even involve weak amino-acceptor interactions (Sponer and Hobza 1994a, 1994b; Luisi et al. 1998; Vlieghe et al. 1999).

12.3.3 Applications of QM Calculations: Molecular Mechanics Force Fields

As noted above, QM calculations play crucial roles as sources of reference data for the validation and reparametrization of molecular modeling force fields (Hobza et al. 1997). For current pair-additive force fields, the largest challenge is the parametrization of the torsion profiles. Bond and angle parameters for force fields can be derived from structural data, IR, and microwave spectroscopy, in combination with high-level QM calculations. To determine intermolecular parameters, relatively straightforward protocols are available. Van der Waals radii and well depths can be derived by matching experimental densities, while atomic charges can be parametrized by fitting to QM-derived electrostatic potentials and energies.

Fitting of the torsional parameters, on the other hand, is difficult because their actual physical meaning is not clearly defined. They do not directly correspond to real electronic structure contributions but rather represent ad hoc functions used to tune the behavior of the force field. Medium-level QM calculations were used to derive the initial torsional profiles of the Cornell et al. (AMBER) force field, which is most widely used to model nucleic acids. Recently, modern electron correlation QM calculations were used to further refine the torsion profiles of the force field. The α/γ torsion profiles were reparametrized in the parmbsc0 force field to prevent the ladder-like degradation of B-DNA in 10+ ns simulations (Perez et al. 2007). This reparametrization is valid for both DNA and RNA. Also, the χ torsion profile was reparametrized for RNA in the parmχOL$_3$ force field to prevent ladder-like degradation of RNA on even longer time scales, as shown in Fig. 12.3 (Banas et al. 2010).

Finally, QM calculations can be combined with empirical force fields, in so-called quantum-chemical/molecular mechanical (QM/MM) "hybrid" approaches, to model molecular and chemical dynamics. In these methods, a small but crucial part of the system, for example, the active site of an enzyme, where chemical

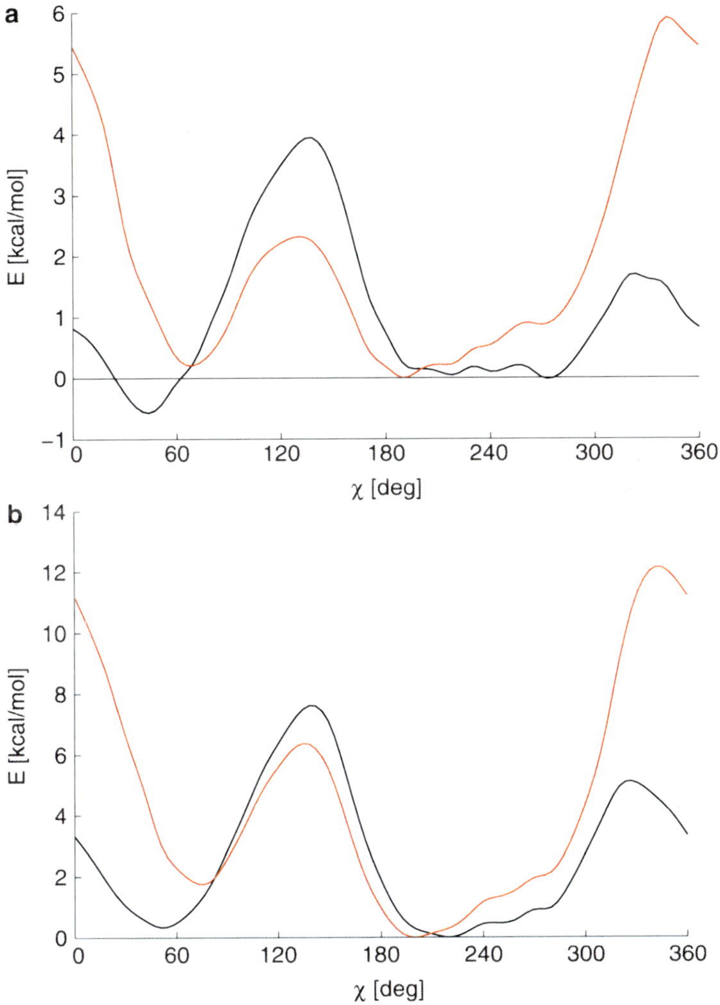

Fig. 12.3 The performance of molecular simulations force fields is substantially affected by dihedral terms, which currently are parametrized using advanced QM methods. The figure shows the full energy profile (calculated by inclusion of a Poisson–Boltzmann water model) of rotation around the glycosidic bond for (**a**) rA and (**b**) rC. The *black curve* is the original Cornell et al. force field (parm94–parmbsc0 versions); the *red curve* is its major refitting using modern QM methods (parmχOL$_3$ version). Note the difference in the balance of the *anti* to *high-anti* regions, which is critical for stabilizing long RNA simulations, and the change of the position of the *syn* minimum, which is critical for stabilizing UNCG tetraloops and eliminating the *syn* vs. *anti* bias of the original force field (Banas et al. 2010)

reactions occur, is treated using QM methods while the rest of the system is treated using a classical force field. For RNA, this approach is especially useful for studies of ribozyme reaction mechanisms [for a recent review, see (Banas et al. 2009)].

12.3.4 Comparison of QM Calculated Energies and Experimental Gas-Phase Energies

The evaluation of the interaction energy corresponds to a hypothetical in vacuo dimerization process at absolute zero (0 K) temperature, during which the interacting monomers are brought from an infinite separation to the specific interaction geometry. Geometries obtained by gradient optimization correspond to local or global electronic energy minima. However, the interaction energy can be calculated for any geometry of interest. In practice, QM methods are used to assess the interaction energy for hundreds or even thousands of configurations, since it is a unique function of the *xyz* coordinates of the molecular complexes under investigation. What is the relation of such calculations to actual experiments carried out in the gas phase? The interaction energy in minimized structures is related to, but not identical with, the gas-phase binding energy, D_0, and the enthalpy of formation of the complex. Direct comparison of the energy calculated by QM would require an experiment conducted at absolute zero. The energy measured in such an experiment would include, however, the zero-point vibrational energy (ZPE). The ZPE can be calculated, usually within the harmonic approximation, by evaluation of second derivatives of the energy at the minimum-energy geometry. For comparison to experiments conducted at nonzero temperatures, the enthalpy and entropy contributions need to be calculated and added to the QM result. This calculation is straightforward when assuming harmonicity and the rigid-rotator approximation (Hobza and Sponer 1999) but becomes quite tedious when it is necessary to consider the anharmonicity of vibrations around minima (Spirko et al. 1997) or competition between several structures on the free energy surface (Kratochvil et al. 1998, 2000).

12.4 QM Calculations of RNA Base Pairs

12.4.1 Methodology for QM Calculations of RNA Base Pairs Involving the Sugar Edge

RNA nucleobases present three edges for base pairing: the Watson–Crick (W), Hoogsteen (H), and sugar (S) edges. Pairing can occur in two relative orientations of the glycosidic bonds, *cis* and *trans*, resulting in 12 basic types or geometric families of base pairs (Leontis and Westhof 2001, Leontis et al. 2002). A unique feature of RNA structure is the occurrence of a large variety of stable base-pairing interactions. Six of the 12 standard geometric base-pair families involve the sugar edge of at least one of the interacting nucleotides. As these interactions are unique to RNA, they are of special interest and all six have been subjected to quantum chemical calculations. When only one sugar edge is involved, it is sufficient for QM calculations to treat model systems consisting of a base and a nucleoside. If the interaction involves two sugar edges, then two nucleosides are needed. Phosphate groups are omitted for the reasons discussed above. All geometry optimizations and subsequent interaction energy calculations are carried out in the gas phase. In some cases, the interaction energies have been reevaluated with inclusion of solvation effects in the form of a continuum dielectric, which compensates for the electrostatic part of the intermolecular stabilization, as this tends to be overestimated by gas-phase calculations compared to the aqueous solution state, as discussed above in Sect. 12.2.7.

12.4.2 Results of QM Computations on RNA Sugar-Edge Base Pairs

We begin this section by explaining how QM results obtained for RNA base pairs are interpreted. It is well established that the biological role and frequency of occurrence of different RNA base pairs are primarily determined by their shapes, at least when their energies are comparable. However, the interaction energy (which is a measure of the intrinsic in vacuo stability of the H-bonding interactions) also plays an important role in determining molecular structure and functional properties. Since the RNA base pairs from different geometric families are very diverse in shape and occur in different structural contexts, we do not expect to observe a direct correlation between interaction energies and frequencies of occurrence of base-pair families per se. However, we suggest that energies can be important as secondary factors, especially when considering mutually interchangeable isosteric and near isosteric base pairs belonging to the same geometric family. In any case, it is useful to know their relative stabilities for a more complete understanding of their roles in functional RNA molecules.

As discussed below, computations also provide insights regarding the nature of these base-pairing interactions. The interaction energy consists of two fundamental parts: the Hartree–Fock (HF) component and the electron correlation component. The electron correlation term consists mainly of the dispersion attraction, which is entirely absent from the HF term, supplemented by some corrections to the other contributions, including electrostatics, exchange repulsion, and induction, that are at least partially accounted for by the HF term. The ratio of the electron correlation component to the HF component thus provides an estimate of the relative contributions of dispersion and electrostatics to base-pair formation. Since the electrostatic component, as discussed above, is counterbalanced by solvent screening, while the dispersion energy is not, it is reasonable to suggest that base pairs with considerable correlation interaction energy behave more hydrophobically than base pairs clearly dominated by the HF term. Although we intentionally do not try to quantify this feature, which can be context dependent, we can say that the biological relevance of this distinction is that base pairs exhibiting more "dispersion" or "hydrophobicity" appear predisposed to form effective tertiary interactions, as these interactions need to compete directly with solvation.

The QM studies we have carried out on RNA base pairs involving sugar–base contacts confirm that the 2′-OH moiety of ribose actively participates in stabilizing these binding patterns. In addition, decomposition of the interaction energies has revealed that electron correlation plays a more significant role in base pairing involving the sugar edge of the nucleotides than in base pairs stabilized purely by base–base contacts. For example, while in the G/C cWW base pair, the correlation energy is only 10% of the total gas-phase interaction energy, in some sugar-edge pairs, it ranges as high as 50%.

The QM calculations also give insights regarding the geometrical features of the base pairs. For those base pairs where there is a good match between computed and observed geometries, we can safely conclude that the observed geometries reflect the intrinsic stabilities of the interactions. However, the more difficult it is to computationally reproduce the geometry observed experimentally, the more likely it is that the observed structures result from an interplay between the interacting nucleosides and the surrounding structural contexts, which can include obligatory interactions with a third nucleoside to form a base triple.

In fact, we find computationally that some sugar-edge base pairs are not stable per se in their experimental geometries. In addition, the discrepancy between computed and observed geometries may indicate the presence of less frequent alternative geometrical substates which occasionally can form in some specific contexts. Of course, all these situations need to be judged case by case as the enormous diversity of the RNA base pairs precludes formulating a universally valid relation between intrinsically preferred and observed structures.

The QM results obtained for base pairs involving the sugar edge are summarized in Table 12.1. The results include intrinsically preferred structures and intrinsic interaction energies and complement structural and frequency-of-occurrence data obtained by structural bioinformatics. This catalogue can be consulted to obtain insights into the intrinsic features of these base pairs. Points of agreement, as well as disagreement, between the computed and observed structures are equally important and provide information about a given geometric base-pair family (type) or a specific base combination forming that base-pair type.

In the following sections, we consider each sugar-edge family in turn. The cited literature can be consulted for more details.

12.4.2.1 The *cis* Watson–Crick/Sugar-Edge (cWS) Base-Pair Family

At this point, all 16 members of the cWS family combinations have been observed in RNA crystal structures (Stombaugh et al. 2009), but when the quantum chemical investigation was carried out (Sponer et al. 2005b), only 13 were known; 12 of these were found by QM methods to be intrinsically stable although two of them required a constrained optimization. By "intrinsically stable," we mean that the geometry obtained for the base pair by unconstrained, gas-phase optimization agreed with the

Table 12.1 Summary of RIMP2/aug-cc-pVDZ interaction energies (*E*, kcal/mol) and occurrence frequencies (*F*, %) of all sequence variants of the six base-pair families involving the sugar edge, cWS, tWS, cHS, tHS, cSS, and tSS

cWS		A	C	G	U
A	E	-17.1	-17.9	-16	-16.5
	F	26.9	26.9	0.4	12.8
C	E	-19.5	-21.8	-17	-20.1
	F	5.0	2.9	1.2	5.4
G	E	-14.6[a]	-15.2[b]	-29.5[c]	-9.4[b]
	F	0.8	2.1	2.1	2.5
U	E	-16.9	-15.3[a]	-16	-17.3[a]
	F	6.6	0.4	3.3	0.8

tWS		A	C	G	U
A	E	-9.9 (-13.8)[d]	-17.2[a]	-16.5 (-16.1)[d]	-16.2[a]
	F	2.9	0.5	49.5	1.4
C	E	-10.8 (-6.6)[d]	-9.7 (-14.4)[d]	-20.9 (-17.0)[d]	-8.1[a]
	F	1.0	3.3	6.2	0.0
G	E		-28.4		-8.1
	F	n.e.	2.4	n.e.	24.8
U	E	-9.9	-15.9	-26.4	-17.7[a]
	F	5.2	1.4	1.4	0.0

cSS		A	C	G	U
A	E	-18.5	-19.0	-23.6	-15.7
	F	6	24.3	11.4	5.8
C	E	-20.0	-21.4[a]	-26.9	-17.2
	F	24.3	0.0	2.2	0.5
G	E	-23.7	-22.9[a]	-24.3	-18.9[a,e]
	F	11.4	2.2	0.7	2.5
U	E	-19.6	-28.2[c]	-23.3	-15.8[a]
	F	5.8	0.5	2.5	0.1

tSS		A	C	G	U
A	E	-16.5	-15.2 (-10.5)[f]	-21.4	-14.3
	F	4.4	15.4	64.6	8.2
C	E				
	F	n.e.	n.e.	n.e.	n.e.
G	E	-21.0[a]	-13.5 (-20.9)[d]	-21.2 (-17.9)[d]	-4.4
	F	0.0	0.6	6.6	0.2
U	E				
	F	n.e.	n.e.	n.e.	n.e.

cHS		A	C	G	U
A	E	-16.9 / -13.6 / -[g]	-15.6 / - / -15.6[g]	-12.1 / - / -15.8[g]	-8.4 / - / -15.2
	F	10.0	0.8	2.3	3.1
C	E	-5.2	-11.7 / -7.7 / -9.5[g]	-6.3	-7.9 / -7.9 / -[g]
	F	5.7	6.5	0.8	15.7
G	E	-6.4		-14.6	
	F	8.4	n.e.	6.5	n.e
U	E	- / -6.3 / -[g]	-5.1	-11.8 / - / -12.4[g]	-11.5 / -10.0 / -[g]
	F	0.0	0.4	34.5	5.4

tHS		A	C	G	U
A	E	-10.2 (-9.9)[d]	-8.3 (7.8)[d]	-15.1 (-15.6)[d]	-8.1 (-8.5)[d]
	F	6.5	2.9	74.7	4.9
C	E	-9.6 (-9.4)[d]	-11.6 (-11.6)[d]		-8.6 (-11.4)[d]
	F	0.9	1.0	n.e.	0.5
G	E			-9.3	
	F	n.e.	n.e.	3.9	n.e.
U	E	-1.7		-4.7	
	F	2.5	n.e.	2.2	n.e.

Interaction energies were obtained from the following references: cWS (Sponer et al. 2005b), tWS (Sponer et al. 2005c), tSS and cSS (Sponer et al. 2005a), cHS (Sharma et al. 2010b), and tHS (Mladek et al. 2009). Occurrence frequencies were calculated using the nonredundant list of RNA 3D structures current as of 10/01/2011. Base combinations that are not expected to form pairs in a given base-pair family are marked "n.e."

[a]Optimized structures with amino-acceptor interactions are in parentheses

[b]No crystallographic data available at the time when the energies were calculated

[c]Constrained optimizations

[d]Water-mediated interaction cannot be directly compared with others

[e]Interaction energy was calculated for the O3′-methylated structure

[f]The first value refers to the fully optimized structure, while the second value was obtained for a partially relaxed geometry with fixed ribose rings

[g]Adjacent/near-adjacent/distant base pair

experimentally observed structure or that, at most, a physically justified constraint sufficed to maintain the interaction geometry during the optimization. Intrinsically unstable base pairs are those for which it is difficult to reproduce the experimentally observed structure by QM calculation in the gas phase, even using constraints. This is taken as evidence that the observed structure is substantially influenced by factors "external" to the studied base pair, which by definition includes only the interacting bases and sugars. This external factor could be an inserted, structural water molecule, an extension of the base pairing by a stabilizing BPh (base-phosphate) interaction (utilizing one of the participating nucleotides), or interaction with a third base to form a base triple. Indeed, the cWS G/rG base pair utilizes water insertion for stabilization. With the exception of cWS G/rU, the base-pairing energy ranges between −16 and −22 kcal/mol, which is comparable to the interaction energy computed at the same theoretical level for the canonical (cWW) A/U base pair (−15.3 kcal/mol) but markedly weaker than that obtained for the G/C cWW base pair (−29.4 kcal/mol). Thus, the majority of the base pairs belonging to the cWS family can be considered as medium strong. The only exception is the G/rU base pair, which is classified as weak (−9.4 kcal/mol), comparable, for example, to cWW and tWW U/U base pairs. In this family, the variations in the sugar–base and base–base contributions to the interaction energy compensate each other, and as a result, with the exception of G/rU and G/rG, the interaction energies fall within a fairly narrow range. Thus, the base pairs of the cWS family are approximately isoenergetic (see Table 12.1). The most frequent base combinations in this family are cWS A/A, A/C, and A/U, which have comparable energies.

12.4.2.2 The *trans* Watson–Crick/Sugar-Edge (tWS) Base-Pair Family

The tWS base-pair family includes 14 members, of which only 10 structures were known crystallographically at the time the QM calculations were carried out. The calculations reproduced their geometries and predicted stable structures for the remaining members of the family (Sponer et al. 2005c), all of which have now been observed in crystal structures (Stombaugh et al. 2009). In some of the crystal structures, the C2′-endo sugar pucker occurs. In these cases, QM calculations were done considering both the C2′-endo and C3′-endo arrangements, and it was found that C3′-endo was in all cases more stable in isolation. Further, for some tWS base pairs, two structural isomers were found, depending on the orientation of the 2′-OH group. In one isomer, the hydroxyl group is involved in a conventional H-bond; in the other, it forms an amino-acceptor contact with C(N4) or A(N6) of the other base as shown in Fig. 12.4. Although the X-ray geometries in all cases support the conventional binding, the amino-acceptor variant cannot be ruled out on the basis of the energy data.

The energies of the tWS base pairs range from −8 to −28 kcal/mol (see Table 12.1), a considerably wider range than observed for cWS pairs (−16 to −22 kcal/mol, see above). Thus, some tWS base pairs are almost as stable as the cWW G/C base pair (−29.4 kcal/mol when evaluated with the same method)

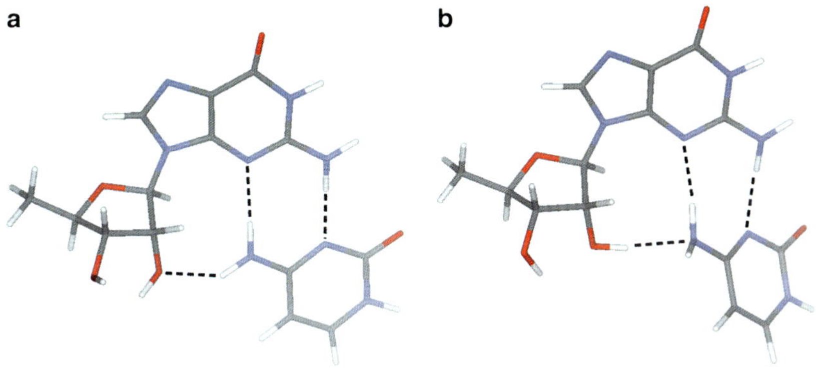

Fig. 12.4 Conventional (**a**) and amino-acceptor (**b**) binding modes in the tWS C/rG base pair. The *dotted lines* indicate key H-bonding contacts. Geometries were obtained from optimization at B3LYP/6-31G** level in gas phase (Sponer et al. 2005a, b, c)

whereas others are markedly weaker than the A/U cWW pair (−15.3 kcal/mol), including A/rA, C/rC, C/rU, and U/rA. The characteristic C1′-N distances adopted by tWS base pairs in experimental structures (ranging from 6.5 to 8.7 Å) as well as in the calculated models (ranging from 6.6 to 8.5 Å) fall within a tighter interval than for the cWS base pairs, which range from 5.2 to 9.3 Å in structures and from 5.3 to 8.7 Å in the gas-phase calculations. The most common base combinations by far in the tWS are A/rG, which usually is part of a base triple, and G/rU, which occurs in the stable UNCG tetraloops, with the G in the syn configuration.

12.4.2.3 The *cis* Sugar-Edge/Sugar-Edge (cSS) Base-Pair Family

The geometries and intrinsic stabilities of the cSS base pairs are dictated by a common structural motif, which includes the 2′-OH groups of both riboses and at least one of the nucleobases. For a majority of these base pairs, we find fair agreement between the computed and experimental geometries, although larger deviations occur between theory and experiment for experimental geometries that were not known at the time the quantum chemical calculations were made (Sponer et al. 2005a). These calculations will need to be revisited in future studies. Interaction energies in this family are substantial, ranging from −15.7 to −26.9 kcal/mol. The most stable pair is rC/rG (−26.9 kcal/mol), followed by rG/rG (−24.3 kcal/mol) and the group rG/rA, rA/rG, rG/rC, and rU/rG (−23 ± 1 kcal/mol). Note that all these interactions involve at least one purine, Y/Y cSS pairs being extremely rare or not observed. Thus, the observed cSS base pairs provide strong interactions. They are found throughout structured RNA molecules, stabilizing tertiary architectures. The most common cSS base combination is rA/rC, which almost invariably occurs as part of a base triple, as in fact do most cSS pairs (see Table 12.2).

Table 12.2 Frequencies (%) of base pairs occurring as components of base triples, by geometric family and by base combination

cWW	A		C		G		U	
	% triples (near-triples)	No.	% triples (near-triples)	No.	% triples (near-triples)	No.	% triples (near-triples)	No.
A	0.0 (31.8)	22	10.2 (44.1)	59	16.6 (53.8)	145	8.7 (38.4)	2929
C			25.0 (25.0)	8	12.5 (38.9)	7718	14.6 (43.9)	41
G					no pair		6.6 (29.8)	1009
U							13.9 (41.0)	244

cHH	A		C		G		U	
	% triples (near-triples)	No.	% triples (near-triples)	No.	% triples (near-triples)	No.	% triples (near-triples)	No.
A	no pair		no pair		100.0 (100.0)	3	no pair	
C			no pair		100.0 (100.0)	1	no pair	
G					100.0 (100.0)	4	no pair	
U							no pair	

tWW	A		C		G		U	
	% triples (near-triples)	No.	% triples (near-triples)	No.	% triples (near-triples)	No.	% triples (near-triples)	No.
A	74.0 (85.6)	104	80.0 (100.0)	5	no pair		29.2 (75.0)	48
C			0.0 (40.0)	10	17.0 (46.8)	47	0.0 (100.0)	2
G					50.0 (50.0)	12	33.3 (100.0)	3
U							0.0 (0.0)	4

tHH	A		C		G		U	
	T (near T)/Bp,%	Bp	T (near T)/Bp,%	Bp	T (near T)/Bp,%	Bp	T (near T)/Bp,%	Bp
A	51.1 (70.2)	188	75.0 (75.0)	4	20.0 (100.0)	5	40.0 (80.0)	5
C			no pair		93.3 (93.3)	15	60.0 (60.0)	5
G					66.7 (100.0)	6	no pair	
U							no pair	

cWH	A		C		G		U	
	% triples (near-triples)	No.	% triples (near-triples)	No.	% triples (near-triples)	No.	% triples (near-triples)	No.
A	no pair		no pair		0.0 (100.0)	3	0.0 (0.0)	0
C	no pair		100.0 (100.0)	4	62.5 (75.0)	8	0.0 (0.0)	0
G	20.0 (40.0)	5	no pair		34.2 (71.2)	111	no pair	
U	61.3 (83.8)	80	no pair		100.0 (100.0)	5	18.2 (54.5)	11

cHS	A		C		G		U	
	% triples (near-triples)	No.	% triples (near-triples)	No.	% triples (near-triples)	No.	% triples (near-triples)	No.
A	50.0 (88.5)	26	0.0 (50.0)	2	66.7 (100.0)	6	62.5 (100.0)	8
C	93.3 (100.0)	15	94.1 (94.1)	17	100.0 (100.0)	2	75.6 (92.7)	41
G	90.9 (100.0)	22	no pair		100.0 (100.0)	17	no pair	
U	0.0 (0.0)	0	100.0 (100.0)	1	78.9 (91.1)	90	28.6 (42.9)	14

tWH	A		C		G		U	
	% triples (near-triples)	No.	% triples (near-triples)	No.	% triples (near-triples)	No.	% triples (near-triples)	No.
A	49.3 (71.0)	69	no pair		100.0 (100.0)	6	no pair	
C	55.3 (75.0)	76	87.5 (100.0)	8	100.0 (100.0)	8	no pair	
G	no pair		no pair		89.2 (94.6)	37	20.0 (40.0)	5
U	45.6 (66.4)	384	no pair		100.0 (100.0)	1	66.7 (91.7)	12

tHS	A		C		G		U	
	% triples (near-triples)	No.	% triples (near-triples)	No.	% triples (near-triples)	No.	% triples (near-triples)	No.
A	24.6 (59.6)	57	48.0 (60.0)	25	25.9 (46.9)	653	48.8 (60.5)	43
C	12.5 (37.5)	8	22.2 (55.6)	9	no pair		0.0 (25.0)	4
G	no pair		no pair		52.9 (67.6)	34	no pair	
U	72.7 (95.5)	22	no pair		52.6 (52.6)	19	no pair	

cWS	A		C		G		U	
	% triples (near-triples)	No.	% triples (near-triples)	No.	% triples (near-triples)	No.	% triples (near-triples)	No.
A	69.2 (83.1)	65	87.7 (92.3)	65	100.0 (100.0)	1	74.2 (87.1)	31
C	75.0 (75.0)	12	85.7 (85.7)	7	100.0 (100.0)	3	92.3 (92.3)	13
G	50.0 (100.0)	2	40.0 (60.0)	5	80.0 (80.0)	5	33.3 (50.0)	6
U	62.5 (68.8)	16	100.0 (100.0)	1	37.5 (75.0)	8	100.0 (100.0)	2

cSS	A		C		G		U	
	% triples (near-triples)	No.	% triples (near-triples)	No.	% triples (near-triples)	No.	% triples (near-triples)	No.
A	84.4 (88.9)	90	99.7 (100.0)	362	93.5 (96.4)	169	88.4 (100.0)	86
C	99.7 (100.0)	362	0.0 (0.0)	0	90.6 (93.8)	32	100.0 (100.0)	7
G	93.5 (96.4)	169	90.6 (93.8)	32	100.0 (100.0)	10	97.3 (100.0)	37
U	88.4 (100.0)	86	100.0 (100.0)	7	97.3 (100.0)	37	100.0 (100.0)	2

tWS	A		C		G		U	
	% triples (near-triples)	No.	% triples (near-triples)	No.	% triples (near-triples)	No.	% triples (near-triples)	No.
A	50.0 (66.7)	6	0.0 (0.0)	1	76.0 (80.8)	104	66.7 (66.7)	3
C	0.0 (50.0)	2	71.4 (100.0)	7	69.2 (100.0)	13	0.0 (0.0)	0
G	no pair		60.0 (60.0)	5	no pair		0.0 (5.8)	52
U	54.5 (81.8)	11	100.0 (100.0)	3	66.7 (100.0)	3	0.0 (0.0)	0

tSS	A		C		G		U	
	% triples (near-triples)	No.	% triples (near-triples)	No.	% triples (near-triples)	No.	% triples (near-triples)	No.
A	84.6 (100.0)	52	100.0 (100.0)	180	92.6 (98.2)	379	93.8 (100.0)	96
C	no pair		no pair		no pair		no pair	
G	92.6 (98.2)	379	100.0 (100.0)	8	62.8 (83.3)	78	0.0 (0.0)	2
U	no pair		no pair		no pair		no pair	

For each base combination, in the left cell, the percentage of base-pair instances that occur as parts of base triples (or near triples) is given, and in the right cell, the number of instances in the nonredundant list of RNA 3D structures selected by FR3D (current as of 10/01/2011). Near pairs lie just outside the FR3D classification limits for base pairs. For example, AA tWW occurs 104 times in the nonredundant list, and 74% of these instances are parts of triples, and 85.6% are parts of triples or near triples (Petrov et al. in preparation)

12.4.2.4 The *trans* Sugar-Edge/Sugar-Edge (tSS) Family

Only eight tSS base pairs have been observed in X-ray structures, and none others are predicted. The fully optimized geometries of *trans* rA/rG, rG/rG, and rG/rC base pairs preserved the main features of the corresponding crystal geometries. The computed interaction energies (Sponer et al. 2005a) for these base pairs are rather substantial, ranging from −14 to −21 kcal/mol. In contrast, upon geometry optimization, a substantial alteration was observed for the less stable tSS base pairs (rA/rA, rA/rC, rA/rU, and rG/rU), which mainly concerned the angle of their C1′-N vectors. When the crystal geometry was constrained during optimization, the intrinsic stabilities of these four base pairs deteriorated substantially to −4 to −10 kcal/mol. This is consistent with the experimental observations that the observed geometries are stabilized by interactions with additional nucleotides. Thus, like other sugar-edge pairs, the tSS base pairs largely do not form independent (self-structured) RNA building blocks but should be considered parts of larger building blocks and motifs. For example, in some A-minor motifs, the A simultaneously forms a cSS and a tSS interaction with two cWW-paired bases, usually C and G. In fact, AC is the most frequent cSS pair and AG the most frequent tSS pair (Stombaugh et al. 2009). Only the G/G tSS base combination occurs to a significant extent as a free-standing base pair (see Table 12.2). G/G, along with A/G, is calculated to be the most stable tSS base combinations, with energies of about −21 kcal/mol each (Table 12.1).

In general, cSS and tSS interactions are primarily stabilized by the correlation component of the interaction energy. This is reflected by the fact that in many SS base pairs, the correlation stabilization exceeds the HF term. We can therefore conclude that the SS interactions are considerably more hydrophobic than standard base pairs, as the dispersion energy provides the leading stabilization force in many SS base pairs. In combination with their structural versatility, this makes the SS base pairs natural candidates for efficient tertiary interactions in RNAs.

12.4.2.5 The *cis* Hoogsteen/Sugar-Edge (cHS) Family

Except for two base combinations, all predicted members of this family have been found in crystal geometries (Stombaugh et al. 2009). Base-pairing strengths in this family vary over a very broad range, from −5.2 to −20.6 kcal/mol (Sharma et al. 2010b). The cHS base pairs are also dominated by the correlation component of the interaction energy and therefore can be considered as more hydrophobic building blocks of RNA architectures. They generally occur between adjacent nucleotides in the sequence, in which case they form "platform" motifs involved in stabilizing RNA architectures and mediating tertiary interactions by forming docking sites for hairpin loops. It is evident that the platform base pairs need to be considered in broader contexts, prompting additional QM computations now are under way (Mladek et al. 2012).

12.4.2.6 The *trans* Hoogsteen/Sugar-Edge (tHS) Family

The tHS base-pair family is sparser, having only ten base combinations that form base pairs. These ten tHS base pairs form two distinct isosteric subfamilies. One family has A or C as the Hoogsteen edge and the other has G or U. Base pairs in the same subfamily are observed to exchange at corresponding positions in homologous RNA molecules (Stombaugh et al. 2009). Base pairs from different subfamilies are not isosteric and do not exchange one for the other. As in the tWS base-pair family, base pairs of the first subfamily can form amino-acceptor contacts between the base and the sugar. Importantly, these amino-acceptor base pairs exhibit geometric parameters similar to those of the parent structures with conventional H-bonds. Further, both the conventional and amino-acceptor variants possess approximately the same intrinsic stabilities. Therefore, similar to the tWS base pairs, tHS base pairing enables a rapid transition between the canonical and the amino-acceptor variants. At the present time, the suggestion that there is some involvement of amino-acceptor interactions in RNA base pairs remains largely speculation. However, it is not clear whether experimentalists would notice such interactions, if in fact they are present.

The three known members of the second subfamily (U/A, U/G, and G/G) do not utilize the ribose in the pairing. They are stabilized by an H-bond donated by the nucleobase of the sugar edge nucleoside to the exocyclic oxo group of the Hoogsteen edge base. Among them, only the G/G pair exhibits considerable gas-phase stability.

In general, tHS base binding provides weak to medium intermolecular stabilization in gas phase, in the range from -1 to -15 kcal/mol (Mladek et al. 2009). The most stable member of the family is the tHS A/rG ("sheared" AG) base pair, with MP2/aug-cc-pVDZ interaction energy of -15.1 kcal/mol (-16.7 kcal/mol with the highest-accuracy method including the CBS extrapolation). Its stability is comparable to that of the A/U cWW base pair (-15.3 and -17.0 kcal/mol with the MP2/aug-cc-pVDZ and CBS reference methods, respectively). The A/G combination is by far the most commonly observed tHS base pair in RNA molecules, and one of the most frequent non-Watson–Crick base pairs regardless of base family (Stombaugh et al. 2009). It is entirely isosteric with the less frequent tHS A/rA base pair, which is intrinsically less stable, having a stabilization of -10.2 kcal/mol. The difference in the frequency of occurrence of the tHS AG and AA base pairs may be largely determined by the difference in their intrinsic base-pair stabilities. The AG base combination constitutes almost 70% of tHS base pairs while AA, the second most common, is only about 9% (Stombaugh et al. 2009).

Other members of the tHS family are considerably less stable. In continuum solvent calculations, the tHS GG pair acquires significant stability, which, like that of tHS AG, is comparable to the stability computed for the AU cWW base pair. The rest of the tHS base pairs are markedly less stable under the same conditions, although AU constitutes about 8% of all tHS base pairs. In conclusion, the most frequently observed member of the tHS base-pair family, also known as the "sheared AG base pair," exhibits prominent stability both in gas phase and in solution calculations.

12.4.3 Implications for Force-Field Parametrizations

QM calculations provide genuine benchmarks for comparing the performance of nucleic acid force fields. Significantly, the QM computations indicate that the AMBER molecular modeling force field performs in a satisfactory manner for base pairs involving the sugar edge (Sponer et al. 2005a, b, c), and this is reflected in the generally good performance of AMBER in simulations of folded RNAs containing these interactions (Ditzler et al. 2010). Sugar-edge base pairs are characteristic features of RNA tertiary structures. Note that the interaction energies reported here and in the original studies were obtained with the RIMP2/aug-cc-pVDZ method, considered a medium-quality method by present-day standards (Sponer et al. 2005a, b, c; Mladek et al. 2009; Sharma et al. 2010a, b). This method, nonetheless, appears to be completely adequate for this purpose. Highest-accuracy CBS energies for some non-Watson–Crick base pairs are also available in the literature (Sponer et al. 2009).

In spite of these successes, the accuracy with which current force fields (including AMBER) can evaluate intermolecular interactions is limited by the approach: Molecular mechanical force fields approximate molecular interactions using Lennard-Jones potentials featuring r^{-6} attractive and r^{-12} repulsive terms, complemented by $1/r$ electrostatic terms calculated with atom-centered point charges, where the atomic charges are derived to approximate the electrostatic potential around the monomers. At H-bond distances and larger separations, the classic coulombic term approximates the QM coulombic term rather well, while all the other, very diverse QM terms must be effectively approximated by the Lennard-Jones term. Note that the force field is inherently not able to capture any electronic redistribution effects associated with intermolecular interactions. In addition, at shorter separations, which are also sampled during MD simulation, the agreement between the QM and classical electrostatic terms breaks down and the repulsive term only crudely approximates the typically e^{-r} behavior of the real contributions due to overlap of the electronic clouds. It has been proposed that a better description can be achieved by combining exponential short-range repulsion with damped dispersion terms (Zgarbova et al. 2010).

12.5 QM Calculations of Tertiary Interactions

QM computations may also be useful in studies of H-bonded base triples and quadruples (quartets). So far, only a limited set of computations has been reported, for key tertiary interactions known as A-minor and packing (P-) interactions. A-minor interactions are formed by adenine bases interacting via their sugar edges in the minor grooves of canonical double helices (Nissen et al. 2001). In an A-minor type I interaction, the adenine, which is generally presented for interaction with the helix by a hairpin or internal loop, simultaneously forms a cSS pair with one base of the cWW base pair and a tSS pair with the other. In the other highly conserved A-minor interaction ("type II"), the A forms a cSS pair with one base of the cWW pair in the helix, usually a cytosine. In A-minor interactions, there is a clear steric preference for adenine as the base interacting in the minor groove over any other base (Nissen et al. 2001). The gas-phase-optimized geometry of the most conserved A-minor interaction ("type I") perfectly matches the crystal geometry (Sponer et al. 2007). In contrast, during gas-phase optimization of the A-minor type II interaction, a significant structural alteration was observed. However, addition of a water molecule to the computational model restored the geometry suggested by X-ray structures. We also conclude that the A-minor type II interaction clearly prefers participation of a water molecule when the primary interaction is between adenine and cytosine (canonical A-minor type II motif). The role of the water molecule is to mediate the contact between adenine and the guanine of the G/C cWW pair. In the significantly less conserved A-minor type 0 and type III contacts, the adenine from the single-stranded segment interacts only with one nucleotide of the Watson–Crick base pair. The A-minor type III interaction is particularly weak and is considered to be the least specific and biologically important. In these cases, structural water molecules had to be included in the computational model to reproduce the crystal geometry in the gas phase. Optimizations of the A-minor type 0 interaction geometry resulted in stable structures both with and without structural water molecules, but both optimized geometries deviated slightly from crystal structures.

The P-interaction brings together a G/U cWW ("wobble") base pair and a cWW base pair, usually, but not exclusively, C/G (Mokdad et al. 2006). The P-interaction is stabilized by an extensive network of H-bonding interactions, and its optimized geometry was almost identical to that observed in the crystal structure.

The direct (without water mediation) A-minor type I and II tertiary contacts are very stable (interaction energies −31 and −26 kcal/mol) and are primarily stabilized by the electron correlation interaction energy term (−17 and −16 kcal/mol). This again indicates that the A-minor type I and II contacts are much more hydrophobic than cWW base pairs, which makes them particularly suitable to form stabilizing tertiary contacts, crucial for RNA architectures. The P-interaction is also very strong (−25 kcal/mol) and is dominated by the correlation component of the intermolecular stabilization (−12.5 kcal/mol). Due to the active participation of the 2′-OH of one or more riboses, the A-minor interactions as well as the P-interaction are stabilized by a remarkably prominent electron correlation component. Thus, the seemingly large energetic contribution of the electron correlation may be one of the key physicochemical features that make the A-minor and P-interactions so prominent in stabilizing RNA architectures.

The calculations provide insight into the physical chemistry of the molecular interactions stabilizing the A-minor type I A/GC interaction, which consists of tSS A/G and cSS A/C sugar-edge base pairs, in addition to the GC canonical pair. While the tSS A/G interaction is substantially stabilized by the dispersion term, the comparably stable cSS A/C interaction is much more electrostatic in nature. Interestingly, a survey of X-ray structures reveals frequent water insertion in the

H-bonded aggregates. As shown in Table 12.2, many base-pair combinations, especially in sugar-edge base pairs, occur wholly or largely as parts of triples. The data in this table were generated using FR3D ("Find RNA 3D") software to search for all annotated instances of each base pair in nonredundant lists of RNA-containing PDB files (Petrov et al. 2011a, b). For each possible base combination in each base-pair family, two searches were performed and collated: first, a two-nucleotide symbolic search for instances of the base pair and then a three-nucleotide search, for instance, in which at least one of the bases in the pair also pairs with a third base, with no restrictions on the type of the second base pair formed. The values in Table 12.2 are the percentage of instances of each base pair in which the base pair is part of a base triple. The values in parentheses include near base pairs with the third base. For example, 99.7% (100% including near pairs) of the 362 instances of cSS AC base pairs are parts of triples. In other words, this pair essentially never exists as an independent base pair. In fact, this is true for most of the cSS base pairs (Table 12.2). By contrast, 62.8% (83.3% including near pairs) of the 78 instances of tSS GG base pairs are parts of triples, indicating this pair can exist independently, whereas the other tSS base combinations are almost always parts of triples.

Finally, a very small number of triples were observed that have intermediate geometries and therefore were not predicted based on the base-pair classification. We anticipate detailed QM calculations will clucidate whether these are simply artifacts of medium- to low-resolution X-ray structures or can, in fact, be expected to recur in other structures and therefore should be included in the RNA triple classification. An online database of base triples is now available that includes observed and predicted triples. See: http://rna.bgsu.edu/Triples/triples.php/.

12.6 Software and Computational Demands

Several quantum chemical program packages are commercially available. The leading ones include Gaussian, ADF, NWChem, MOLPRO and Turbomole. Gaussian has an excellent optimizer, while Turbomole provides highly accurate MP2 energies with relatively low computational costs, by making use of the resolution of identity (RI) procedure. Typical runtimes for the optimization of a base pair consisting of two nucleosides (i.e., ca. 60 atoms) at the DFT level are 1–2 days on a cluster of four parallel, coupled Opteron 2.6 GHz processors. For the same system and using the same computer platform, equipped with about 0.5 GB dynamic memory, computations to obtain the interaction energy at the RIMP2/aug-cc-pVDZ level can be executed within 2–3 days using the Turbomole code.

Acknowledgments The authors thank Anton I. Petrov for assistance in organizing data and preparing tables. This work was supported by grants 203/09/1476, P208/12/1878, and P208/11/1822 from the Grant Agency of the Czech Republic, grants AV0Z50040507 and AV0Z50040702 from the Ministry of Education of the Czech Republic, and "CEITEC - Central European Institute of

Technology" (CZ.1.05/1.1.00/02.0068) from the European Regional Development Fund and by grants from the National Institutes of Health to NBL (grant numbers 1R01GM085328-01A1 and 2R15GM055898-05).

References

Abu Almakarem AS, Petrov AI, Stombaugh J, Zirbel CL, Leontis NB (2011) Comprehensive survey and geometric classification of base triples in RNA structures. Nucleic Acids Res. doi:10.1093/nar/gkr810

Banas P, Jurecka P, Walter NG, Sponer J, Otyepka M (2009) Theoretical studies of RNA catalysis: hybrid QM/MM methods and their comparison with MD and QM. Methods 49:202–216. doi:10.1016/j.ymeth.2009.04.007

Banas P, Hollas D, Zgarbova M, Jurecka P, Orozco M, Cheatham TE, Sponer J, Otyepka M (2010) Performance of molecular mechanics force fields for RNA simulations: stability of UUCG and GNRA hairpins. J Chem Theory Comput 6:3836–3849. doi:10.1021/ct100481h

Blas JR, Luque FJ, Orozco M (2004) Unique tautomeric properties of isoguanine. J Am Chem Soc 126:154–164

Brandl M, Meyer M, Suhnel J (2000) Water-mediated base pairs in RNA: a quantum-chemical study. J Phys Chem A 104:11177–11187. doi:10.1021/jp002022d

Brandl M, Meyer M, Suhnel J (2001) Quantum-chemical analysis of C-H center dot center dot center dot O and C-H center dot center dot center dot N interactions in RNA base pairs – H-bond versus anti-H-bond pattern. J Biomol Struct Dyn 18:545–555

Bugg CE, Thomas JM, Rao ST, Sundaral M (1971) Stereochemistry of nucleic acids and their constituents. 10. Solid-state base-stacking patterns in nucleic acid constituents and polynucleotides. Biopolymers 10:175–219

Cieplak P, Cornell WD, Bayly C, Kollman PA (1995) Application of the multimolecule and multiconformational RESP methodology to biopolymers – charge derivation for DNA, RNA, AND proteins. J Comput Chem 16:1357–1377

Cieplak P, Dupradeau FY, Duan Y, Wang JM (2009) Polarization effects in molecular mechanical force fields. J Phys Condens Matter 21:333102. doi:33310210.1088/0953-8984/21/33/333102

Colominas C, Luque FJ, Orozco M (1996) Tautomerism and protonation of guanine and cytosine. Implications in the formation of hydrogen-bonded complexes. J Am Chem Soc 118:6811–6821

Cornell WD, Cieplak P, Bayly CI, Gould IR, Merz KM, Ferguson DM, Spellmeyer DC, Fox T, Caldwell JW, Kollman PA (1995) A 2nd generation force-field for the simulation of proteins, nucleic-acids, and organic-molecules. J Am Chem Soc 117:5179–5197

Dey M, Moritz F, Grotemeyer J, Schag EW (1994) Base-pair formation of free nucleobases and mononucleosides in the gas-phase. J Am Chem Soc 116:9211–9215

Ditzler MA, Otyepka M, Sponer J, Walter NG (2010) Molecular dynamics and quantum mechanics of RNA: conformational and chemical change we can believe in. Acc Chem Res 43:40–47. doi:10.1021/ar900093g

Dong F, Miller RE (2002) Vibrational transition moment angles in isolated biomolecules: a structural tool. Science 298:1227–1230

Dudev T, Lim C (2003) Principles governing Mg, Ca, and Zn binding and selectivity in proteins. Chem Rev 103:773–787. doi:10.1021/cr020467n

Dudev T, Lim C (2008) Metal binding affinity and selectivity in metalloproteins: insights from computational studies. Annu Rev Biophys 37:97–116. doi:10.1146/annurev.biophys.37.032807.125811

Egli M, Gessner RV (1995) Stereoelectronic effects Of deoxyribose O4′ on DNA conformation. Proc Natl Acad Sci USA 92:180–184

Ennifar E, Yusupov M, Walter P, Marquet R, Ehresmann B, Ehresmann C, Dumas P (1999) The crystal structure of the dimerization initiation site of genomic HIV-1 RNA reveals an extended duplex with two adenine bulges. Structure 7:1439–1449

Feyereisen MW, Feller D, Dixon DA (1996) Hydrogen bond energy of the water dimer. J Phys Chem 100:2993–2997

Florian J, Sponer J, Warshel A (1999) Thermodynamic parameters for stacking and hydrogen bonding of nucleic acid bases in aqueous solution: ab initio/Langevin dipoles study. J Phys Chem B 103:884–892

Grimme S (2011) Density functional theory with London dispersion corrections. Wiley Interdiscip Rev Comput Mol Sci 1:211–228. doi:10.1002/wcms.30

Hammond NB, Tolbert BS, Kierzek R, Turner DH, Kennedy SD (2010) RNA internal loops with tandem AG pairs: the structure of the 5′G(UG)U/3′U(GA)G loop can be dramatically different from others, including 5′A(AG)U/3′U(GA)A. Biochemistry 49:5817–5827. doi:10.1021/bi100332r

Hanus M, Ryjacek F, Kabelac M, Kubar T, Bogdan TV, Trygubenko SA, Hobza P (2003) Correlated ab initio study of nucleic acid bases and their tautomers in the gas phase, in a microhydrated environment and in aqueous solution. Guanine: surprising stabilization of rare tautomers in aqueous solution. J Am Chem Soc 125:7678–7688. doi:10.1021/ja034245y

Hobza P, Sponer J (1999) Structure, energetics, and dynamics of the nucleic acid base pairs: nonempirical ab initio calculations. Chem Rev 99:3247–3276

Hobza P, Sponer J, Polasek M (1995) H-bonded and stacked 2nd-base pairs – cytosine dimer – an ab-initio 2nd-order Moller-Plesset study. J Am Chem Soc 117:792–798

Hobza P, Selzle HL, Schlag EW (1996) Potential energy surface for the benzene dimer. Results of ab initio CCSD(T) calculations show two nearly isoenergetic structures: T-shaped and parallel displaced. J Phys Chem 100:18790–18794

Hobza P, Kabelac M, Sponer J, Mejzlik P, Vondrasek J (1997) Performance of empirical potentials (AMBER, CFF95, CVFF, CHARMM, OPLS, POLTEV), semiempirical quantum chemical methods (AM1, MNDO/M, PM3), and ab initio Hartree-Fock method for interaction of DNA bases: comparison with nonempirical beyond Hartree-Fock results. J Comput Chem 18:1136–1150

Hunter CA (1993) Sequence-dependent DNA-structure – the role of base stacking interactions. J Mol Biol 230:1025–1054

Jurecka P, Sponer J, Hobza P (2004) Potential energy surface of the cytosine dimer: MP2 complete basis set limit interaction energies, CCSD(T) correction term, and comparison with the AMBER force field. J Phys Chem B 108:5466–5471. doi:10.1021/jp049956c

Jurecka P, Sponer J, Cerny J, Hobza P (2006) Benchmark database of accurate (MP2 and CCSD(T) complete basis set limit) interaction energies of small model complexes, DNA base pairs, and amino acid pairs. Phys Chem Chem Phys 8:1985–1993. doi:10.1039/b600027d

Katz AK, Glusker JP, Beebe SA, Bock CW (1996) Calcium ion coordination: a comparison with that of beryllium, magnesium, and zinc. J Am Chem Soc 118:5752–5763

Klamt A, Mennucci B, Tomasi J, Barone V, Curutchet C, Orozco M, Luque FJ (2009) On the performance of continuum solvation methods. A comment on "Universal approaches to solvation modeling". Acc Chem Res 42:489–492. doi:10.1021/ar800187p

Koller AN, Bozilovic J, Engels JW, Gohlke H (2010) Aromatic N versus aromatic F: bioisosterism discovered in RNA base pairing interactions leads to a novel class of universal base analogs. Nucleic Acids Res 38:3133–3146

Kollman PA, Massova I, Reyes C, Kuhn B, Huo SH, Chong L, Lee M, Lee T, Duan Y, Wang W, Donini O, Cieplak P, Srinivasan J, Case DA, Cheatham TE (2000) Calculating structures and free energies of complex molecules: combining molecular mechanics and continuum models. Acc Chem Res 33:889–897. doi:10.1021/ar000033j

Kopitz H, Zivkovic A, Engels JW, Gohlke H (2008) Determinants of the unexpected stability of RNA fluorobenzene self pairs. ChemBioChem 9:2619–2622. doi:10.1002/cbic.200800461

Kratochvil M, Engkvist O, Sponer J, Jungwirth P, Hobza P (1998) Uracil dimer: potential energy and free energy surfaces. Ab initio beyond Hartree-Fock and empirical potential studies. J Phys Chem A 102:6921–6926

Kratochvil M, Sponer J, Hobza P (2000) Global minimum of the adenine center dot center dot center dot thymine base pair corresponds neither to Watson-Crick nor to Hoogsteen structures.

Molecular dynamic/quenching/AMBER and ab initio beyond Hartree-Fock studies. J Am Chem Soc 122:3495–3499

Leontis NB, Westhof E (2001) Geometric nomenclature and classification of RNA base pairs. RNA 7:499–512

Leontis NB, Stombaugh J, Westhof E (2002) The non-Watson-Crick base pairs and their associated isostericity matrices. Nucleic Acids Res 30:3497–3531

Luisi B, Orozco M, Sponer J, Luque FJ, Shakked Z (1998) On the potential role of the amino nitrogen atom as a hydrogen bond acceptor in macromolecules. J Mol Biol 279:1123–1136

Mathews DH, Turner DH (2006) Prediction of RNA secondary structure by free energy minimization. Curr Opin Struct Biol 16:270–278. doi:10.1016/j.sbi.2006.05.010

Mathews DH, Sabina J, Zuker M, Turner DH (1999) Expanded sequence dependence of thermodynamic parameters improves prediction of RNA secondary structure. J Mol Biol 288:911–940

Miller JL, Kollman PA (1996) Solvation free energies of the nucleic acid bases. J Phys Chem 100:8587–8594

Mladek A, Sharma P, Mitra A, Bhattacharyya D, Sponer J, Sponer JE (2009) Trans Hoogsteen/sugar edge base pairing in RNA. Structures, energies, and stabilities from quantum chemical calculations. J Phys Chem B 113:1743–1755. doi:10.1021/jp808357m

Mladek A, Sponer JE, Jurecka P, Banas P, Otyepka M, Svozil D, Sponer J (2010) Conformational energies of DNA sugar-phosphate backbone: reference QM calculations and a comparison with density functional theory and molecular mechanics. J Chem Theory Comput 6:3817–3835. doi:10.1021/ct1004593

Mladek A, Sponer JE, Kulhanek P, Lu XJ, Olson WK, Sponer J (2012) Understanding the sequence preference of recurrent RNA building blocks using quantum chemistry: the intrastrand RNA dinucleotide platform. J Chem Theory Comput 8:335–347. doi:10.1021/ct200712b

Mokdad A, Krasovska MV, Sponer J, Leontis NB (2006) Structural and evolutionary classification of G/U wobble basepairs in the ribosome. Nucleic Acids Res 34:1326–1341. doi:10.1093/nar/gkl025

Morgado CA, Jurecka P, Svozil D, Hobza P, Sponer J (2009) Balance of attraction and repulsion in nucleic-acid base stacking: CCSD(T)/complete-basis-set-limit calculations on uracil dimer and a comparison with the force-field description. J Chem Theory Comput 5:1524–1544. doi:10.1021/ct9000125

Nir E, Kleinermanns K, de Vries MS (2000) Pairing of isolated nucleic-acid bases in the absence of the DNA backbone. Nature 408:949–951

Nissen P, Ippolito JA, Ban N, Moore PB, Steitz TA (2001) RNA tertiary interactions in the large ribosomal subunit: the A-minor motif. Proc Natl Acad Sci USA 98:4899–4903

Oliva R, Cavallo L, Tramontano A (2006) Accurate energies of hydrogen bonded nucleic acid base pairs and triplets in tRNA tertiary interactions. Nucleic Acids Res 34:865–879. doi:10.1093/nar/gkj491

Oliva R, Tramontano A, Cavallo L (2007) Mg^{2+} binding and archaeosine modification stabilize the G15-C48 Levitt base pair in tRNAs. RNA 13:1427–1436. doi:10.1261/rna.574407

Perez A, Sponer J, Jurecka P, Hobza P, Luque FJ, Orozco M (2005) Are the hydrogen bonds of RNA (A U) stronger than those of DNA (A T)? A quantum mechanics study. Chem Eur J 11:5062–5066. doi:10.1002/chem.200500255

Perez A, Marchan I, Svozil D, Sponer J, Cheatham TE, Laughton CA, Orozco M (2007) Refinenement of the AMBER force field for nucleic acids: improving the description of alpha/gamma conformers. Biophys J 92:3817–3829. doi:10.1529/biophysj.106.097782

Petrov AI, Zirbel CL, Leontis NB (2011a) WebFR3D–a server for finding, aligning and analyzing recurrent RNA 3D motifs. Nucleic Acids Res 39:W50–W55. doi:10.1093/nar/gkr249

Petrov AS, Bowman JC, Harvey SC, Williams LD (2011b) Bidentate RNA-magnesium clamps: On the origin of the special role of magnesium in RNA folding. RNA 17:291–297. doi:10.1261/rna.2390311

Prive G, Heinemann U, Chandrasegaran S, Kan L, Kopka M, Dickerson R (1987) Helix geometry, hydration, and G.A mismatch in a B-DNA decamer. Science 238:498–504. doi:10.1126/science.3310237

Rappoport D, Crawford NRM, Furche F, Burke K (2009) Approximate density functionals: Which should I choose? In: Solomon EI, Scott RA, King RB (eds) Computational inorganic and bioinorganic chemistry. Wiley-VCH, New York, pp 159–172

Razga F, Koca J, Sponer J, Leontis NB (2005) Hinge-like motions in RNA kink-turns: the role of the second A-minor motif and nominally unpaired bases. Biophys J 88:3466–3485. doi:10.1529/biophysj.104.054916

Reblova K, Strelcova Z, Kulhanek P, Besscova I, Mathews DH, Van Nostrand K, Yildirim I, Turner DH, Sponer J (2010) An RNA molecular switch: intrinsic flexibility of 23S rRNA helices 40 and 68 5′-UAA/5′-GAN internal loops studied by molecular dynamics methods. J Chem Theory Comput 6:910–929. doi:10.1021/ct900440t

Reblova K, Sponer JE, Spackova N, Besseova I, Sponer J (2011) A-minor tertiary interactions in RNA kink-turns. Molecular dynamics and quantum chemical analysis. J Phys Chem B 115:13897–13910

Roy A, Panigrahi S, Bhattacharyya M, Bhattacharyya D (2008) Structure, stability, and dynamics of canonical and noncanonical base pairs: quantum chemical studies. J Phys Chem B 112:3786–3796. doi:10.1021/jp076921e

Shankar N, Kennedy SD, Chen G, Krugh TR, Turner DH (2006) The NMR structure of an internal loop from 23S ribosomal RNA differs from its structure in crystals of 50S ribosomal subunits. Biochemistry 45:11776–11789. doi:10.1021/bi0605787

Sharma P, Mitra A, Sharma S, Singh H, Bhattacharyya D (2008) Quantum chemical studies of structures and binding in noncanonical RNA base pairs: the trans Watson-Crick: Watson-Crick family. J Biomol Struct Dyn 25:709–732

Sharma P, Sharma S, Chawla M, Mitra A (2009) Modeling the noncovalent interactions at the metabolite binding site in purine riboswitches. J Mol Model 15:633–649. doi:10.1007/s00894-008-0384-y

Sharma P, Chawla M, Sharma S, Mitra A (2010a) On the role of Hoogsteen:Hoogsteen interactions in RNA: Ab initio investigations of structures and energies. RNA 16:942–957. doi:papers://8F282AF1-C00B-4965-8450-641CADBEB600/Paper/p1255

Sharma P, Sponer JE, Sponer J, Sharma S, Bhattacharyya D, Mitra A (2010b) On the role of the cis Hoogsteen:sugar-edge family of base pairs in platforms and triplets-quantum chemical insights into RNA structural biology. J Phys Chem B 114:3307–3320. doi:10.1021/jp910226e

Siegfried NA, Metzger SL, Bevilacqua PC (2007) Folding cooperativity in RNA and DNA is dependent on position in the helix. Biochemistry 46:172–181. doi:10.1021/bi0613751

Špacková N, Cheatham TE, Ryjácek F, Lankaš F, van Meervelt L, Hobza P, Šponer J (2003) Molecular dynamics simulations and thermodynamics analysis of DNA–drug complexes. Minor groove binding between 4′,6-diamidino-2-phenylindole and DNA duplexes in solution. J Am Chem Soc 125:1759–1769. doi:10.1021/ja025660d

Spirko V, Sponer J, Hobza P (1997) Anharmonic and harmonic intermolecular vibrational modes of the DNA base pairs. J Chem Phys 106:1472–1479

Sponer J, Hobza P (1994a) Bifurcated hydrogen-bonds in DNA crystal-structures – an ab-initio quantum-chemical study. J Am Chem Soc 116:709–714

Sponer J, Hobza P (1994b) Nonplanar geometries of DNA bases – ab-initio 2nd-order Moller-Plesset study. J Phys Chem 98:3161–3164

Sponer J, Lankas F (eds) (2006) Computational studies of RNA and DNA. Challenges and advances in computational chemistry and physics. Springer, Dordrecht

Sponer J, Leszczynski J, Hobza P (1996a) Nature of nucleic acid-base stacking: nonempirical ab initio and empirical potential characterization of 10 stacked base dimers. Comparison of stacked and H-bonded base pairs. J Phys Chem 100:5590–5596

Sponer J, Leszczynski J, Hobza P (1996b) Structures and energies of hydrogen-bonded DNA base pairs. A nonempirical study with inclusion of electron correlation. J Phys Chem 100:1965–1974

Sponer J, Leszczynski J, Vetterl V, Hobza P (1996c) Base stacking and hydrogen bonding in protonated cytosine dimer: the role of molecular ion-dipole and induction interactions. J Biomol Struct Dyn 13:695–706

Sponer J, Gabb HA, Leszczynski J, Hobza P (1997) Base-base and deoxyribose-base stacking interactions in B-DNA and Z-DNA: a quantum-chemical study. Biophys J 73:76–87

Sponer J, Sabat M, Gorb L, Leszczynski J, Lippert B, Hobza P (2000) The effect of metal binding to the N7 site of purine nucleotides on their structure, energy, and involvement in base pairing. J Phys Chem B 104:7535–7544. doi:10.1021/jp001711m

Sponer J, Leszczynski J, Hobza P (2001) Electronic properties, hydrogen bonding, stacking, and cation binding of DNA and RNA bases. Biopolymers 61:3–31. doi:10.1002/bip.10048

Sponer J, Mokdad A, Sponer JE, Spackova N, Leszczynski J, Leontis NB (2003) Unique tertiary and neighbor interactions determine conservation patterns of cis Watson-Crick A/G base-pairs. J Mol Biol 330:967–978. doi:10.1016/s0022-2836(03)00667-3

Sponer J, Jurecka P, Hobza P (2004) Accurate interaction energies of hydrogen-bonded nucleic acid base pairs. J Am Chem Soc 126:10142–10151. doi:10.1021/ja048436s

Sponer JE, Leszczynski J, Sychrovsky V, Sponer J (2005a) Sugar edge/sugar edge base pairs in RNA: stabilities and structures from quantum chemical calculations. J Phys Chem B 109:18680–18689. doi:10.1021/jp053379q

Sponer JE, Spackova N, Kulhanek P, Leszczynski J, Sponer J (2005b) Non-Watson-Crick base pairing in RNA. Quantum chemical analysis of the cis Watson-Crick/sugar edge base pair family. J Phys Chem A 109:2292–2301. doi:10.1021/jp050132k

Sponer JE, Spackova N, Leszczynski J, Sponer J (2005c) Principles of RNA base pairing: structures and energies of the trans Watson-Crick/sugar edge base pairs. J Phys Chem B 109:11399–11410. doi:10.1021/jp051126r

Sponer J, Jurecka P, Marchan I, Luque FJ, Orozco M, Hobza P (2006) Nature of base stacking: reference quantum-chemical stacking energies in ten unique B-DNA base-pair steps. Chem Eur J 12:2854–2865. doi:10.1002/chem.200501239

Sponer JE, Reblova K, Mokdad A, Sychrovsky V, Leszczynski J, Sponer J (2007) Leading RNA tertiary interactions: structures, energies, and water insertion of a-minor and p-interactions. A quantum chemical view. J Phys Chem B 111:9153–9164. doi:10.1021/jp0704261

Sponer J, Riley KE, Hobza P (2008) Nature and magnitude of aromatic stacking of nucleic acid bases. Phys Chem Chem Phys 10:2595–2610. doi:10.1039/b719370j

Sponer J, Zgarbova M, Jurecka P, Riley KE, Sponer JE, Hobza P (2009) Reference quantum chemical calculations on RNA base pairs directly involving the 2′-OH group of ribose. J Chem Theory Comput 5:1166–1179. doi:10.1021/ct800547k

Sponer J, Sponer JE, Petrov AI, Leontis NB (2010) Quantum chemical studies of nucleic acids can we construct a bridge to the RNA structural biology and bioinformatics communities? J Phys Chem B 114:15723–15741. doi:10.1021/jp104361m

Stombaugh J, Zirbel CL, Westhof E, Leontis NB (2009) Frequency and isostericity of RNA base pairs. Nucleic Acids Res 37:2294–2312. doi:10.1093/nar/gkp011

Svozil D, Hobza P, Sponer J (2010) Comparison of intrinsic stacking energies of ten unique dinucleotide steps in A-RNA and B-DNA duplexes. Can we determine correct order of stability by quantum-chemical calculations? J Phys Chem B 114:1191–1203. doi:10.1021/jp910788e

Swart M, Guerra CF, Bickelhaupt FM (2004) Hydrogen bonds of RNA are stronger than those of DNA, but NMR monitors only presence of methyl substituent in uracil/thymine. J Am Chem Soc 126:16718–16719. doi:10.1021/ja045276b

Tomasi J, Mennucci B, Cammi R (2005) Quantum mechanical continuum solvation models. Chem Rev 105:2999–3093. doi:10.1021/cr9904009

Vlieghe D, Sponer J, Van Meervelt L (1999) Crystal structure of d(GGCCAATTGG) complexed with DAPI reveals novel binding mode. Biochemistry 38:16443–16451

Vokacova Z, Sponer J, Sponer JE, Sychrovsky V (2007) Theoretical study of the scalar coupling constants across the noncovalent contacts in RNA base pairs: the cis- and trans-Watson-Crick/sugar edge base pair family. J Phys Chem B 111:10813–10824. doi:10.1021/jp072822p

Yanson IK, Teplitsky AB, Sukhodub LF (1979) Experimental studies of molecular-interactions between nitrogen bases of nucleic-acids. Biopolymers 18:1149–1170

Yildirim I, Turner DH (2005) RNA challenges for computational chemists. Biochemistry 44:13225–13234. doi:10.1021/bi051236o

Yildirim I, Stern HA, Sponer J, Spackova N, Turner DH (2009) Effects of restrained sampling space and nonplanar amino groups on free-energy predictions for RNA with imino and sheared tandem GA base pairs flanked by GC, CG, iGiC or iCIG base pairs. J Chem Theory Comput 5:2088–2100. doi:10.1021/ct800540c

Zgarbova M, Otyepka M, Sponer J, Hobza P, Jurecka P (2010) Large-scale compensation of errors in pairwise-additive empirical force fields: comparison of AMBER intermolecular terms with rigorous DFT-SAPT calculations. Phys Chem Chem Phys 12:10476–10493

Zgarbova M, Jurecka P, Banas P, Otyepka M, Sponer JE, Leontis NB, Zirbel CL, Sponer J (2011a) Noncanonical hydrogen bonding in nucleic acids. Benchmark evaluation of key base-phosphate interactions in folded RNA molecules using quantum-chemical calculations and molecular dynamics simulations. J Phys Chem A 115:11277–11292. doi:10.1021/jp204820b

Zgarbová M, Otyepka M, Sponer J, Mládek A, Banáš P, Cheatham TE, Jurečka P (2011b) Refinement of the Cornell et al. nucleic acids force field based on reference quantum chemical calculations of glycosidic torsion profiles. J Chem Theory Comput 7:2886–2902

Zhao Y, Truhlar DG (2008) Density functionals with broad applicability in chemistry. Acc Chem Res 41:157–167. doi:10.1021/ar700111a

Zirbel CL, Sponer JE, Sponer J, Stombaugh J, Leontis NB (2009) Classification and energetics of the base-phosphate interactions in RNA. Nucleic Acids Res 37:4898–4918. doi:10.1093/nar/gkp468

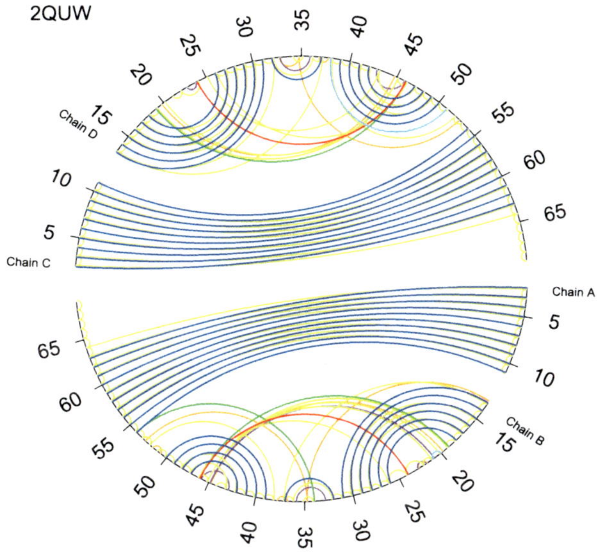

Fig. 13.1 *Circular diagram* indicating the pairwise interactions in PDB file 2QUW.pdb (Chi et al. 2008) as annotated by FR3D. *Dark blue chords* indicate the 42 nested Watson–Crick base pairs, *red chords* the 2 nonnested Watson–Crick base pairs, *cyan* the 2 nested non-Watson–Crick base pairs, *green* the 3 nonnested non-Watson–Crick base pairs, *yellow* the 146 stacking interactions, *magenta* the 7 base–phosphate interactions, and *orange* the 10 base–ribose interactions. The *circular diagrams* are reproduced from the entry for 2QUW (see http://rna.bgsu.edu/FR3D/AnalyzedStructures/2QUW)

We treat PDB files that contain more than one biological unit as follows: if all biological units in the file are structurally equivalent, we simply take the first one listed. "Structurally equivalent" means they have the same number of nucleotides, and the nucleotides make the same interactions, as captured by our annotations. We annotate base pairs according to the Leontis/Westhof system of base-pair classification (Leontis and Westhof 2001), using the FR3D program suite, which has been carefully fine-tuned to identify the base pairs annotated manually (Sarver et al. 2008). When the biological units differ, we establish how they differ and choose the one that is most suitable for the NR dataset. We structurally align all pairs of biological units to determine whether there are significant conformational differences, as measured by the geometric discrepancy between aligned nucleotides, which we calculate as previously described (Sarver et al. 2008). If the geometric discrepancy between two biological units exceeds 0.4, it is likely that each unit has valuable, nonredundant information, and so they are not labeled redundant. Among redundant biological units, one may be better modeled. As a proxy for modeling quality, we use the number of annotated base pairs per nucleotide to choose the representative biological unit. A special case, illustrated in the left panel of Fig. 13.2, is when two of the biological units interact directly by base-pairing. In this case, we keep both units in order to capture the additional interactions. For each NR dataset file, we list the chains comprising the biological unit chosen for inclusion (see below).

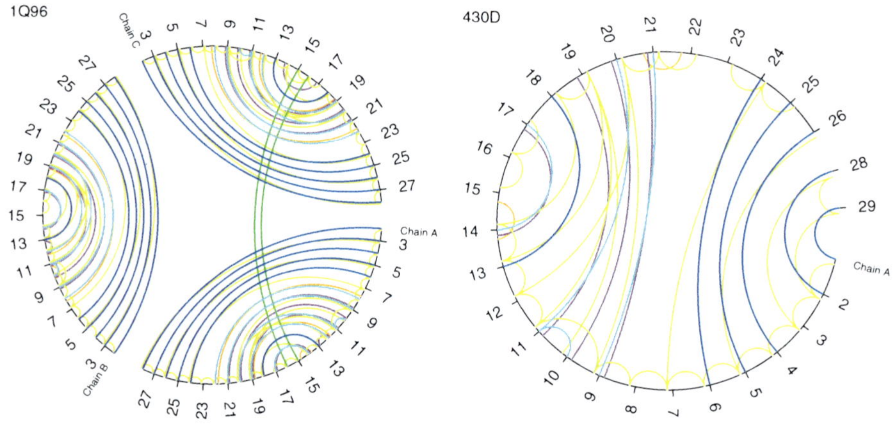

Fig. 13.2 Different crystal forms of essentially the same RNA motif, here the Sarcin–Ricin loop (SRL) from rat 28S rRNA. Circular interaction diagrams for PDB files 1Q96 (*left panel*) and 430D (*right panel*). File 1Q96 contains three copies (chains A–C) of the biological unit, with two base pairs between chains A and C. File 430D has one biological unit and a modified nucleotide at residue 27 (currently not read by FR3D). For the NR set, 1Q96 chains A and C are chosen to represent this class; both chains are kept to include the quaternary base pairs between them

13.2.2 Redundancy Between PDB Files

The PDB/NDB database serves as the international repository of biological macromolecular structure investigations. As such, the database contains every published protein, DNA, or RNA 3D structure. Consequently, for a given macromolecule, the database may contain more than one file, representing essentially the same 3D structure. This occurs for a number of reasons. First of all, more than one research group may have solved essentially the same structure. An example is the cyclic-di-G type 1 riboswitch, PDB files 3IRW and 3IWN (Kulshina et al. 2009; Smith et al. 2009). Secondly, once the wild-type structure has been solved, investigators may be interested to solve structures containing various mutations, to test functional and structural hypotheses. Each of the resulting structures constitutes a separate entry in the PDB/NDB. For the c-di-G riboswitch, the database contains structures for the G20A and C92U mutations, as well as the double mutation, in different files, 3MUM, 3MUR, and 3MUT (Smith et al. 2010). The mutated residues were designed to change the specificity of the riboswitch from c-di-G (3MUT) to c-di-A (3MUV). Except for the mutated residues and the bound analyte, all the RNA structures are essentially identical to the wild type (3IRW) and thus largely redundant for our purposes. Below we discuss future plans to capture the informative differences between structures that are mostly, but not entirely, redundant.

A third source of redundancy stems from the interest in seeing how a macromolecule interacts with alternative ligands. For example, the lysine riboswitch has been solved bound to different analytes, in addition to lysine, and with different metal ions, besides the physiologically relevant magnesium (Garst et al. 2008;

Serganov et al. 2008). Reported structural studies show that the 3D structure of the lysine riboswitch changes very little when bound to different analytes or ions, and again, the different PDB/NDB entries for this RNA are largely redundant for our purposes. In other cases, the differences may be significant and so must be examined case by case.

A fourth situation pertains to structures of ribozymes. To elucidate the mechanisms of enzymes, including ribozymes, crystallographers try to capture "snapshots" of the progress of the chemical reactions they catalyze. For example, structures of a hammerhead ribozyme, with catalytic residue G12 mutated to A to slow the reaction, were obtained before (2QUS) and after (2QUW) strand cleavage (Chi et al. 2008). The cleavage reaction changes the number of distinct chains in the file. Except for the presence of the cleavage site, the 3D structures of the RNA in 2QUS and 2QUW are also very similar and largely redundant for our purposes. Similarly for ribosomes "caught" at different stages of translation.

Fifthly, the PDB/NDB contains structures of homologous RNA molecules. These are RNA molecules related by evolution via the processes of speciation or gene duplication and thus share similar functions and 3D structures, while differing in sequence to variable degrees. Large portions of homologous molecules can be identical or very similar in structure, if not in sequence. However, when there is sufficient sequence variation between homologous RNA molecules, the structural redundancy is actually very interesting, as it documents sequence changes which are structurally neutral and likely to occur frequently during evolution. So unlike the other sources of redundancy, which we try to identify and exclude in constructing NR datasets, we seek to retain structural redundancy due to the presence of different homologs of the same molecule. Our aim, therefore, becomes to identify and place in our NR datasets the best copy of each distinct homolog in the PDB/NDB. Thus, our NR datasets contain representative 16S rRNA structures from *Escherichia coli*, *T. thermophilus*, and *D. radiodurans*, as well as the homologous 18S rRNA from yeast. We also retain representatives of all tRNAs that differ by species or codon specificity.

A sixth situation arises from the fact that the same motif may be crystallized in two or more different crystal forms, usually as a consequence of using different RNA constructs to promote crystallization or to introduce heavy atoms for phasing. For example, PDB files 430D and 1Q96 both contain structures of the Sarcin–Ricin Loop (SRL) motif from rat 28S ribosomal RNA. The asymmetric unit in 1Q96 contains three biological units, each a distinct, but very similar, copy of the SRL, solved to 1.8 Å (Correll et al. 2003). File 430D contains one SRL motif, solved to 2.1 Å (Correll et al. 1998). Figure 13.2 contains circular diagrams showing the contents of these files. There are small differences between the structures, but they are largely equivalent. For example, the structure in 430D contains bromocytidine, which is not currently read by our programs, resulting in a gap at position 27 in the diagram. More significantly, two of the SRL motifs in 1Q96 (chains A and C) form an interaction comprising two non-WC base pairs (green chords). We choose file 1Q96 and retain both chains in the NR data, to sample the interaction between them.

13.3 Identifying Redundant Files in the RNA 3D Structure Database

Many of the over 2,000 RNA-containing 3D structure files in the PDB/NDB are largely redundant, for one or more of the reasons described above. In this section, we describe the procedure that we use to identify and cluster redundant files in the RNA 3D structure database. Our aim is to flag uninteresting redundancy for removal while retaining interesting sequence variation, such as that between RNA homologs that are sufficiently different in sequence to justify retaining them. We first flag possible redundancy between PDB files by sequence comparisons. We then verify redundancy by structural superposition and geometric analysis. Once we have clustered all redundant files in their respective classes, we select one file to represent each class.

Because PDB structure files may contain more than one RNA molecule (chain), we simplify the procedure by focusing on the longest chains present in each file. For example, if a file contains a 16S rRNA and one or more tRNA molecules, we focus on the 16S rRNA and ignore the tRNAs. However, we focus on the tRNA when it is the largest RNA in the file. In Sect. 13.8.1, we discuss plans for considering also the shorter chains in making choices of data to include. For NMR files, we use the first model of the longest chain.

For each file, the longest chain is identified; in case of ties, the first chain is chosen. While the biological source of this chain is usually provided in the SOURCE record, in cases where the RNA sequence has been reengineered to facilitate crystallization, the source is only listed in the PDB as "synthetic." When considering redundancy between two 3D structure files, if both files have biological source information indicating that the longest RNA chains are from different species, the structures are declared to be nonredundant to each other, and no further analysis is carried out, as detailed in the following paragraphs.

Next, we proceed to sequence comparisons. For a pair of files with longest chains X and Y, the number, N, of identical bases in the alignment is counted, and the smaller of the two lengths, L, is recorded. We make a provisional assignment of redundancy between the files in a way that depends on N and L and the lengths of the chains X and Y. The idea is to assign provisional redundancy between chains when more than 95% of the nucleotides are identical. For chains of length 80 or less, we loosen this criterion and allow up to four base differences, as 95% sequence identity allows for very few differences. However, for chains shorter than 19 nucleotides (corresponding to the sizes of the shortest micro-RNAs), we insist on identical sequences; many of the structures at this length are high-resolution structures of duplexes containing individual non-Watson–Crick base pairs or 3D motifs. Here, small differences involving just one or two bases are of interest. Thus, up to length 18, the longest chains are only compared to other longest chains of exactly the same length. Chains of length 19 and longer are compared to longest chains in other files up to twice their length. This is to prevent clustering structural fragments (e.g., domains of ribosomal RNAs) with the intact structures (e.g., complete 16S or 23S rRNA). The sequences of each pair of chains that meet the top of the table. Within each equivalence class, the chosen representative is the first structure listed, and the other structures are listed in decreasing order of the number of base pairs per nucleotide. Each line of the table provides the PDB ID, linked to the corresponding PDB resource page, a link to FR3D annotations for the file, the number of nucleotides and base pairs detected in the file, the NR chains chosen to represent the file, and the nominal resolution. Also included are metadata fields including author, deposition date and biological source, where available, and at the end of each line, the PDB identifier of the structure which currently represents the equivalence class. Thus, the representative file for each equivalence class is associated with all members of the class, and the equivalence classes can be reconstructed from this listing.

Separate links provide tables with NR lists up to resolution thresholds 20, 4.0, 3.5, 3.0, 2.5, 2.0, 1.5, and 1.0 Å. Each line in the NR tables corresponds to a distinct equivalence class and contains information pertaining to the file representing that class. The last column of each row lists all the other structures belonging to the equivalence class, at all resolutions, sorted by decreasing number of base pairs per nucleotide. To facilitate database searches for recurrent motifs, we have integrated the NR lists into WebFR3D, the FR3D Web server (see: http://rna.bgsu.edu/WebFR3D/).

13.8 Issues for Future Work

We anticipate continued improvements in the procedures for maintaining and refining NR sets for RNA structural analysis. Improvements can be made in (1) the construction of the equivalence classes, (2) the choice of biological unit within individual files, and (3) the choice of file to represent the class. We discuss some ideas to improve each of these steps in turn. In addition, there is the issue of how to include structurally interesting variation between files belonging to the same equivalence class, something that we currently overlook. We will conclude by discussing some ideas to tap this resource as well.

13.8.1 Improving the Construction of Equivalence Classes

The construction of equivalence classes is a problem in clustering and can be improved by (1) identifying structures that are currently assigned to different clusters but which essentially represent the same RNAs and which therefore should be placed in the same equivalence class and (2) identifying structures that are currently placed in the same class but which really should be separated. This is an ongoing task that requires further study and then automation. To facilitate the process, we are developing additional analysis and visualization tools to compare members of clusters and representatives of different clusters.

Structures that should be assigned to the same equivalence class. After identifying two structures as provisionally redundant based on sequence similarity, we check that they are also geometrically similar using rigid-body superposition. If by this criterion they are deemed geometrically dissimilar, they are placed in different equivalence classes. However, we have found problems with this approach. For example, files 3IRW and 3IWN (Kulshina et al. 2009; Smith et al. 2010) are essentially identical structures of the wild-type cyclic-di-GMP riboswitch, but they are currently placed by the procedure in different equivalence classes. This is because, by rigid-body rotation, they have an average geometric discrepancy of 0.67 Å/nucleotide, which exceeds the 0.5 Å cutoff. The reason for this is that the structures were solved in different laboratories, and while both research groups used the same riboswitch from the same organism and even the same kind of RNA engineering to facilitate crystallization, there is a small difference in the constructions that is amplified during structure comparison. Although both groups attached the U1A protein-binding hairpin loop to the same peripheral helical stem to facilitate crystallization, two extra base pairs were included in the stem in the construction of the molecule in 3IWN. Consequently, the main body of the riboswitches and the protein-binding hairpin loops cannot be superposed simultaneously, even though individually these domains do superpose well. This results in an average geometric discrepancy exceeding the threshold.

We have developed a local structural alignment algorithm which we have implemented in our R3DAlign software and Web application that is able to overcome this problem (Rahrig et al. 2010), but we have not yet incorporated it in the NR pipeline. R3DAlign determines a nucleotide-to-nucleotide alignment between two 3D structures by making 4-nucleotide neighborhoods around each nucleotide and choosing the alignment that maximizes the number of aligned neighborhoods that superimpose well. The localized nature of the neighborhoods means that R3DAlign can align nucleotides even when the global 3D structure does not superimpose under a single rigid transformation. Applying R3DAlign to 3IWR and 3IWN, we find that the nucleotides that are common to both structures in the main body of the riboswitch and in the U1A-binding hairpin loop both superimpose well locally, giving a small overall geometric discrepancy per nucleotide. All the structural differences are concentrated at the site where the two base pairs are inserted. By using R3DAlign in the pipeline, these two files will be placed in the same equivalence class, where they belong. A similar situation occurs with the lysine riboswitch, again solved by two different groups using slightly different constructs and consequently placed in different equivalence classes by the current algorithm (Garst et al. 2008; Serganov et al. 2008).

Structures that should be assigned to different equivalence classes. With regard to the second issue, the current approach focuses on the longest chain in each structure file and does not consider unique shorter chains that may be present. For example, a file containing a ribosome may also contain a bound tRNA or mRNA fragment that may be unique in the database. Thus, under our current procedures, a tRNA in a ribosome structure file is not being compared to other tRNAs. Likewise an mRNA fragment containing, for example, an IRES element, is not being

Chapter 14
Ions in Molecular Dynamics Simulations of RNA Systems

Pascal Auffinger

Abstract Ions and water molecules are intricately associated with biomolecular systems and play important structural and functional roles that are still not well understood. For RNA systems, the functions of these ions are not limited to the neutralization of the charges carried by the polyanionic backbone, since they also bind to very specific locations of the RNA 3D fold. Hence, it is essential to include them with the greatest possible accuracy in 3D structural models and especially in molecular dynamics (MD) simulations. This review aims at describing some of the successes achieved in the modeling of monovalent and divalent ions in RNA systems, as well as to highlight important deficiencies of current force fields and MD techniques that represent important challenges for future development.

Keywords Molecular dynamics simulation • Crystallography • RNA • DNA • Solvation • Hydration • Monovalent cation • Divalent cation • Sodium • Potassium • Magnesium • Na^+ • K^+ • Mg^{2+}

14.1 Introduction

In addition to water, ions are an integral part of nucleic acid systems and play a crucial role in RNA stability and folding (Takamoto et al. 2002; Woodson 2005; Auffinger and Hashem 2007; Auffinger et al. 2011). It is now well documented that structural and functional properties of nucleic acid systems can be strongly altered by the type and concentration of the surrounding ionic species. Hence, it is important to better understand this interplay, both from an experimental and a theoretical point of view. The aim of this review is to address various issues

P. Auffinger (✉)
Architecture et réactivité de l'ARN, Université de Strasbourg, CNRS, IBMC, 15 rue René Descartes, 67084 Strasbourg, France
e-mail: p.auffinger@ibmc-cnrs.unistra.fr

N. Leontis and E. Westhof (eds.), *RNA 3D Structure Analysis and Prediction*, Nucleic Acids and Molecular Biology 27, DOI 10.1007/978-3-642-25740-7_14,

associated with modeling ions in molecular dynamics (MD) simulations of nucleic acids and, more specifically, of RNA systems.

14.2 Modeling Monovalent Cations (Na^+, K^+, NH_4^+, . . .)

14.2.1 Is a Neutralizing Ionic Atmosphere Sufficient?

Ions were first introduced in MD simulations for the purpose of neutralizing the negative charge carried by the anionic backbone of these biopolymers. For convenience, co-ions were generally omitted and assumed to be unimportant. Such an assumption seemed reasonable in the early stages of the development of MD techniques and led to trajectories acceptable at the time. However, this approximation does not allow for accurate modeling of subtle effects associated with the presence of co-ions in different concentration ranges, such as the sequence-dependent increase of helical twist and reduction of groove width of RNA duplexes documented by recent MD simulations (Besseova et al. 2009). Differences between simulations using minimal Na^+ or excess KCl salt conditions were also recently reported (Reblova et al. 2010b). Without doubt, minimal salt models represent a distant approximation of physiological ionic conditions, not to mention most experimental in vitro conditions, and do not allow to model recently described "anion/nucleic acid" interactions (Auffinger et al. 2004b).

Moreover, one has to consider that simulations using minimal salt conditions suffer from finite size artifacts and from much slower and inappropriate ionic relaxation times (Chen et al. 2009b). Insidiously, minimal salt conditions were originally recommended to avoid the formation of NaCl or KCl salt clusters (Fig. 14.1) observed in MD simulations using some unrefined force-field parameters (Chen et al. 2009b). With the development of improved ionic force fields (Chen and Pappu 2007b; Joung and Cheatham 2008, 2009; Lopes et al. 2009; Yu et al. 2010; Zhang et al. 2010), such salt clustering artifacts (Vaiana et al. 2006; Auffinger et al. 2007; Chen and Pappu 2007b; Noy et al. 2009) are no longer an issue and the use of minimal salt models should now definitely be relegated to the past.

It is worth noting that ionic aggregation occurs in most cases above a specific transition point located usually between 0.1 and 0.2 M in NaCl or KCl (Auffinger et al. 2007). Consequently, it is possible to "conceal" this phenomenon by choosing excess monovalent ion concentrations below this limit. Of course, this is not advised, since ion/RNA and especially ion/phosphate interactions are still affected by the use of "deficient" ion parameters, even at low excess salt concentration (Noy et al. 2009). The preceding issue represents a strong incentive for continuing our efforts to improve nucleic acid force fields and associated ionic parameters.

Fig. 14.1 View of a KCl aggregate formed in the vicinity of an RNA/antibiotic complex after 4 ns of MD simulation using the K^+ van der Waals parameters implemented in earlier AMBER force fields. The largest of these clusters is marked by a *yellow arrow*. AMBER 10 and 11 versions (Case et al. 2010) include improved ionic parameters (Joung and Cheatham 2008, 2009) [reproduced by permission from (Vaiana et al. 2006)]

14.2.2 *Finite Size Artifacts*

A recent study documented finite size artifacts that take place when the box size surrounding the simulated RNA is too small and too few ions are present. Remarkably, one observes significantly different phosphate/Na^+ radial distribution function profiles (Fig. 14.2) in a box of 80 and 120 Å at a constant excess salt concentration of 800 mM NaCl (Chen et al. 2009b). Another study reported that aqueous solutions of $MgCl_2$ attain their bulk properties only at a distance larger than 18 Å from the RNA. Surprisingly, in NaCl, the RNA charges extend their influence on the much longer 25 Å scale (Kirmizialtin and Elber 2010), setting lower limits to the minimal solvation shell size that considerably exceed those generally in use. Such effects have to be taken into consideration in future developments and applications of MD techniques.

14.2.3 *Choice of Monovalent Cations: K^+ Versus Na^+ or NH_4^+*

The monovalent Na^+, K^+, and NH_4^+, and marginally, Cs^+ cations (Chen et al. 2009a), have been used in MD simulations of RNA systems. It is worth noting that Na^+ cations are used in simulations much more often than K^+ or NH_4^+ (89 with Na^+, 14 with K^+, and 8 MD simulations with NH_4^+ cations were listed in a 2007 survey (Hashem et al. 2008). Authors prefer to use Na^+ cations because MD

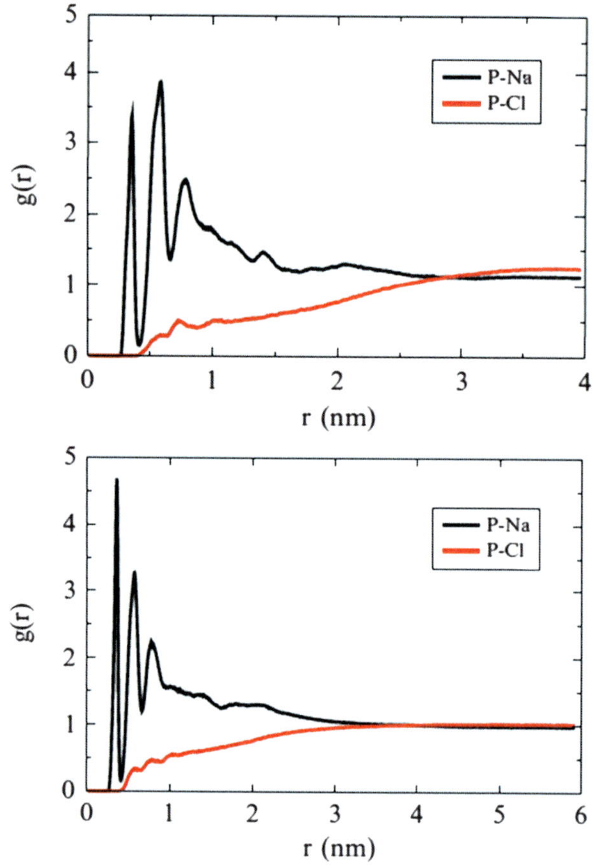

Fig. 14.2 RNA phosphate-counterion radial distribution functions for a Tar–Tar* complex in an (**a**) 80 Å and (**b**) 120 Å box of 800 mm NaCl that illustrate finite size artifacts. Note the differences in the short-range P–Na^+ profiles and the fact that the anion concentration is ≈15% different at large separation [reproduced by permission from (Chen et al. 2009b)]

simulations are mostly based on NMR or crystallographic structures and most of these are determined in Na^+-containing buffers. The preference for Na^+ over K^+ cations in experimental work seems completely at odds with conditions prevailing in vivo, since K^+ is the major cation found in cells, while Na^+ dominates in *extracellular* fluids (Auffinger et al. 2011). Hence, for most MD simulations of intracellular systems, K^+ should be the monovalent cation of choice.

NH_4^+ cations were sometimes included in MD simulations because they are considered to favor the crystallization of RNA systems and to stabilize RNA folds. Yet, there is no documented justification for using NH_4^+ over other monovalent cations. Moreover, NH_4^+ cations are cytotoxic and are converted into less toxic compounds such as urea in mammals. NH_4^+ cations are consequently irrelevant to RNA functions under most in vivo conditions. In bacteria, NH_4^+ toxicity must be evaluated with different criteria (Muller et al. 2006). However, it would be surprising to uncover ammonium cations contacting RNA systems in in vivo conditions.

14.2.4 *Locating Binding Pockets for Monovalent Cations*

Monovalent cations are often undetectable by crystallographic methods given their resemblance to water molecules (Das et al. 2001). Hence, the most common MD strategy for locating monovalent cation-binding pockets consists in randomly placing, at a certain distance from the RNA, an appropriate number of monovalent cations and anions (Hashem and Auffinger 2009). In the course of a short period of time, generally during the equilibration phase, monovalent ions track down the most electronegative (for cations) or electropositive (for anions) binding pockets and settle there for extended periods of time, after which they exchange with nearby cations or water molecules.

A first success of this strategy, and of nucleic acid MD simulations in general, is associated with early and relatively short simulations (≈1.5 ns) of a DNA duplex that suggested that ions might intrude into DNA grooves (Young et al. 1997). This conclusion was reached before clear experimental evidence of specific ion binding to nucleic acids was provided and changed our conception of the structure of the ionic atmosphere surrounding nucleic acids (Auffinger and Hashem 2007).

14.2.5 *Where Do Monovalent Cations Bind?*

Monovalent cations are no longer considered to function solely as components of the diffuse ionic cloud that neutralizes the negative charge carried by the nucleic acid polyanionic backbone (Manning 1978). Rather, it is now well appreciated that they can also intrude in a sequence specific manner into nucleic acid grooves (Auffinger and Westhof 2000, 2001; Kirmizialtin and Elber 2010) where they can bind to electronegative pockets created by particular RNA folds.

For regular $r(CG)_{12}$ and $r(AU)_{12}$ duplexes, r(GpC) and r(ApU) steps were identified as efficient cation-binding pockets while r(CpG) and r(UpA) steps, given specific steric and electrostatic factors, are repulsive to cations (Fig. 14.3) (Auffinger and Westhof 2000, 2001). As observed in numerous crystal structures (Auffinger et al. 2011) and unpublished MD studies (Hashem and Auffinger), r(GpG) steps are also important cation-binding sites since cations display a clear propensity to bind to N7 and O6 guanine sites. All these cation-binding sites are located in the major groove which can be considered as an "ionophilic groove." In contrast, the RNA minor groove exhibits a rather "ionophobic" character as observed in MD simulations of the 5S ribosomal loop E motif (Reblova et al. 2003b; Auffinger et al. 2004a).

Monovalent-binding sites around complex RNA folds were also described in following studies (Csaszar et al. 2001; Reblova et al. 2003a, b, 2007; Auffinger et al. 2004a; Krasovska et al. 2006; Razga et al. 2006). When experimentally detected Mg^{2+} cations are removed from the starting model, one or two

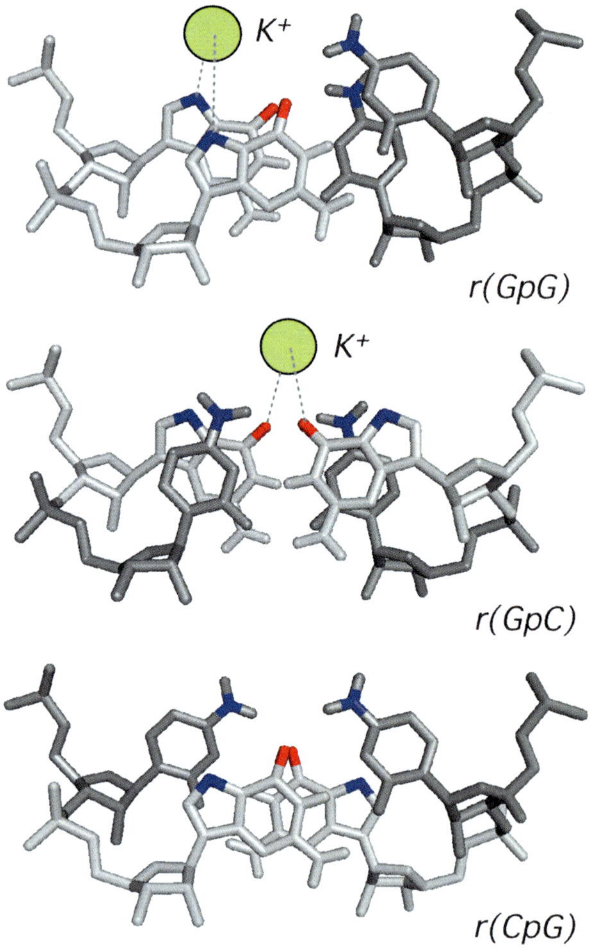

Fig. 14.3 Schematic representation of K^+ ion binding features for the r(GpG), r(GpC) and r(CpG) steps [top figure: Hashem and Auffinger, unpublished data; middle and bottom figures adapted from (Auffinger and Westhof 2000)]. rC and rG residues are shown in *dark* and *light gray*, respectively. Deep groove O, N and K^+ atoms are shown in *red*, *blue* and *yellow*, respectively

monovalent cations quickly occupy the vacant site during MD simulations (Huang et al. 2009), but when Mg^{2+} cations are left in place, adjacent monovalent cation-binding sites are more weakly occupied (Reblova et al. 2004). For example, monovalent cations were found to bind close to experimentally characterized divalent-binding pockets (Auffinger et al. 2004a; Ditzler et al. 2009) and to associate with the aminoglycoside-binding pocket of the ribosomal A-site at locations where charged ammonium groups of the aminoglycoside attach (Romanowska et al. 2008). In the 5S ribosomal loop E motif, it was observed that a monovalent binding site match the binding site of a lysine ammonium group (Auffinger et al. 2004a). Hence, MD simulations using monovalent cations are an efficient tool for detecting potential binding sites of charged functional groups belonging to amino acids or drugs.

14.2.6 Cation Dynamics

Long monovalent cation residency times were reported for several RNA systems (Reblova et al. 2003b, 2004; Auffinger et al. 2004a; Krasovska et al. 2006; Spackova et al. 2010). Estimating ion residency times is particularly difficult because of: variations due to (1) force field approximations, (2) the nature of the chosen anions, (3) the ionic concentration dependance (4) or the choice of the methods used for calculating the residency times themselves (Auffinger et al. 2004a; Krasovska et al. 2006; Eargle et al. 2008; Chen et al. 2009a; Kirmizialtin and Elber 2010). For the kissing loop Tar–Tar* complex, mean counterion residency times of 56, 38, and 35 ps were reported for Na^+, K^+, and Cs^+ (Chen et al. 2009a). Earlier studies on RNA duplexes proposed maximum residency times for K^+ around 500 ps (Auffinger and Westhof 2000, 2001). For the ribosomal 5S loop E motif that displays a highly "ionophilic" major groove, monovalent cation residency times exceeding 5 ns were noted in two independent sets of MD simulations (Reblova et al. 2003b; Auffinger et al. 2004a). For more complex binding pockets, residency times in the 6–13 ns range were reported (Krasovska et al. 2006). Yet, in most instances, when one ion dissociates from the RNA, it is rapidly replaced by another, and so occupancies of binding pockets ranging from 80 to 100% are sometimes reported. This latter fact strongly supports the concept that monovalent ions are integral components of RNA structures.

14.2.7 K^+ Versus Na^+ in MD Simulations

A small number of studies have been undertaken comparing the effects of Na^+ versus K^+ cations on RNA structure. Some of them reported that simulations with net-neutralizing Na^+ and 0.2 M excess salt conditions appear in all aspects equivalent (Razga et al. 2006; Besseova et al. 2010; Spackova et al. 2010), although more insightful simulations demonstrated that a $\approx$0.65 M K^+ excess caused a modest sequence-dependent compaction of canonical A-RNA double helices (Besseova et al. 2009). Simulations of the ribosomal A-site finger in the presence of K^+ revealed a slightly larger propensity for more "closed" structures. The authors concluded, however, that such an observation might not be significant given the size of the simulated system and its unusual flexibility (Reblova et al. 2010a). However, in a study describing a set of simulations of the smaller ribosomal UAA/GAA internal loop structural element, the same authors reported similar narrowing of the major groove in the presence of excess KCl (Reblova et al. 2010b). They suggested that "these results are explained by better screening of phosphate groups with higher ionic strength which allows their closer approach across the groove" and concluded that "the stability of the functional H40 conformation may be affected by ionic conditions or other interactions reducing the interphosphate repulsion." Surprisingly, in a third study by this group, RNA simulations of hairpin ribozyme structures led to the artifactual generation of irreversible "ladder-like," underwound A-RNA structures in one of the external helices in NaCl but not in KCl excess salt conditions (Mlynsky et al. 2010).

Indeed, if effects induced by the choice of monovalent ions exist, as suggested by some biophysical experiments (Gluick et al. 1997; Heddi et al. 2007; Vieregg et al. 2007; Lambert et al. 2009), they are likely to be quite subtle and possibly out of reach of current MD techniques. In all cases they deserve further in-depth investigations. For instance, the inversely proportional stability of the Tar–Tar* complex to the crystallographic radius of the monovalent counterion (Lambert et al. 2009) has been reproduced by MD simulations and is associated with a more effective calculated condensation of Na^+ with respect to K^+ cations around RNA systems (Chen et al. 2009b).

14.3 Modeling Divalent Cations

14.3.1 Magnesium Cations (Mg^{2+})

Despite the importance of Mg^{2+} cations for RNA structure and function, few MD simulations of RNA systems have been carried out using Mg^{2+} cations (see Table 14.1). A survey of MD simulations of RNA systems (up to September 2007) revealed that only 22 out of a total of 113 simulations of RNA systems included Mg^{2+} cations [(Hashem et al. 2008); the present list comprises 14 additional references]. The tendency to avoid inclusion of Mg^{2+} when simulating RNA is a consequence of the difficulties that persist in modeling these cations. First, water molecules bound to Mg^{2+} cations display very long residence times (2–10 ms) that are several orders of magnitude longer than the residence times of water molecules bound to monovalent cations (Ohtaki 2001). Hence, current MD techniques cannot simulate the desolvation process of Mg^{2+} cations required to form inner-sphere complexes. Moreover, Mg^{2+} cations display very slow diffusion rates and therefore have poor sampling properties. Force-field concerns have also been raised (McDowell et al. 2007) necessitating the development parameters reproducing subtle polarization effects (Jiao et al. 2006; Yu et al. 2010) based on accurate experimental and high-level theoretical data (Markham et al. 2002; Petrov et al. 2002, 2005; Bock et al. 2006; Harding 2006; Ikeda et al. 2007; Rao et al. 2008; Oliva and Cavallo 2009; Callahan et al. 2010).

Note that in some instances, it has been reported that outer-sphere bound Mg^{2+} cations quickly lose a water molecule from their inner coordination sphere and directly chelate to RNA nucleotides, an observation consistent with the known bias toward inner-shell binding of Mg^{2+} cations in RNA simulations due to force-field approximations (Reblova et al. 2003b, 2006) and in agreement with a recent MD simulation of solvated Mg^{2+} cations using a polarizable force field that reported first solvation shell lifetimes in the order of only hundreds of picoseconds (Jiao et al. 2006).

Table 14.1 List of RNA systems (up to March 2011) that have been studied by MD simulations using at least one Mg^{2+} ionic condition

System	References
RNA hairpins	Sorin et al. (2005)
RNA duplex	Kirmizialtin and Elber (2010)
Hiv-1 dimerization initiation site (DIS)	Reblova et al. (2007)
RNA three way junction	Besseova et al. (2010)
RNA kink-turns	Razga et al. (2006)
Ribosomal 5S loop E motif	Auffinger et al. (2003, 2004a) and Reblova et al. (2003b)
Ribosomal 5S loop E in complex with L25	Reblova et al. (2004)
Ribosomal A-site	Romanowska et al. (2008)
Ribosomal16S helix 44	Reblova et al. (2006)
Ribosomal L1-stalk/tRNA complex	Trabuco et al. (2010)
GluRS/tRNA complex	Black Pyrkosz et al. (2010)
tRNA/EF1u complex	Eargle et al. (2008)
Full ribosome	Sanbonmatsu and Tung (2006)
Hammerhead ribozyme	Hermann et al. (1997, 1998), Torres and Bruice (2000), Torres et al. (2003), Lee et al. (2007, 2009), Lee and York (2008) and Martick et al. (2008)
Hepatitis delta virus ribozyme	Krasovska et al. (2005, 2006, Sefcikova et al. (2007) and Veeraraghavan et al. (2010)
Hepatitis C virus IRES IIId domain	Golebiowski et al. (2004)
Hairpin ribozyme	Rhodes et al. (2006) and Ditzler et al. (2009)
Guanine riboswitch	Villa et al. (2009)
Add-A riboswitch	Sharma et al. (2009)
SAM riboswitch	Huang et al. (2009), Priyakumar (2010)
L1 ligase molecular switch	Giambasu et al. (2010)

Yet, most other MD settings report no first solvation shell water exchange on current MD time scales (Kirmizialtin and Elber 2010). This point has consequently to be addressed with great concern given the lack of precise experimental data on the water exchange mechanism.

Given above-mentioned limitations, only a few studies among those mentioned in Table 14.1, have addressed the binding properties of Mg^{2+} cations to RNA and their influence on RNA structure, function, and dynamics while the others mentioned no more than the presence of Mg^{2+} in their settings without providing further details.

The roles played by Mg^{2+} cations on the structure and catalytic mechanism of the hammerhead ribozyme were addressed several times in attempts to shed some light on the still elusive roles of these cations (Hermann et al. 1997, 1998; Torres and Bruice 2000; Torres et al. 2003; Lee et al. 2007, 2009; Lee and York 2008; Martick et al. 2008; Banas et al. 2009; Park and Boero 2010). Interestingly, a catalytic mechanism without direct participation of metal ions has recently been suggested based on crystallographic and MD data (Martick et al. 2008). Yet, much remains to be understood regarding the mechanism of this most intensively studied RNA ribozyme (Leclerc 2010).

Different effects related to the presence of Mg^{2+} cations on the structure and dynamics of two closely related subtypes of the HIV-1 DIS were reported. For subtype B, the conversion from an open to a closed conformation occurred in the presence of Mg^{2+} while the absence of divalent cations led to an increased conformational dynamics. By contrast, the stability of the active site architecture of subtype A was not affected by the presence or absence of Mg^{2+} cations (Reblova et al. 2007).

In line with this observation, several studies concluded that Mg^{2+} cations have no significant effects on the structure and dynamics of RNA systems, at least on the investigated time scales. A study on tetraloops reported that explicit representations of cations are not necessary to model the folding of these RNA fragments (Sorin et al. 2005). Simulations of ribosomal RNA kink-turns carried out in the presence of Mg^{2+} or K^+ cations suggested that changes in ionic conditions do not affect the flexibility of the RNA (Razga et al. 2006). But such conclusions are certainly structure dependent and should not be generalized. In a rare occurrence, it was reported that Mg^{2+} cations, introduced in an MD model of the ribosomal A-site, destabilized the RNA structure (Romanowska et al. 2008) suggesting that great care has to be taken in the initial placement of these cations.

Indeed Mg^{2+} cations play a significant role in the stabilization and folding of RNA systems (Draper et al. 2005), but much has still to be learned about the mechanisms involved. The very specific binding geometry of hydrated Mg^{2+} cations was investigated (Auffinger et al. 2003). Mg^{2+} cations are known to display a high affinity for specific anionic oxygen atoms belonging to phosphate groups, to which they bind by losing a water molecule from their hexacoordinated hydration shell. Contrary to the $Mg(H_2O)_6^{2+}$ form that displays significant tumbling motions when bound to RNA, the pentahydrated form establishes a large array of water-mediated contact with distant RNA residues that stabilizes the entire structure by acting as a freezing agent (Fig. 14.4). This coordination-clamp mechanism is probably of universal significance.

14.3.2 *Other Divalent Cations (Mn^{2+}, Ca^{2+}, Sr^{2+}, Ba^{2+}, ...)*

Mn^{2+}, Ca^{2+}, Sr^{2+}, and Ba^{2+} cations are found in crystallographic structures of RNA systems and are supposed to mimic some properties of Mg^{2+} cations (Auffinger et al. 2011). Yet, to the best of our knowledge, no MD simulations of RNA systems with divalent cations other than Mg^{2+} have been published. The challenges associated with an accurate modeling of the binding specificities of the softer Mn^{2+} cations are particularly delicate and are probably out of reach of classical MD techniques (Bock et al. 1999). Although both ions display similar ionic radii and charge, Mn^{2+} prefers to bind to softer atoms (nitrogen, sulfur) while Mg^{2+} has a greater affinity for oxygen atoms. The only MD study referring to Mn^{2+} cations is that of a hammerhead

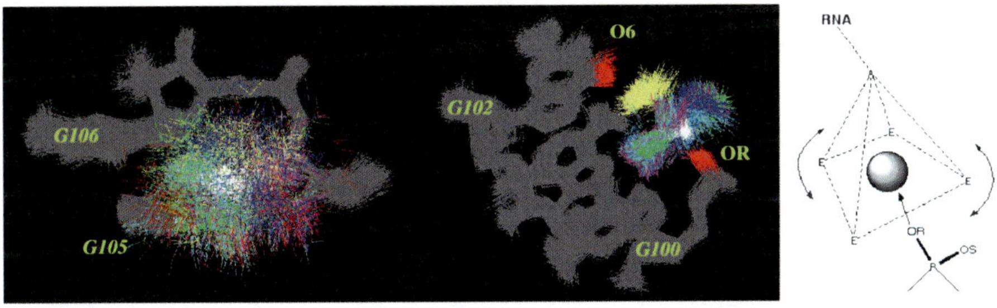

Fig. 14.4 Dynamics of RNA–Mg^{2+} complexes derived from MD simulations of the 5S rRNA loop E motif (Auffinger et al. 2003). (*right*) High mobility of a hexahydrated cation bound to a Watson–Crick GpG step. (*middle*) Reduced mobility of a pentahydrated cation interacting with two guanines of the internal loop. (*left*) Scheme describing a coordination clamp mechanism through which pentahydrated Mg^{2+} cations stabilize RNA systems. *A*: apical water molecule; *E*: equatorial water molecules [reproduced by permission from (Auffinger and Hashem 2007)]

ribozyme in its crystal environment, where all Mn^{2+} cations were replaced by Mg^{2+} cations (Martick et al. 2008). In other studies, crystallographic Sr^{2+} cations (Priyakumar 2010), Ca^{2+} cations, (Rhodes et al. 2006) or cobalt(III) hexammine cations (Ditzler et al. 2009) were replaced by Mg^{2+} cations and in some occurrences, divalent Mg^{2+} cations were ignored and replaced by monovalent Na^{+} cations or water molecules (Reblova et al. 2003b; Banas et al. 2010). Indeed, crystallographers often use saturating ionic conditions that lead to an excess of Mg^{2+} cations bound to the biopolymeric system and some of these cations, involved in lattice contacts, do not contribute to the stabilization of the RNA system in solution (Auffinger et al. 2003, 2004a; Auffinger 2006).

Note that a further difficulty in modeling Mg^{2+} cations is related to the fact that electron densities are sometimes incorrectly assigned. For instance, in at least two documented cases, electron densities generated by crystallographic SO_4^{2-} or Cl^{-} anions were assigned to Mg^{2+} cations, which resulted in inaccurate starting models for MD simulation (Auffinger et al. 2004b; Reblova et al. 2004, 2007; Hashem and Auffinger 2009; Kieft et al. 2010). Calculated barrier heights for the chemical reaction of the hepatitis delta virus were shown to be particularly sensitive to the precise positioning of Mg^{2+} cations (Banas et al. 2008). Hence, one has to examine with great care the starting models used for MD studies (Auffinger 2006).

14.4 Modeling Other Cations [$Co(NH_3)_6^{3+}$ and Polyamines]

Given the difficulty of the task (related to the scarcity of reliable experimental data and the lack of accurate force-field parameters), only a few attempts related to modeling multivalent cations (cobalt hexamine and polyamines) bound to DNA have been published.

MD simulations investigating the stabilization of A-DNA by $Co(NH_3)_6^{3+}$, reproduced a spontaneous transition from B-DNA to A-DNA forms for a short duplex in agreement with experimental work and NMR studies. The large hexamine cation coordinated mainly in the major groove of GpG pockets to promote the transition to A-forms (Cheatham and Kollman 1997, 2000).

The effects of natural polyamines, including spermine (Spm^{4+}), spermidine (Spd^{3+}), and putrescine (Put^{2+}) and the synthetic diaminopropane (DAP^{2+}), on DNA systems was investigated in several studies. Spermine dehydrates the DNA minor groove by binding to the phosphate groups delimiting the groove and the hydrophobic methylene groups reduce the organization of water at the positions of spermine binding (Korolev et al. 2002). But spermine molecules, on account of their flexibility, do not appear to form long-lived and structurally well-defined complexes with nucleic acids, which hinders their straightforward detection by crystallographic methods (Korolev et al. 2001). Other polyamines exhibit significant binding differences. For instance, DAP^{2+} is able to form bridges connecting neighboring phosphate groups along the DNA strand and a small fraction of DAP^{2+} and Put^{2+} localizes to the major groove while Spd^{3+} does not (Korolev et al. 2003). It has been suggested that the higher structuring potential of the synthetic DAP^{2+} compared to the more dynamic character of natural polyamines might explain the occurrence of the latter in cells, in preference to DAP^{2+} (Korolev et al. 2004).

14.5 Modeling Anions (Cl^-, SO_4^{2-}, ...)

To approximate physiological ionic conditions, it is necessary to take into account anions neutralizing the excess co-ions found around nucleic acid systems. Anions are commonly considered to have a minimal effect on nucleic acid structure and function. Yet, anions establish, in specific structural contexts, direct contacts with electropositive atoms of nucleic acid. In a survey of nucleic acid structures extracted from the PDB, anion-binding sites were mapped and found to match binding sites of nucleic acid phosphate groups and side chains of the two negatively charged aspartic and glutamic amino acids (Auffinger et al. 2004b). Most of these anion-binding sites were recently identified and characterized in RNA crystal structures by selenate (SeO_4^{2-}) soaking techniques at high but also near biologically relevant ionic strengths (Fig. 14.5) (Kieft et al. 2010).

Given such recent recognition of their binding potential, anions have rarely been included in MD models and only a few studies have described direct binding of chlorides to nucleic acid fragments (Makarov et al. 1998; Feig and Pettitt 1999; Auffinger et al. 2004b; Kirmizialtin and Elber 2010). Given their binding potential and their unexplored effects on nucleic acid structure and function, it is obvious that anions should be included in theoretical models. For that, great care has to be paid to the choice of force-field parameters to avoid artifacts like ion clustering (Auffinger et al. 2007; Chen and Pappu 2007a). Recent parameterization studies (Chen and Pappu 2007b; Joung and Cheatham 2008, 2009) should help alleviate

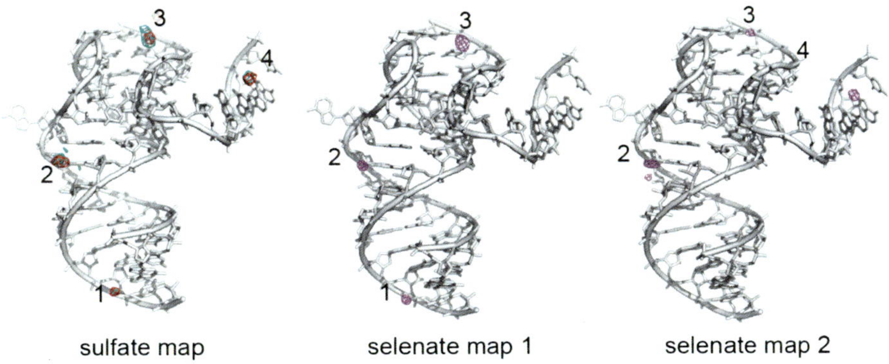

Fig. 14.5 Characterization of sulfate (SO_4^{2-}) and selenate (SeO_4^{2-}) anions bound to the cricket paralysis virus intergenic region of the internal ribosome entry site (IRES; pdb code: 3MJ3; 3MJA; 3MJB) domain 3 RNA (Kieft et al. 2010). (*right*) SO_4^{2-} crystallographic difference map and (*middle* and *left*) SeO_4^{2-} difference maps at different ionic strengths [reproduced by permission from (Kieft et al. 2010)]

such issues and facilitate the investigation of subtle ionic effects on structure and dynamics. In specific instances, some MD studies demonstrated their ability to locate known anion binding sites (Feig and Pettitt 1999; Auffinger et al. 2004b; Kirmizialtin and Elber 2010).

14.6 Toward More Complex Ionic Models?

As usual in science, there is a strong incentive for simplifying models. The first reported MD simulations of biomolecular systems neglected to take into account water molecules until we were computationally able to deal with them, incidentally assessing their fundamental role in structure and function of macromolecular systems. Similarly, the effects associated with the presence of the surrounding ionic atmosphere were ignored for a long time and MD was carried out in early work with minimal salt models. We are now able to acknowledge that the structural and dynamical effects of these ions are certainly as important, although different, than those induced by water molecules. Henceforth, we have to design the most accurate possible "all solvent (water, cations, anions) models", aiming to approach in vivo conditions as closely as possible. This is best done using a mixture of K^+ and Mg^{2+} cations, balanced electrostatically with Cl^- anions. It is further advised, to disregard when possible, nonphysiological conditions (no ions, minimal cationic environment, high Mg^{2+} conditions, ...) that are sometimes reported in the literature. For example, the crystal structure of the ribosomal 5S loop E motif displays five contacting Mg^{2+} and no monovalent cations (Correll et al. 1997). The observation of five Mg^{2+} cations in the vicinity of this 24-nucleotide RNA fragment results most probably from the specific ionic conditions needed for crystallization. As MD simulations suggested (Auffinger et al. 2003, 2004a), only one or two of the deep groove electropositive sites of loop E are occupied by Mg^{2+} cations while the others are occupied by K^+ cations. Indeed, saturating so many electropositive RNA sites with Mg^{2+} cations, as observed in crystals, probably never occurs in vivo.

14.7 Conclusions

It is now well appreciated that ions are an integral part of nucleic acids. Consequently, great care has to be taken in modeling them. The widespread use of MD simulation of nucleic acid systems has brought to light some imperfections of current force fields and prompted significant efforts to improve them. Through this tedious trial and error process significant knowledge related to the binding of ions to nucleic acids has been gained, on the basis of which the following conclusions can be drawn and recommendations made:

1. Minimal salt conditions had their usefulness but should now be replaced by the use of excess neutralizing salt conditions, to allow modeling more subtle ionic effects, including those associated with anions. Specific ion-induced effects will definitely come to light with longer MD simulations using refined force fields. Furthermore, the use of excess ions should significantly improve the sampling of the available configurational space.
2. Great care should be paid to recently described finite size artifacts that alter the distribution of ions around charged groups (i.e., phosphate groups). Simulations exhibiting finite-size artifacts should not be used to calibrate ionic parameters in force fields.
3. It is time to recognize that Na^+ is not a biologically relevant cation; inside cells, Na^+ is found in low concentration while K^+ is found in high concentration. Consequently, the default choice should be to use K^+ (or KCl), even when no clear differences in dynamics are observed. This conclusion is reinforced by important studies reporting observable differences between MD simulation in the presence of K^+ versus Na^+.
4. As noted above, other studies report a small to insignificant sensitivity of RNA dynamics on the choice of counterion. Yet, given the fact that biopolymeric systems exhibit nonlinear behavior and are chaotic by nature, the effects of specific ionic conditions on a given RNA system are difficult to extrapolate. In most cases, it is therefore wiser to simply select ionic conditions that match in vivo conditions.
5. In cells, mixtures of ions of different charges (K^+ and Mg^{2+}) are generally necessary for ensuring biological function. Ionic models should reproduce an appropriate proportion of both species. Note that crystallographic conditions include often a level of divalent ions far above the level prevailing in vivo.
6. Inclusion of Mg^{2+} cations in MD models is considerably more difficult than inclusion of monovalent cations. Yet, RNA simulations will only realistically

Gluick TC, Gerstner RB, Draper DE (1997) Effects of Mg^{2+}, K^+, and H^+ on an equilibrium between alternative conformations of an RNA pseudoknot. J Mol Biol 270:451–463
Golebiowski J, Antonczak S, Di-Giorgio A, Condom R, Cabrol-Bass D (2004) Molecular dynamics simulation of hepatitis C virus IRES IIId domain: structural behavior, electrostatic and energetic analysis. J Mol Model 10:60–68
Harding MH (2006) Small revisions to predicted distances around metal sites in proteins. Acta Crystallogr D62:678–682
Hashem Y, Auffinger P (2009) A short guide to molecular dynamics simulations of RNA systems. Methods 47:187–197
Hashem Y, Westhof E, Auffinger P (2008) Milestones in molecular dynamics simulations of RNA systems. In: Schwede T, Peitsch MC (eds) Computational structural biology, vol 13. World Scientific, London, pp 363–399
Heddi B, Foloppe N, Hantz E, Hartmann B (2007) The DNA structure responds differently to physiological concentrations of K(+) or Na(+). J Mol Biol 368:1403–1411
Hermann T, Auffinger P, Scott WG, Westhof E (1997) Evidence for a hydroxide ion bridging two magnesium ions at the active site of the hammerhead ribozyme. Nucleic Acids Res 25:3421–3427
Hermann T, Auffinger P, Westhof E (1998) Molecular dynamics investigations of the hammerhead ribozyme RNA. Eur J Biophys 27:153–165
Huang W, Kim J, Jha S, Aboul-ela F (2009) A mechanism for S-adenosyl methionine assisted formation of a riboswitch conformation: a small molecule with a strong arm. Nucleic Acids Res 37:6528–6539
Ikeda T, Boero M, Terakura K (2007) Hydration properties of magnesium and calcium ions from constrained first principles molecular dynamics. J Chem Phys 127:074503
Jiao D, King C, Grossfield A, Darden TA, Ren P (2006) Simulation of Ca^{2+} and Mg^{2+} solvation using polarizable atomic multipole potential. J Phys Chem B 110:18553–18559
Joung IS, Cheatham TE III (2008) Determination of alkali and halide monovalent ion parameters for use in explicitly solvated biomolecular simulations. J Phys Chem B 112:9020–9041
Joung IS, Cheatham TE III (2009) Molecular dynamics simulations of the dynamic and energetic properties of alkali and halide ions using water-model-specific ion parameters. J Phys Chem B 113:13279–13290
Kieft JS, Chase E, Costantino DA, Golden BL (2010) Identification and characterization of anion binding sites in RNA. RNA 16:1118–1123
Kirmizialtin S, Elber R (2010) Computational exploration of mobile ion distributions around RNA duplex. J Phys Chem B 114:8207–8220
Korolev N, Lyubartsev AP, Nordenskiold L, Laaksonen A (2001) Spermine: an "invisible" component in the crystals of B-DNA. A grand canonical Monte Carlo and molecular dynamics simulation study. J Mol Biol 308:907–917
Korolev N, Lyubartsev AP, Laaksonen A, Nordenskiold L (2002) On the competition between water, sodium ions, and spermine ion binding to DNA: a molecular dynamics computer simulation study. Biophys J 82:2860–2875
Korolev N, Lyubartsev AP, Laaksonen A, Nordenskiold L (2003) A molecular dynamics simulation study of oriented DNA with polyamine and sodium counterions: diffusion and averaged binding of water and cations. Nucleic Acids Res 31:5971–5981
Korolev N, Lyubartsev AP, Laaksonen A, Nordenskiold L (2004) Molecular dynamics simulation study of oriented polyamine- and Na-DNA: sequence specific interactions and effects on DNA structure. Biopolymers 73:542–555
Krasovska MV, Sefcikova J, Spackova N, Sponer J, Walter NG (2005) Structural dynamics of precursor and product of the RNA enzyme from the hepatitis delta virus as revealed by molecular dynamics simulations. J Mol Biol 351:731–748
Krasovska MV, Sefcikova J, Reblova K, Schneider B, Walter NG, Sponer J (2006) Cations and hydration in catalytic RNA: molecular dynamics of the hepatitis delta virus ribozyme. Biophys J 91:626–638

Lambert D, Leipply D, Shiman R, Draper DE (2009) The influence of monovalent cation size on the stability of RNA tertiary structures. J Mol Biol 390:791–804

Leclerc F (2010) Hammerhead ribozymes: true metal or nucleobase catalysis? Where is the catalytic power from? Molecules 15:5389–5407

Lee TS, York DM (2008) Origin of mutational effects at the C3 and G8 positions on hammerhead ribozyme catalysis from molecular dynamics simulations. J Am Chem Soc 130:7168–7169

Lee TS, Silva-Lopez C, Martick M, Scott WG, York DM (2007) Insight into the role of Mg^{2+} in hammerhead ribozyme catalysis from X-ray crystallography and molecular dynamics simulation. J Chem Theor Comput 3:325–327

Lee TS, Giambasu GM, Sosa CP, Martick M, Scott WG, York DM (2009) Threshold occupancy and specific cation binding modes in the hammerhead ribozyme active site are required for active conformation. J Mol Biol 388:195–206

Lopes PE, Roux B, Mackerell AD (2009) Molecular modeling and dynamics studies with explicit inclusion of electronic polarizability. Theory and applications. Theor Chem Acc 124:11–28

Makarov VA, Feig M, Andrews BK, Pettitt MM (1998) Diffusion of solvent around biomolecular solutes: a molecular dynamics simulation study. Biophys J 75:150–158

Manning GS (1978) The molecular theory of polyelectrolyte solutions with applications to the electrostatic properties of polynucleotides. Q Rev Biophys 11:179–246

Markham GD, Glusker JP, Bock CW (2002) The arrangement of first and second-sphere water molecules in divalent magnesium complexes: results from molecular orbital and density functional theory and from structural crystallography. J Phys Chem B 106:5118–5134

Martick M, Lee TS, York DM, Scott WG (2008) Solvent structure and hammerhead ribozyme catalysis. Chem Biol 15:332–342

McDowell SE, Spackova N, Sponer J, Walter NG (2007) Molecular dynamics simulations of RNA: an in silico single molecule approach. Biopolymers 85:169–184

Mlynsky V, Banas P, Hollas D, Reblova K, Walter NG, Sponer J, Otyepka M (2010) Extensive molecular dynamics simulations showing that canonical G8 and protonated A38H+ forms are most consistent with crystal structures of hairpin ribozyme. J Phys Chem B 114:6642–6652

Muller T, Walter B, Wirtz A, Burkovski A (2006) Ammonium toxicity in bacteria. Curr Microbiol 52:400–406

Noy A, Soteras I, Luque FJ, Orozco M (2009) The impact of monovalent ion force field model in nucleic acids simulations. Phys Chem Chem Phys 11:10596–10607

Ohtaki H (2001) Ionic solvation in aqueous and nonaqueous solutions. Monatshefte für Chemie 132:1237–1268

Oliva R, Cavallo L (2009) Frequency and effect of the binding of Mg^{2+}, Mn^{2+}, and Co^{2+} ions on the guanine base in Watson-Crick and reverse Watson-Crick base pairs. J Phys Chem B 113:15670–15678

Park JM, Boero M (2010) Protonation of a hydroxide anion bridging two divalent magnesium cations in water probed by first-principles metadynamics simulation. J Phys Chem B 114:11102–11109

Petrov AS, Lamm G, Pack GR (2002) Water-mediated magnesium-guanine interactions. J Phys Chem B 106:3294–3300

Petrov AS, Lamm G, Pack GR (2005) Calculation of the binding free energy for magnesium-RNA interactions. Biopolymers 77:137–154

Priyakumar UD (2010) Atomistic details of the ligand discrimination mechanism of S(MK)/SAM-III riboswitch. J Phys Chem B 114:9920–9925

Rao JS, Dinadayalane TC, Leszczynski J, Sastry GN (2008) Comprehensive study on the solvation of mono- and divalent metal cations: Li^+, Na^+, K^+, Be^{2+}, Mg^{2+} and Ca^{2+}. J Phys Chem A 112:12944–12953

Razga F, Zacharias M, Reblova K, Koca J, Sponer J (2006) RNA kink-turns as molecular elbows: hydration, cation binding, and large-scale dynamics. Structure 14:825–835

Chapter 15
Modeling RNA Folding Pathways and Intermediates Using Time-Resolved Hydroxyl Radical Footprinting Data

Joshua S. Martin, Paul Mitiguy, and Alain Laederach

Abstract The analysis of time-resolved hydroxyl radical (•OH) footprinting data can reveal the complex and rugged folding landscape of an RNA molecule. This analysis requires the identification and subsequent optimization of a kinetic model and its parameters. The number of possible kinetic models increases factorially with the complexity of the molecule, complicating the modeling process. We detail here a computational approach that allows complex models involving up to five kinetic intermediates to be run on a desktop computer. Our approach involves an initial "model-free" analysis of the data, which reduces the computational complexity of the subsequent kinetic parameter optimization. Our method is able to systematically identify the best fitting kinetic model and reveals the underlying folding mechanism of an RNA.

15.1 Introduction

Understanding and predicting the process by which an RNA molecule adopts its native and active structure remains a contemporary challenge in the biophysical sciences (Woodson 2002; Vicens et al. 2007; Thirumalai and Hyeon 2005; Talkington et al. 2005; Takamoto et al. 2004). Ribozymes, including the L-21 *Tetrahymena thermophila* group I intron, adopt a specific conformation to achieve their catalytic function in the cell (Laederach et al. 2006, 2007). Many RNAs have

J.S. Martin • A. Laederach (✉)
Department of Biology, University of North Carolina at Chapel Hill, Chapel Hill, NC 27599, USA
e-mail: jsmartin@bio.unc.edu; alain@unc.edu

P. Mitiguy
Department of Mechanical Engineering, Stanford University, Stanford, CA 94305, USA
e-mail: mitiguy@stanford.edu

N. Leontis and E. Westhof (eds.), *RNA 3D Structure Analysis and Prediction*, Nucleic Acids and Molecular Biology 27, DOI 10.1007/978-3-642-25740-7_15,

been identified as changing conformations to regulate various cellular processes (Tucker and Breaker 2005). The active conformation is achieved by navigating the complex folding landscape and is highly dependent on many environmental factors (Russell and Herschlag 2001; Russell et al. 2006; Laederach et al. 2007).

The fact that the RNA is able to traverse this landscape on the timescale of seconds in vitro illustrates the extent to which it is susceptible to misfolding (Shcherbakova et al. 2008). The rate-determining steps in RNA folding depend on many factors, including the electrostatic environment, temperature, and exogenous molecule binding (Russell and Herschlag 2001; Russell et al. 2006; Laederach et al. 2007). We have shown that changes in the folding conditions (such as variation of the counterion concentration and mutation) have profound effects on the observed rate constants, suggesting an intricate relationship between the sequence, structure, environment, and folding dynamics of an RNA molecule (Laederach et al. 2006, 2007).

Chemical and enzymatic mapping techniques are particularly well suited for the study of RNA structure and kinetics because they can probe kinetic details with single-nucleotide resolution (Wilkinson et al. 2005, 2008; Mitra et al. 2008). Coupled with novel benchtop approaches to collect kinetic data with millisecond resolution (Shcherbakova et al. 2006), these experimental approaches produce large data sets that require advanced modeling. The interpretation and modeling of RNA kinetic data originally required a distributed computing approach to identify and optimize the kinetic model and its parameters (Laederach et al. 2006). Due to the computational complexity of the problem, systems involving more than three intermediates could not be resolved with this approach.

This chapter outlines algorithmic developments for determining the underlying kinetic model that best describes the folding of an RNA molecule based on the analysis of time-resolved hydroxyl radical ($^{\bullet}$OH) footprinting data (Laederach et al. 2006, 2007). Our algorithm has reduced the number of CPU hours by a factor of 2,000 in comparison to the original KinFold software for a two intermediate system (Laederach et al. 2006) and has allowed the analysis of larger molecules with up to five intermediates on a desktop computer (Martin et al. 2009).

15.1.1 Software Availability

The algorithms described in this chapter are implemented in the KinFold software (version 2.2), which is freely available for download at https://simtk.org/home/KinFold. The software was created using MathWorks' Matlab software (version 7.5.0.338) and Python (version 2.5.1) under the OS X operating system and is compatible on other systems for which Matlab and Python are available (Windows and Linux). The downloadable zip archive contains the necessary scripts, example data set, basic instructions, and a graphical user interface (GUI) wrapper for running KinFold. The GUI consists of four major sections (Fig. 15.1); each section

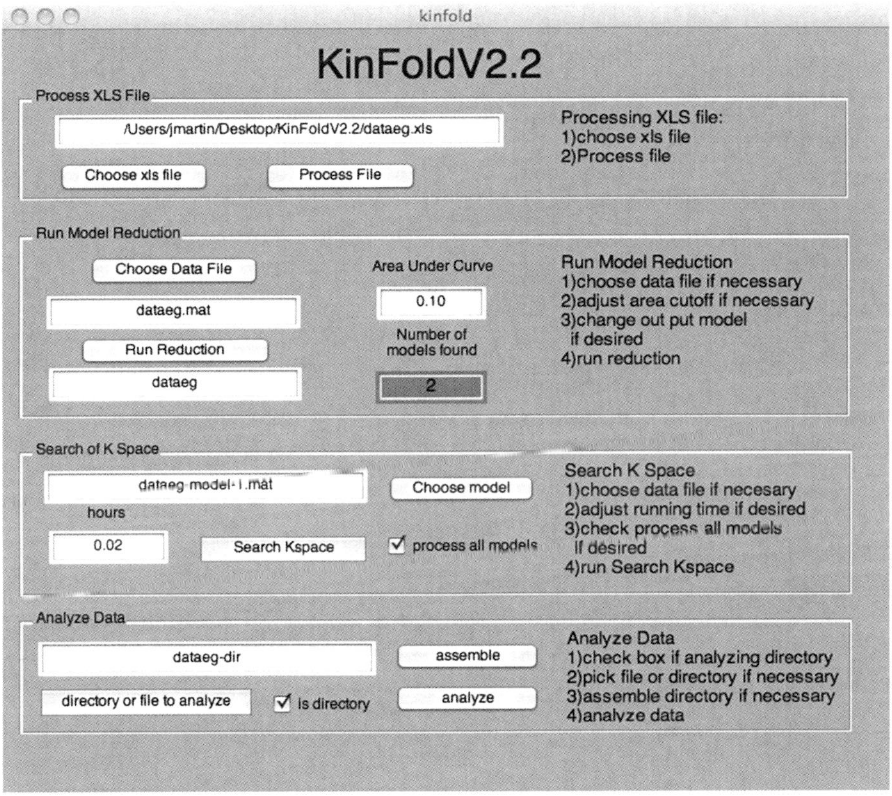

Fig. 15.1 Screen shot of the KinFold GUI run using the example data, dateeg.xls, found with the software package. The GUI is broken down into four distinct independent sections that can be run in sequence or individually. The programs that are used to carry out the calculations can be run without the GUI and directly though Matlab's command line

can be run individually in Matlab's command input without invoking the GUI, as described in the readme file.

15.1.2 *Kinetic Models Describe the Folding Reaction*

RNA secondary structure is very stable and is formed on the microsecond timescale (Woodson 2000, 2002; Pan et al. 1997; Heilman-Miller et al. 2001). In this chapter, we focus on the rate-limiting step in RNA folding, the formation of the catalytically active tertiary structure. We describe this process using a kinetic model, illustrated in Fig. 15.2. The RNA folds from the unfolded state, U, through multiple, long-lived intermediates, I, to reach the final folded state, F (Russell et al. 2006). The concentration of the unfolded state as a function of time is related to the rate constants and concentrations of the other states being converted to and away from U. For the

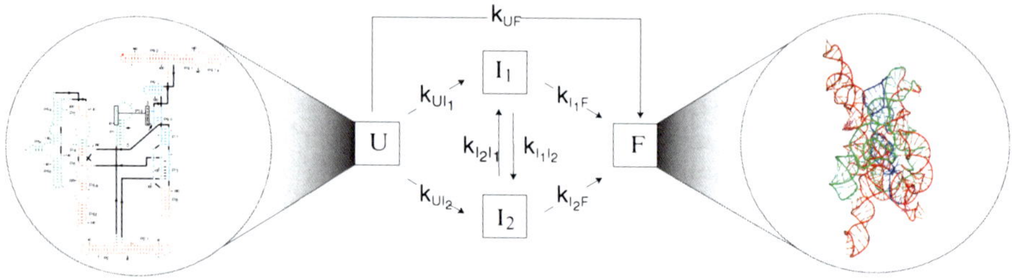

Fig. 15.2 The unfolded RNA (*U* state) with only secondary structure elements formed (secondary structure diagram shown in the *left-hand circle*) must reach a final folded state (*F*) which has the full complement of tertiary interactions(shown in the *right-hand circle*). The domains of the RNA are colored for ease of identification, so *green* corresponds to the P4P6 subdomain, *red* to the periphery, and *blue* to the catalytic core of the molecule. The folding of the RNA has the possibility to go through multiple intermediates *I* that populate the pathways from the *U* state to the *F* state putting the folding reaction on the order of hours to go to completion. The transition rate from state *i* to state *j* is given by k_{ij} and is indicated with an *arrow* in the diagram. For clarity, we only show the major transitions between states which usually correspond to increased folding since the reverse rates are on average much lower (Figure modified from Martin et al. 2009)

example shown in Fig. 15.2 for the L-21 *T. thermophila* group I intron with two intermediates, the change in concentration of the unfolded state is described by

$$\frac{\mathrm{d}U(t)}{\mathrm{d}t} = k_{\mathrm{I1U}}I_1(t) - k_{\mathrm{UI1}}U(t) + k_{\mathrm{I2U}}I_2(t) - k_{\mathrm{UI2}}U(t) + k_{\mathrm{FU}}F(t) - k_{\mathrm{UF}}U(t). \quad (15.1)$$

When (15.1) is written out for every state in a folding process with *N* intermediates, we obtain the following set of *N* + 2-coupled differential equations:

$$\begin{aligned}
\frac{\mathrm{d}U(t)}{\mathrm{d}t} &= k_{\mathrm{I1U}}I_1(t) - k_{\mathrm{UI1}}U(t)\ldots + k_{\mathrm{FU}}F(t) - k_{\mathrm{UF}}U(t),\\
\frac{\mathrm{d}I_1(t)}{\mathrm{d}t} &= k_{\mathrm{UI1}}U(t) - k_{\mathrm{I1U}}I_1(t)\ldots + k_{\mathrm{FI1}}F(t) - k_{\mathrm{I1F}}I_1(t),\\
&\vdots\\
\frac{\mathrm{d}I_N(t)}{\mathrm{d}t} &= k_{\mathrm{UIN}}U(t) - k_{\mathrm{INU}}I_N(t)\ldots + k_{\mathrm{FIN}}F(t) - k_{\mathrm{INF}}I_N(t),\\
\frac{\mathrm{d}F(t)}{\mathrm{d}t} &= k_{\mathrm{UF}}U(t) - k_{\mathrm{FU}}F(t)\ldots + k_{\mathrm{INF}}I_N(t) - k_{\mathrm{FIN}}F(t).
\end{aligned} \quad (15.2)$$

Equation (15.2) can be written in a more compact matrix form by defining the state vector

$$\vec{x}(t) = \begin{pmatrix} U(t) \\ I_1(t) \\ I_2(t) \\ \vdots \\ I_N(t) \\ F(t) \end{pmatrix} \tag{15.3}$$

and the **K** and **D** matrices as:

$$\mathbf{K} = \begin{bmatrix} 0 & k_{\mathrm{UI}_1} & k_{\mathrm{UI}_2} & \cdots & k_{\mathrm{UF}} \\ k_{\mathrm{I}_1\mathrm{U}} & 0 & k_{\mathrm{I}_1\mathrm{I2}} & \cdots & k_{\mathrm{I}_1\mathrm{F}} \\ k_{\mathrm{I}_2\mathrm{U}} & k_{\mathrm{I}_2\mathrm{I}_1} & 0 & \cdots & k_{\mathrm{I}_2\mathrm{F}} \\ \vdots & \vdots & \vdots & \ddots & \vdots \\ k_{\mathrm{FU}} & k_{\mathrm{FI}_1} & k_{\Gamma\mathrm{I}_{21}} & \cdots & 0 \end{bmatrix} \tag{15.4}$$

$$D_{ij} = \begin{cases} i \neq j; & K_{ji} \\ i = j; & -\sum_{i=1}^{n} K_{ji} \end{cases}, \tag{15.5}$$

resulting in the equation

$$\frac{\mathrm{d}\vec{x}(t)}{\mathrm{d}t} = \boldsymbol{D}\vec{x}(t). \tag{15.6}$$

The kinetic model for any RNA folding reaction such as the one illustrated in Fig. 15.2 can therefore be written in the same form as (15.6) using the proper values for $\vec{x}(t)$ and **D**.

The solution to (15.6) is equivalent to that of finding eigenvalues and eigenvectors for **D**. The values of λ that satisfies the condition

$$(\boldsymbol{D} - \lambda \mathrm{I})\vec{x} = 0 \tag{15.7}$$

are the eigenvalues of this problem. Equation (15.7) can only be true for nonzero values of $\vec{x}(t)$when the determinant of $D - \lambda \mathrm{I}$ is zero. The corresponding eigenvector $\vec{\Lambda}_i$ for the eigenvalue λ_i is found by solving

$$D\vec{\Lambda}_i = \lambda_i \vec{\Lambda}_i. \tag{15.8}$$

By writing the matrix **D** in the bases of the eigenvectors, we decouple (15.6) and reduce it to a set of linear first-order differential equations of the form

$$\frac{\mathrm{d}x(t)}{\mathrm{d}t} = \lambda x(t). \tag{15.9}$$

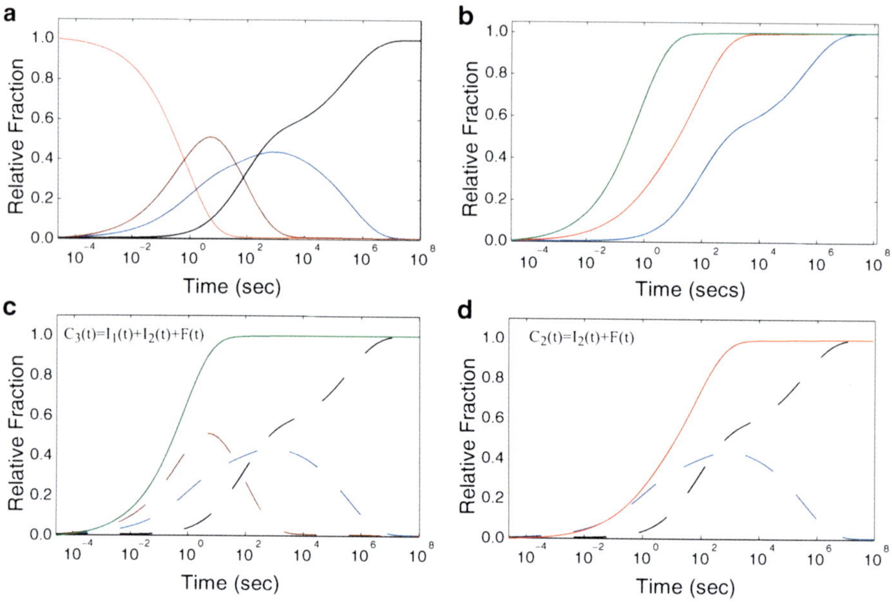

Fig. 15.3 (**a**) Resulting state curves $\vec{x}(t)$ for *U* (*orange*), I_1 (*magenta*), I_2 (*purple*), and *F* (*black*) that describe the relative fraction of each species as a function of time for the folding of the L-21 *T. thermophila* group I intron in the presence of 10 mM $MgCl_2$ (Laederach et al. 2006). These curves are obtained from (15.10). (**b**)The corresponding $^{\bullet}$OH footprinting curves ($\vec{C}_P(t)$) for the folding of the *T. thermophila* group I intron in the presence of 10 mM $MgCl_2$. The *green curve* corresponds to the P4P6 subdomain, *red* to the periphery, and *blue* to the catalytic core of the molecule matching the color scheme chosen for the structures in Fig. 15.2. (**c**) The $^{\bullet}$OH footprinting curve for the P4P6 subdomain as a linear combination of the $I_1(t)$, $I_2(t)$, and $F(t)$ state curves. (**d**) The $^{\bullet}$OH footprinting curve for the periphery is a linear combination of the $I_2(t)$ and $F(t)$ state curves (figure modified from Martin et al. 2009)

The general solution to a differential equation of the form of (15.9) is $x(t) = c \times \exp(\lambda t)$, where c is determined by the initial conditions. By collecting and rewriting all the solutions to the individual uncoupled differential equations in the original basis of **D**, we get the final solution:

$$\vec{\boldsymbol{x}}(t) = \sum_i c_i \vec{\Lambda}_i \exp(\lambda_i t). \tag{15.10}$$

The constants c are found by solving the set of linear equations represented by (15.10) with the initial conditions of $\vec{\boldsymbol{x}}(t = 0)$.

This solution offers a computationally simple solution in comparison to numerical methods used to solve (15.6) as previously implemented (Laederach et al. 2006). The curves shown in Fig. 15.3a for the L-21 *T. thermophila* were generated using this solution and the reported **K** matrix (Laederach et al. 2006).

15.1.3 Hydroxyl Radical Footprinting Measures the Folding Reaction

Hydroxyl radical footprinting involves generating a burst of $^{\bullet}$OH radicals that selectively cleave the RNA backbone at residues that are exposed to the solution and not buried within the molecule (Shcherbakova et al. 2006). Hydroxyl radicals are generated using either synchrotron radiation (Sclavi et al. 1997, 1998a) or the Fenton reaction (Shcherbakova et al. 2006). As the molecule folds into a compact structure, some regions become more buried and are thus increasingly protected from cleavage as folding progresses. Kinetic data is obtained by measuring the change in accessibility as a function of time as the RNA folds. The RNA fragments resulting from $^{\bullet}$OH radical cleavage are then identified using gel electrophoresis (Das et al. 2005) or a capillary sequencer (Mitra et al. 2008) yielding a time-dependent change in accessibility for each nucleotide in the RNA. A more detailed description is beyond the scope of this chapter but can be found in the literature (Shcherbakova et al. 2006; Sclavi et al. 1997, 1998a, b; Brenowitz et al. 1986, 2002; Shcherbakova and Brenowitz 2008).

The accessibility changes measured for each nucleotide as the molecule folds are illustrated in Fig. 15.3b for the L-21 *T. thermophila*. We label these time progress curves as $\vec{C}(t)$. The $^{\bullet}$OH footprinting curves shown in Fig. 15.3b correspond to individual subdomains of the molecule; in this case, the green curve is the average change in accessibility of nucleotides in the P4P6 subdomain, while the red curves correspond to the peripheral helices, and the blue curves correspond to nucleotides in the catalytic core (Laederach et al. 2006). The subdomain coloring schematic is consistent throughout this chapter (e.g., between Figs. 15.2 and 15.3).

The progress curves, $\vec{C}(t)$, are created from linear combinations of the state curves, $\vec{x}(t)$. It can be seen in Fig. 15.3c that the green progress curve from Fig. 15.3b is made up of the addition of $I_1(t)$, $I_2(t)$, and $F(t)$, while in Fig. 15.3d, the red curve is made up of $I_2(t)$ and F(t). These progress curves can be written as a group of equation as

$$\begin{aligned} C_1(t) &= 0U(t) + 0I_1(t) + 0I_2(t) + 1F(t), \\ C_2(t) &= 0U(t) + 0I_1(t) + 0I_2(t) + 1F(t), \\ C_3(t) &= 0U(t) + 0I_1(t) + 0I_2(t) + 1F(t). \end{aligned} \tag{15.11}$$

This group of equations can be simplified by writing them in matrix form as

$$\vec{C}_P(t) = \boldsymbol{P}\vec{\boldsymbol{x}}(t), \tag{15.12}$$

where

$$\mathbf{P} = \begin{bmatrix} 0 & 0 & 0 & 1 \\ 0 & 0 & 1 & 1 \\ 0 & 1 & 1 & 1 \end{bmatrix}. \tag{15.13}$$

The matrix **P** represents the linear combinations of state curves, $\vec{x}(t)$, that enumerate the progress curves, $\vec{C}(t)$. It should be noted that we simplify the possible number of **P** matrices by realizing that the final folded state will have all the nucleotides folded into their final structure, while the unfolded state will have no nucleotides folded into the final state. Mathematically, this means that the last column in **P** will consist of all 1's, while the first column will consist of all 0's. This property of **P** makes it possible to extract the sub-matrix **P** consisting of everything but the first column of **P** without losing any information. For example, the **P** matrix derived from the **P** from (15.13) is

$$\mathbf{P} = \begin{bmatrix} 0 & 0 & 1 \\ 0 & 1 & 1 \\ 1 & 1 & 1 \end{bmatrix}. \tag{15.14}$$

A priori, neither **P** nor **K** is known for a given RNA folding reaction, since only the experimental progress curves can be measured. The problem of identifying the proper kinetic model for a particular folding reaction thus becomes that of finding the best **P** and **K** matrices that describe the $^{\bullet}$OH footprinting data.

15.2 Implementation

15.2.1 Experimental Progress Curves

The scaled experimental time progress curves are then clustered which serves the duel purposes of reducing the number of progress curves and averages out the noise in the data (Laederach et al. 2006). The number of clusters is determined by the Gap statistic (Tibshirani et al. 2001). The Gap score is calculated by generating 100 different sets of random time progress curves from a normal distribution of random, single exponential curves. These curves are then clustered using k-means clustering to determine the within cluster dispersion (W_k^*) for the random set and compared to the clustered data (W_k) as a function of increasing k. The Gap score is computed using

$$\text{Gap}(k) = \frac{1}{B}\sum_b \log(W_k^*) - \log(W_k), \tag{15.15}$$

where B is the number of random sets of time progress curves (Laederach et al. 2006). The optimal value of k was chosen such that $\text{Gap}(k) \geq \text{Gap}(K+1) - s_k + 1$ where $s_k + 1$ is the standard deviation of the Gap parameter for the random time progress curves (Laederach et al. 2006).

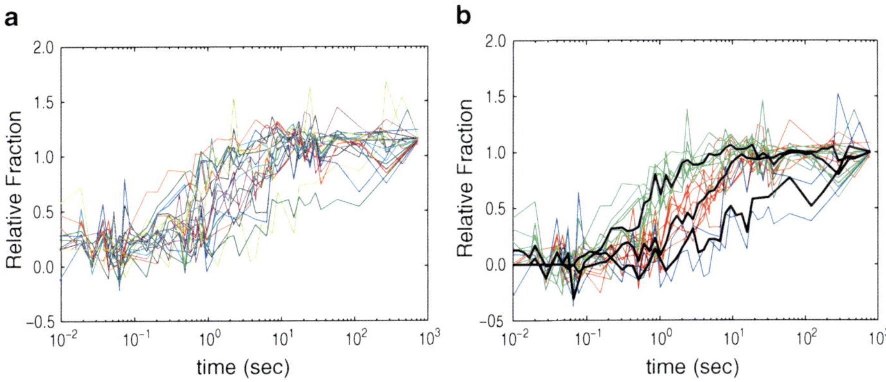

Fig. 15.4 Example of the clustering for the L-21 *T. thermophila* in 10 mM $MgCl_2$ (Laederach et al. 2006). (**a**) The time progress curves from hydroxyl radical footprinting after they have been scaled between zero and unity. The Gap statistic is used to calculate an optimal cluster number of three (Laederach et al. 2006). (**b**) The resulting clusters shown as *red*, *green*, and *blue curves* corresponding to the same colors in the structures shown in Fig. 15.2. The cluster centroids are shown as *heavy black lines* and are the time progress curves that are analyzed using the KinFold algorithm

An example of this clustering is shown in Fig. 15.4 for experimental data for the L-21 *T. thermophila* collected at 10 mM $MgCl_2$ (Laederach et al. 2006). Figure 15.4a shows the hydroxyl radical curves after they have been scaled between zero and one. The Gap statistic determines that three clusters are the optimal, and we color these green, red, and blue (Fig. 15.4b). The cluster centroids are shown in black in Fig. 15.4b and correspond to the experimental curves that are used when performing kinetic modeling.

15.2.2 Factorial Explosion of P

To calculate the state curves $\vec{x}(t)$ from the progress curves $\vec{C}(t)$, we make use of the fact that the two sets of curves are related through the **P** matrix as given by (15.12). Since the rows in the sub-matrix **P** are linearly independent, we can solve for all but $U(t)$ in $\vec{x}(t)$ by inverting **P**:

$$\sum_{i=2}^{n} x_i(t) = \boldsymbol{P}^{-1}\vec{C}(t). \tag{15.16}$$

The mass-balance equation allows us to generate $U(t)$ by subtracting the other state curves from unity,

$$U(t) = 1 - I_1(t) - I_2(t) \cdots - I_N(t) - F(t), \tag{15.17}$$

and allows the determination of $\vec{x}(t)$. In other words, given the experimental data $\vec{C}_{\mathrm{E}}(t)$, we are able to obtain the time progress of the different species in solution using only the **P** matrix and mass balance.

The number of different **P** matrices is related to the number of different ways one can combine the intermediate curves to generate the state curves. These combinations are equivalent to the number of ways you can pick *j* items, out of a set of *i* items which is described by

$$\binom{I}{j} = \frac{I!}{j!(I-j)!}. \tag{15.18}$$

In this situation, the number of ways of selecting 0, 1, 2, ..., *N* intermediates must be summed to obtain the final number of ways of selecting the intermediates:

$$\sum_{j=0}^{I}\binom{I}{j} = \binom{I}{0} + \binom{I}{1} + \ldots + \binom{I}{I}. \tag{15.19}$$

These factors are the binomial expansion of order *I* written as

$$\binom{I}{0}X^{I} + \binom{I}{1}x^{I-1}\,y + \ldots + \binom{I}{1}y^{I} = (x+y)^{I}. \tag{15.20}$$

Equation (15.19) is equal to (15.20) when $x = y = 1$ resulting in the number of possible combinations being $(x+y)^{I} = 2^{I}$. The number of ways to arrange these 2^{I} vectors into an $I + 1$ matrix where order matters is also a combinatorial problem given by

$$n = \binom{2^{I}}{I+1} = \frac{2^{I}!}{(2^{I}-I-1)!}. \tag{15.21}$$

This combinatorial explosion in the number of **P** matrices is shown as the solid blue curve in Fig. 15.5. The original implementation of KinFold reduced the number of **P** matrices (as shown in red in Fig. 15.5) using several simplifying assumptions (Laederach et al. 2006).

15.2.3 Testing P Without Fitting K

Section 15.2.2 described how the number of **P** matrices increases combinatorially with the number of intermediates. This increase makes it impractical to find the best fitting **K** matrix for every one of these **P** matrices on a desktop computer for any

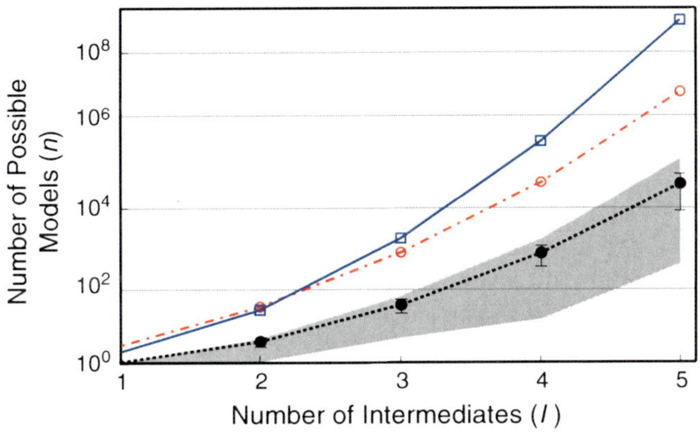

Fig. 15.5 Illustration of the combinatorial explosion in the enumeration of all possible **P** matrices as a function of the number of intermediates I. In *blue*, we plot the total number of possible models, n, as given by (15.21). The *red curve* represents the number of models (**P** matrices) that are tested when using the previous implementation of KinFold by nonlinear least-squares optimization (Laederach et al. 2006). The *black curve* is the average number of models that now need to be test based on a sampling of 100 random data sets using our new approach. Error bars represent three standard deviations, and the *light gray* shadows the maximum and minimum values of n for each I for the random data set used. The actual number of models is highly dependent on the data set (figure from Martin et al. 2009)

system with more than two intermediates. One can drastically reduce the number of models that must be fitted by recognizing that not all state curves generated from the inverse of a given **P** matrix for a given set of progress curves are physically possible. This reduction is done by carefully inspecting all the **P** matrices and sets of state curves, $\vec{x}(t)$, generated for a given set of time progress curves, $\vec{C}(t)$. Any set of $\vec{x}(t)$ which contains a curve with negative values of the relative fraction of molecules is not physically realistic and can be disregarded in the analysis. The only way an individual combination of state curves is eliminated is when all the **P** matrices containing that combination are eliminated. This is illustrated in Fig. 15.6 where all the possible **P** matrices are generated and used to calculate all the possible state curves for our example system. Visual inspection of Fig. 15.6 reveals which **P** matrices are not physically viable solutions of the system. For example, the matrix-labeled $\mathbf{P}_1$ generates a $U(t)$ state curve that has values that are less than zero, which is unphysical, so it can be rejected before fitting is attempted. The matrix-labeled $\mathbf{P}_4$ is also an unphysical case because it generates an $F(t)$ state curve that has values less than zero. The two remaining matrices $\mathbf{P}_2$ and $\mathbf{P}_3$ are possible models; however, they are degenerate, meaning that for all practical purposes, they are the same model. In this particular case, the labeling of I_1 and I_2 is switched between the two surviving models. This degeneracy is the result of the interchangeability of the columns of the **P** matrix for the different combinations of the state curves. Consequently, for a given set of time progress curves, there are $I!$ possible degenerate matrices.

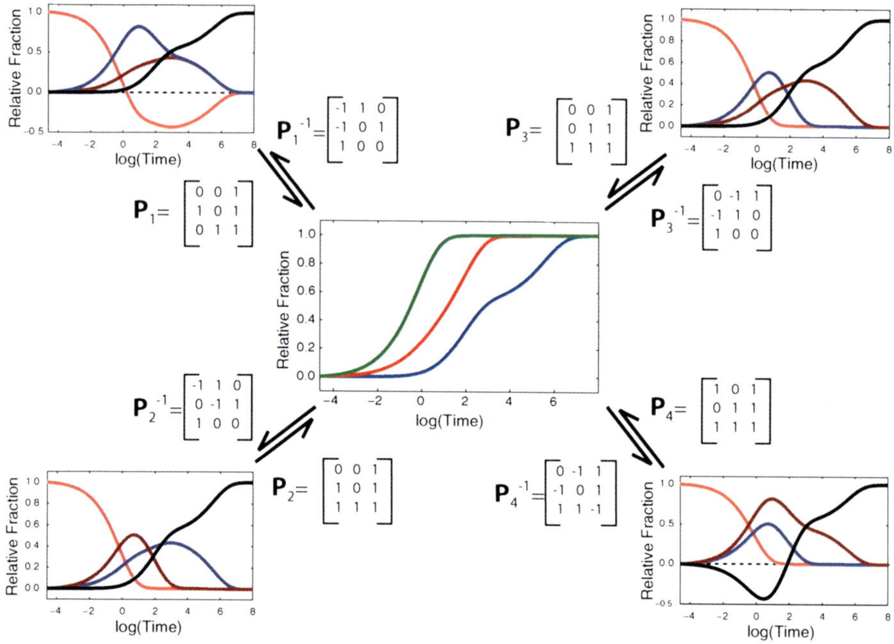

Fig. 15.6 Illustration of the application of (15.16) to $C_{\mathrm{E}}(t)$ (center panel) to generate possible $\vec{x}(t)$ state curves (external panels); colors are identical to those used in Fig. 15.2. Both $\mathbf{P}_{-1}$ and $\mathbf{P}_{-1}$ matrices generate negative state curves allowing us to eliminate them without the need to optimize **K** with nonlinear least squares. $\mathbf{P}_{-1\ 2}$ and $\mathbf{P}_{-1}$ yield identical curve shapes, but *I*1 and *I*2 are inverted. In this case, these two matrices yield degenerate models that are equivalent, such that a single kinetic model describes the RNA folding reaction (figure from Martin et al. 2009)

15.3 Validation and Results

15.3.1 Experimentally Acquired $^{\bullet}$OH Data

To demonstrate the analysis described above on real data, we apply it to the folding of the L-21 *T. thermophila* in 10 mM $MgCl_2$, which was previously analyzed (Laederach et al. 2006). When we apply (15.16) for the four possible **P** matrices to the example data shown in Fig. 15.7a, all sets of the resulting $\vec{x}(t)$ curves have a curve that dips below zero. However, closer examination of the curves reveals that certain solutions only result in minor excursions below zero which are easily accounted for by experimental error, as shown in Fig. 15.7. This can be taken into account by adjusting the criteria to allow up to 10% of the area under the curve (AUC) to be negative before we eliminate the corresponding **P** matrix as a potential model. This results in the identification of a single set of unique state curves (Fig. 15.7b). If we now optimize **K** using nonlinear least square regression on the

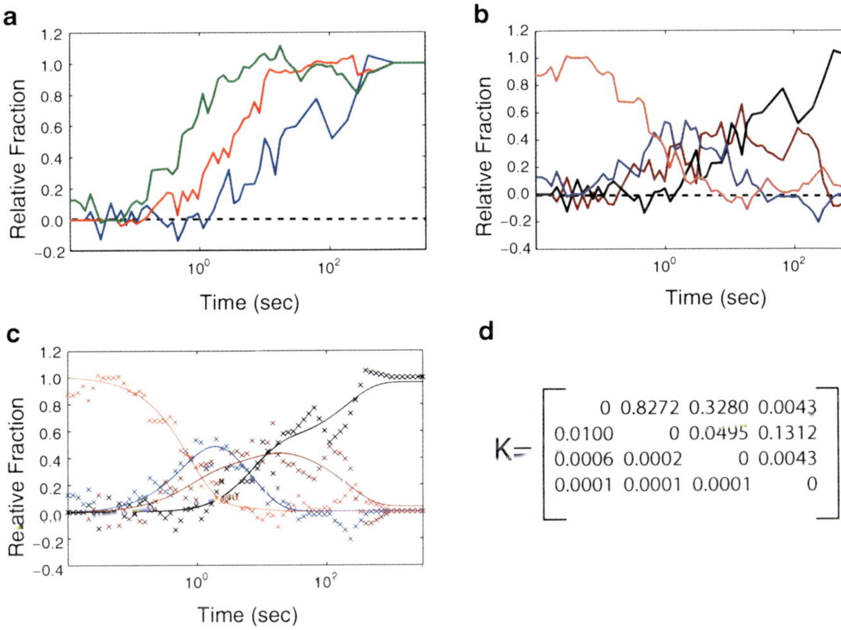

Fig. 15.7 Illustration of current algorithm when applied to actual experimentally obtained $^{\bullet}$OH footprinting data; curve colors are consistent with Fig. 15.2. (**a**) Time progress curves with experimental noise. (**b**) Resulting state curves $\vec{x}(t)$ determined by applying (15.16) to the raw data and only selecting the **P** matrices that satisfy the AUC criteria. (**c**) Optimized fit using nonlinear least-squares minimization of **K** of state curves. (**d**) **K** matrices obtained using the old and new approaches to determining **P** and **K**, which yield identical results within error. Error on the rate values vary between 5 and 20% (figure from Martin et al. 2009)

resulting $\vec{x}(t)$, we are able to fit these data accurately (Fig. 15.7c). Furthermore, the values we obtain for **K** (Fig. 15.7d) are equivalent within error to those obtained using the original KinFold algorithm (Laederach et al. 2006; Martin et al. 2009).

Adjusting the AUC criteria allows us to fine tune the sensitivity of our approach. In this particular example, we chose 10% as this correctly identified the set of state curves corresponding to the folding model. Had we chosen less stringent criteria, such as 50% AUC, we would have identified additional models (**P** matrices) that require testing using nonlinear optimization of the **K** matrix. In essence, raising the AUC criteria results in having to test more models with least-squares optimization and thus makes the problem more computationally intensive. When using our approach with a novel data set, AUC criteria can be selected such that a minimal number of models need to be tested. In practice however, it is preferable to test several models to evaluate the significance in the difference in root-mean-square error (RMSE) of the fit. The AUC criteria allow users of the algorithm to balance computational cost and desire to comprehensively fit all models.

15.3.2 Large Systems

As can be seen in Fig. 15.5, the number of possible **P** matrices increases factorially with the number of intermediates. For systems with four and five intermediates, very large numbers of **P** matrices must therefore be tested. Fortunately, this approach allows one to first efficiently test all these combinations to determine which **P** matrices require nonlinear least-squares optimization to get **K**. The black line in Fig. 15.5 reports on the number of surviving models (**P** matrices) that require testing for computationally generated data as a function of the number of intermediates (Martin et al. 2009).

For all *I*, it is clear that the approach described here offers a significant reduction in the number of models that need to be tested, making this approach computationally tractable for systems with large numbers of intermediates like the ribosome. Interestingly, the number of models that need to be tested is highly dependent on the curves, as evidenced by the large standard deviation over the 100 tested models (and the even larger spread in the min and max values, gray shadow Fig. 15.5). It is therefore difficult to *a priori* predict the total number of **P** matrices that will produce only positive curves for a given data set. The ability to identify a single kinetic model that best fits the experimental data will ultimately depend on the quality of the experimental data.

15.4 Discussion

Our analysis of larger systems with up to five intermediates shows that our approach will scale and remain computationally tractable even for the largest experimentally known systems. These results also illustrate one fundamental limitation of the approach: it may not always be possible for these large systems to identify a single combination of **P** and **K** that fits the data better than all others. This suggests that the information content of the data is not sufficient and that other sources of data will be required. In the case of ribosome assembly, methods like pulse-chase mass spectrometry (Talkington et al. 2005) reveal the protein's perspective on the RNA folding reaction and can provide the additional kinetic information to identify a single model. Furthermore, time-resolved small angle X-ray scattering can provide global compaction measures (Pollack et al. 2001; Russell et al. 2002), while catalytic activity measurements indicate the rate of appearance of the native molecule (Russell et al. 2006). Taken together, these varied sources of experimental data have the potential to accurately describe the folding reaction of very large RNA molecules.

Kinetic modeling, such as the approach we describe here, will be critical in laying the foundation for addressing many of the unanswered questions that remain in RNA folding. These include the identification of conserved themes, the role of counterion concentration, and the role of sequence and kinetic traps in the folding.

The rate-determining steps in RNA folding depend on many factors including the electrostatic environment, temperature, and exogenous molecule binding (Russell and Herschlag 2001; Russell et al. 2006; Laederach et al. 2007). Kinetic models and new experimental approaches will allow for a better understanding of the mechanism for which the RNA changes conformation in response to regulatory elements (Tucker and Breaker 2005) and mutations of the sequence that effect the kinetics of folding.

Our kinetic models provide quantitative and mechanistic insight into the folding of large RNAs. It is important to be aware of the fact that these models describe the in vitro folding reaction in the absence of proteins and other cofactors that act as folding chaperones. Furthermore, we fold fully transcribed RNAs by adding counterions. In reality, RNA is folded co-transcriptionally in the cell. It is likely that the presence of chaperones and the co-transcriptional folding process may simplify and/or accelerate the folding process by eliminating some of the pathways that lead to long-lived, misfolded intermediates. In fact, there is evidence that the *T. thermophila* group I intron folds an order of magnitude faster in vivo than in vitro (Woodson 2002). It is therefore critical to remember that the folding models we develop represent possible folding pathways, but do not represent the actual biological folding mechanism.

Acknowledgments We thank Michael Brenowitz and Joerg Schlatterer for their insightful discussions and comments during the preparation of this chapter. This work is supported by the US National Institutes of Health, NIGMS R00 079953 grant to A.L. Source code, and example data sets can be downloaded from https://simtk.org/home/KinFold.

References

Brenowitz M, Senear DF, Shea MA, Ackers GK (1986) "Footprint" titrations yield valid thermodynamic isotherms. Proc Natl Acad Sci USA 83(22):8462–8466

Brenowitz M, Chance MR, Dhavan G, Takamoto K (2002) Probing the structural dynamics of nucleic acids by quantitative time-resolved and equilibrium hydroxyl radical "footprinting". Curr Opin Struct Biol 12(5):648–653

Das R, Laederach A, Pearlman SM, Herschlag D, Altman RB (2005) Safa: semiautomated footprinting analysis software for high-throughput quantification of nucleic acid footprinting experiments. RNA 11(3):344–354

Heilman-Miller SL, Thirumalai D, Woodson SA (2001) Role of counterion condensation in folding of the tetrahymena ribozyme. I. Equilibrium stabilization by cations. J Mol Biol 306 (5):1157–1166

Laederach A, Shcherbakova I, Liang M, Brenowitz M, Altman RB (2006) Local kinetic measures of macromolecular structure reveal partitioning among multiple parallel pathways from the earliest steps in the folding of a large RNA molecule. J Mol Biol 358(358):1179–1190

Laederach A, Shcherbakova I, Jonikas MA, Altman RB, Brenowitz M (2007) Distinct contribution of electrostatics, initial conformational ensemble, and macromolecular stability in RNA folding. Proc Natl Acad Sci USA 104(17):7045–7050

Martin JS, Simmons K, Laederach A (2009) Exhaustive enumeration of kinetic model topologies for the analysis of time-resolved RNA folding. Algorithms 2(1):2000–2214

Mitra S, Shcherbakova IV, Altman RB, Brenowitz M, Laederach A (2008) High-throughput single-nucleotide structural mapping by capillary automated footprinting analysis. Nucleic Acids Res 36(11):e63

Pan J, Thirumalai D, Woodson SA (1997) Folding of RNA involves parallel pathways. J Mol Biol 273(1):7–13

Pollack L, Tate MW, Finnefrock AC, Kalidas C, Trotter S, Darnton NC, Lurio L, Austin RH, Batt CA, Gruner SM, Mochrie SG (2001) Time resolved collapse of a folding protein observed with small angle X-ray scattering. Phys Rev Lett 86(21):4962–4965

Russell R, Herschlag D (2001) Probing the folding landscape of the tetrahymena ribozyme: commitment to form the native conformation is late in the folding pathway. J Mol Biol 308 (5):839–851

Russell R, Millett IS, Tate MW, Kwok LW, Nakatani B, Gruner SM, Mochrie SG, Pande V, Doniach S, Herschlag D, Pollack L (2002) Rapid compaction during RNA folding. Proc Natl Acad Sci USA 99(7):4266–4271

Russell R, Das R, Suh H, Travers KJ, Laederach A, Engelhardt MA, Herschlag D (2006) The paradoxical behavior of a highly structured misfolded intermediate in RNA folding. J Mol Biol 363(2):531–544

Sclavi B, Woodson S, Sullivan M, Chance MR, Brenowitz M (1997) Time-resolved synchrotron X-ray "footprinting", a new approach to the study of nucleic acid structure and function: application to protein-DNA interactions and RNA folding. J Mol Biol 266(1):144–159

Sclavi B, Sullivan M, Chance MR, Brenowitz M, Woodson SA (1998a) RNA folding at millisecond intervals by synchrotron hydroxyl radical footprinting. Science 279(5358):1940–1943

Sclavi B, Woodson S, Sullivan M, Chance M, Brenowitz M (1998b) Following the folding of RNA with time-resolved synchrotron X-ray footprinting. Methods Enzymol 295:379–402

Shcherbakova I, Brenowitz M (2008) Monitoring structural changes in nucleic acids with single residue spatial and millisecond time resolution by quantitative hydroxyl radical footprinting. Nat Protoc 3(2):288–302

Shcherbakova I, Mitra S, Beer RH, Brenowitz M (2006) Fast Fenton footprinting: a laboratory-based method for the time-resolved analysis of DNA, RNA and proteins. Nucleic Acids Res 34 (6):e48

Shcherbakova IV, Mitra S, Laederach A, Brenowitz M (2008) Energy barriers, pathways, and dynamics during folding of large, multidomain RNAs. Curr Opin Chem Biol 12(6):655–666

Takamoto K, Das R, He Q, Doniach S, Brenowitz M, Herschlag D, Chance MR (2004) Principles of RNA compaction: insights from the equilibrium folding pathway of the p4-p6 RNA domain in monovalent cations. J Mol Biol 343(5):1195–1206

Talkington MW, Siuzdak G, Williamson JR (2005) An assembly landscape for the 30s ribosomal subunit. Nature 438(7068):628–632

Thirumalai D, Hyeon C (2005) Rna and protein folding: common themes and variations. Biochemistry 44(13):4957–4970

Tibshirani R, Walther G, Hastie T (2001) Estimating the number of clusters in a data set via the gap statistic. J R Stat Soc Series B Stat Methodol 63(2):411–423

Tucker BJ, Breaker RR (2005) Riboswitches as versatile gene control elements. Curr Opin Struct Biol 15(3):342–348

Vicens Q, Gooding AR, Laederach A, Cech TR (2007) Local RNA structural changes induced by crystallization are revealed by shape. RNA 13(4):536–548

Wilkinson KA, Merino EJ, Weeks KM (2005) RNA shape chemistry reveals nonhierarchical interactions dominate equilibrium structural transitions in tRNA(asp) transcripts. J Am Chem Soc 127(13):4659–4667

Wilkinson KA, Gorelick RJ, Vasa SM, Guex N, Rein A, Mathews DH, Giddings MC, Weeks KM (2008) High-throughput shape analysis reveals structures in hiv-1 genomic RNA strongly conserved across distinct biological states. PLoS Biol 6(4):e96

Woodson SA (2000) Recent insights on RNA folding mechanisms from catalytic RNA. Cell Mol Life Sci 57(5):796–808

Woodson SA (2002) Folding mechanisms of group I ribozymes: role of stability and contact order. Biochem Soc Trans 30(Pt 6):1166–1169

Chapter 16
A Top-Down Approach to Determining Global RNA Structures in Solution Using NMR and Small-Angle X-ray Scattering Measurements

Yun-Xing Wang, Jinbu Wang, and Xiaobing Zuo

Abstract RNA plays important roles in many biological processes. RNA functions are embedded in its structures and dynamics. Structure elucidation of RNA, using experimental, computational, or combined approaches, remains a major research challenge and focus of interest in contemporary biology. In this chapter, we present a method that uses global orientation and shape restraints, which are derived from experimental NMR measurements and small-angle X-ray scattering (SAXS) data, to determine global structures of sizable RNAs in solution. The global structures may be used as initial structures for high-resolution structure determination by computational or experimental approaches. This chapter outlines the theory and procedures and presents experimental examples to demonstrate the method. A Web page is also included for readers to download the program toolkit, calculation scripts, and examples.

16.1 Introduction

The discovery of the critical roles that RNA plays in the regulation of gene expression at various levels is one of greatest advances in modern biology. For example, RNA is an active participant in the regulation of gene expression by

Y.-X. Wang (✉) • J. Wang
Protein-Nucleic Acid Interaction Section, Structural Biophysics Laboratory, Center for Cancer Research, National Cancer Institute, National Institutes of Health, Frederick, MD 21702, USA
e-mail: wangyunx@mail.nih.gov; wangjinb@mail.nih.gov

X. Zuo
Protein-Nucleic Acid Interaction Section, Structural Biophysics Laboratory, Center for Cancer Research, National Cancer Institute, National Institutes of Health, Frederick, MD 21702, USA

Current address: X-ray Science Division, Advanced Photon Source, Argonne National Laboratory, 9700 South Cass Avenue, Argonne, IL 60439, USA
e-mail: zuox@anl.gov

N. Leontis and E. Westhof (eds.), *RNA 3D Structure Analysis and Prediction*, Nucleic Acids and Molecular Biology 27, DOI 10.1007/978-3-642-25740-7_16,

interference (Fire et al. 1998) or by riboswitches (Tucker and Breaker 2005), in the processing of RNA introns (Kruger et al. 1982), in the maintenance of chromosome ends by telomerase (Blackburn 1992), and in protein synthesis by the ribosome (Moore and Steitz 2002). RNA function is encoded in its dynamics and structure (Zhang et al. 2006, 2007; Cruz and Westhof 2009), and determining RNA structures remains a major goal in contemporary biology. In the past decade, despite significant advances in determining RNA structure using X-ray crystallography and solution NMR, as well as in structure prediction (Martinez et al. 2008; Parisien and Major 2008; Jonikas et al. 2009b), the number of validated bona fide RNA structures is dwarfed by both the number of protein structures and the growing number of known functional RNAs. This disparity is due to the difficulty of growing crystals and/or obtaining phase information and to the size limitations of structure determination by solution NMR spectroscopy. Clearly, a new strategy for determining RNA structure is critically needed.

A survey of the current RNA structure databases reveals that RNA structures consist mainly of duplexes that form the major building blocks punctuated by loops (Leontis et al. 2006; Wang et al. 2009). Cruz and Westhof recently pointed out that RNA architectures are dominated by duplexes, arranged through coaxial stacking and packed in parallel or orthogonal to one another (Cruz and Westhof 2009). The underlying forces that "glue" various building blocks together are tertiary Watson–Crick base pairings, such as those observed in kissing loops or pseudoknots and tertiary non-Watson–Crick base pairings, as observed in loop-receptor and A-minor interactions and within helical junctions (Leontis et al. 2006; Lescoute and Westhof 2006). However, due to the lack of a clear propensity correlation between a sequence and tertiary interactions, these interactions, especially noncanonical base pairings, are difficult to predict based on a sequence alone using a pure computation modeling approach. One of the links between an RNA primary sequence and an atomic resolution structure is the global structure of the RNA. We have developed an experimental method for global structure determination that may make it feasible to reliably predict intricate tertiary interactions (Leontis and Westhof 2001, 2002, 2003; Leontis et al. 2006; Jonikas et al. 2009a). Moreover, this method may also open the door to high-resolution structure determination of sizable RNAs through the combined use of solution-based NMR spectroscopy and small-angle X-ray scattering (SAXS) methods.

Since RNA architectures are dominated by duplexes that tend to pack approximately parallel/antiparallel or orthogonal to each other, and/or to stack coaxially (Leontis et al. 2006; Cruz and Westhof 2009), determining the global structure may begin with determining the relative orientations and relative positions of the duplexes. Both the relative orientation and position of the duplexes can be experimentally measured, as demonstrated in the following section. Since both the orientation and position are global measurements, we have called the method global measurements to global structure (G2G).

16.2 Theory

16.2.1 Duplex Orientation

Residual dipolar coupling (RDC) in solution is an orientation-dependent, geometrical physical property that can be measured in a weakly aligned medium and magnetic field using NMR. The application for studying macromolecule structures was pioneered by Bothner (Gayathri et al. 1982; Bothner 1996). Tolman et al. were first to measure the orientation-dependent ^{1}H–^{15}N splitting in paramagnetic myoglobin (Tolman et al. 1995), and Bolton and his colleagues observed that the magnetic field induced natural alignment of DNA molecules (Kung et al. 1995). The RDC of spin pairs in a repetitive and periodical structure is wave-like when it is plotted against the residue numbers. This wave is called a dipolar wave or more explicitly the RDC–structure periodicity correlation. It was first reported for α-helices in membrane proteins (Mesleh et al. 2002) and later for duplexes of RNA molecules (Walsh et al. 2004). The shape of a dipolar wave depends on the orientation of the repetitive and periodical structure relative to a reference axis (Fig. 16.1a, b). Therefore, the dipolar wave contains orientation information and can be used to extract the orientation information of periodical structural elements in macromolecular structure determination (Walsh and Wang 2005; Wang et al. 2007).

The dipolar coupling splitting, D_{AB}, between two near spin-½ nuclei A and B, is expressed by (16.1):

$$D_{AB} = -\frac{\gamma_A \gamma_B \hbar}{4\pi^2 r_{AB}^3} <3\cos^2\theta - 1>, \tag{16.1}$$

where γ_i is the gyromagnetic ratio of spin A or B, r_{AB} is the distance between spins A and B, and θ is the angle of the AB internuclear vector with respect to the magnetic field (Ernst et al. 1987). The angular brackets "< >" indicate averaging due to overall molecular tumbling and local dynamics. In an isotropic solution, $<3\cos^2\theta - 1> = 0$. In a weakly aligned rigid system in solution, the equation simplifies when rewritten in spherical coordinates, as shown in (16.2):

$$D_{AB}(\theta, \varphi) = D_a[(3\cos^2\theta - -1) + 3/2R\sin^2\theta\cos 2\varphi], \tag{16.2}$$

where D_a and R are the axial and rhombic components of alignment tensors and (θ, φ) are the azimuthal and polar angles in spherical coordinates describing the orientation of the internuclear vector. For periodically repetitive structures, such as α-helixes, β-strands, and DNA or RNA duplexes, D_{AB} can be expressed in terms of both the orientation (Θ, Φ) of a structural element in the alignment tensor frame and the orientation (δ, ρ) of the bond vector within this structural element (Fig. 16.1c–e). (Θ, Φ) are the orientation of the periodical structure axis relative to the global reference axis (Fig. 16.1e), as demonstrated in the following discussion (Walsh et al. 2004).

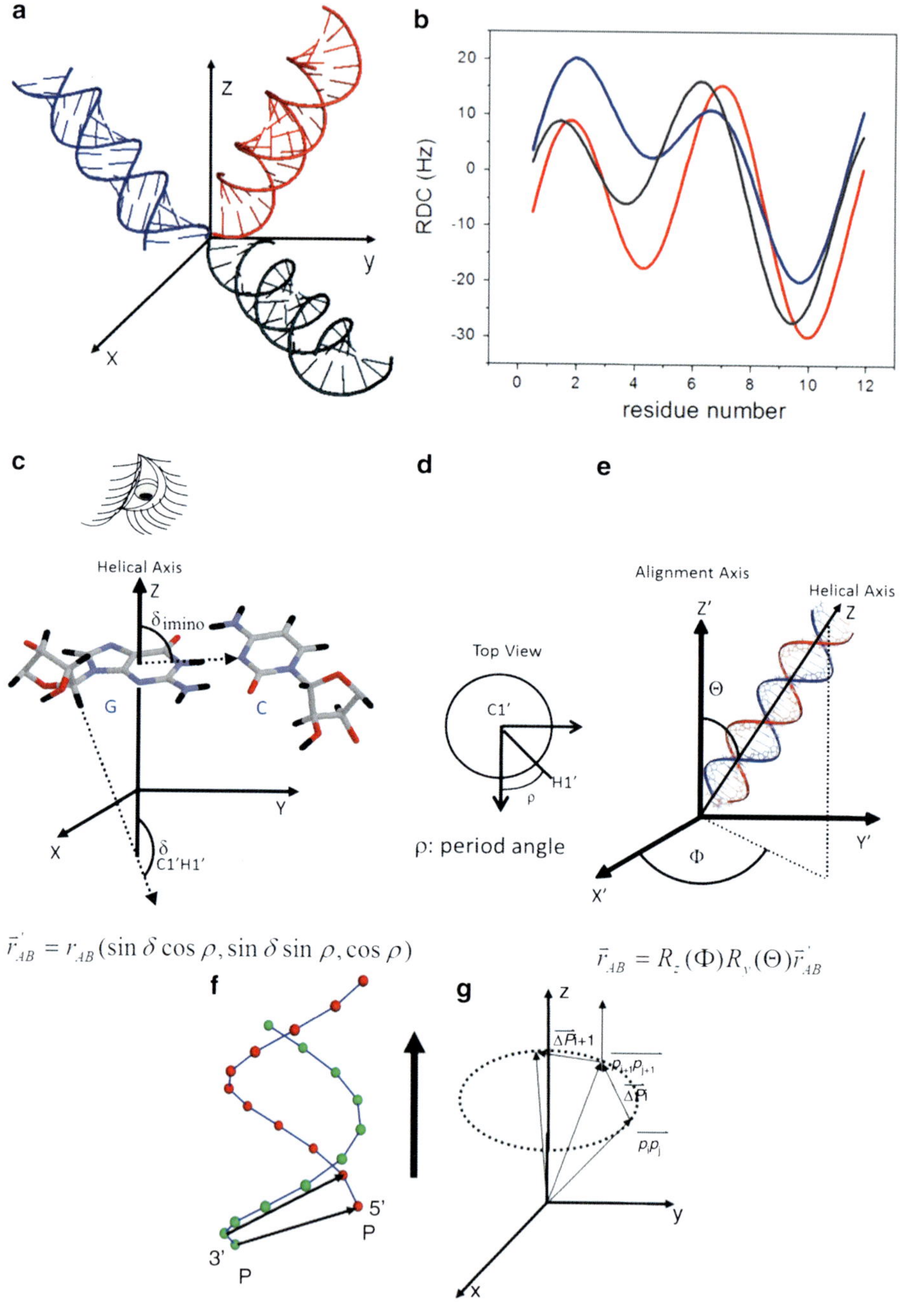

Fig. 16.1 Simulated RDC–structural periodicity correlation of duplexes and the pictorial definitions of duplex orientation (Φ, Θ) and phase ρ_0. (**a**) Three different orientations; (**b**) their corresponding RDC–structural periodicity correlation curves; (**c**) the imino and C1′H1′ slant angles (δ) with respect to the helix with a vector expression; (**d**) the C1′–H1′ period angle ρ; (**e**) the definition of the

In spherical coordinates, the bond vector of the nth residue of a periodically and repeating structure, such as a nucleic acid duplex, can be expressed in terms of the bond length, r_{AB}, angle δ_n to the duplex axis, and the angle ρ_n with the x-axis (in a coordinate system where Z is along the helix axis and the X–Y plane is perpendicular to it, see Fig. 16.1c). The angle $\rho_n = (\alpha_n + \rho_0)$ is given by the phase of the first bond vector in the duplex, ρ_0, plus a phase offset (or "phase," for short), which is characteristic of duplex periodicity: $\alpha_n = 2\pi(n - 1)/T$, $n = 1, 2, 3 \ldots$; T is the period of the duplex (A-RNA: $T = 11$; B-DNA: $T = 10$; α-helix: $T = 3.6$). The bond vector in the duplex reference frame is then given by (16.3):

$$\vec{r}'_{AB} = r_{AB}(\sin\delta_n \cos\rho_n, \sin\delta_n \sin\rho_n, \cos\delta_n). \tag{16.3}$$

Rotating this bond vector by the helix orientation angles (Θ, Φ) gives its orientation in the alignment tensor reference frame.

$$\begin{aligned}\vec{r}_{AB} &= R_z(\Phi)R_y(\Theta)\vec{r}'_{AB}\\ &= \begin{pmatrix}\cos\Phi & -\sin\Phi & 0\\ \sin\Phi & \cos\Phi & 0\\ 0 & 0 & 1\end{pmatrix}\begin{pmatrix}\cos\Theta & 0 & \sin\Theta\\ 0 & 1 & 0\\ -\sin\Theta & 0 & \cos\Theta\end{pmatrix}\begin{pmatrix}d_{AB}\sin\delta_n\cos\rho_n\\ d_{AB}\sin\delta_n\sin\rho_n\\ d_{AB}\cos\delta_n\end{pmatrix}\\ &= r_{AB}\begin{pmatrix}\cos\Phi(\cos\Theta\sin\delta_n\cos\rho_n + \sin\Theta\cos\delta_n) - \sin\Phi\sin\delta_n\sin\rho_n\\ \sin\Phi(\cos\Theta\sin\delta_n\cos\rho_n + \sin\Theta\cos\delta_n) + \cos\Phi\sin\delta_n\sin\rho_n\\ -\sin\Theta\sin\delta_n\cos\rho_n + \cos\Theta\cos\delta_n\end{pmatrix}.\end{aligned} \tag{16.4}$$

Equation 16.3 can be recast in Cartesian coordinates,

$$D_{AB} = \frac{D_a}{r_{AB}}\left\{\left(\frac{3}{2}R - 1\right)x^2 - \left(\frac{3}{2}R + 1\right)y^2 + 2z^2\right\} \tag{16.5}$$

and substituted in the Cartesian coordinates parameterized in terms of duplex parameters from (16.4); then the relationship between the RDC of an internuclear vector, D_{AB}, and the orientation of a repetitive and periodical structural element can be explicitly expressed by (16.6) (Walsh et al. 2004):

Fig. 16.1 (continued) orientation of a duplex (Θ, Φ) and the relationship between a tilted duplex and a reference axis. Simulated RDC–structural periodicity correlation curves (**b**) are color-coded as in (**a**). The shapes of the RDC–structural periodicity correlation curves (RDC waves) depend on the orientation (Θ, Φ) and the phase ρ_0 of duplexes. The equation in (**e**) shows the interconvention of the bond vector in the alignment tensor reference frame and the duplex reference frame, with an orientation of (Θ, Φ) with respect to the alignment tensor reference frame (Walsh et al. 2004). (**f**) Each ball represents a phosphor atom (P). A bp PP vector stands for the vector from 3′ P to 5′ P in the same bp, *Black arrow at right* indicates the duplex orientation. (**g**) The vector $\overrightarrow{\Delta P_i} = \overrightarrow{p_{i+1}p_{j+1}} - \overrightarrow{p_ip_j}$ belongs to the plane perpendicular to the duplex axis. The cross product $n_i = \Delta P_i \otimes \Delta P_{i+1}$ is the normal of the plane perpendicular to the duplex axis, which is parallel to the duplex axis. The duplex orientation is calculated as an average of the normals

$$D_{AB} = C_1(\Theta, \Phi, \delta_n)\cos 2\rho_n + C_2(\Theta, \Phi, \delta_n)\sin 2\rho_n + C_3(\Theta, \Phi, \delta_n)\cos \rho_n + C_4(\Theta, \Phi, \delta_n)\sin \rho_n + C_5(\Theta, \Phi, \delta_n), \quad (16.6)$$

where C_i ($i = 1, 2, \ldots 5$) are functions of (Φ, Θ, ρ_n) and are given by (16.7):

$$C_1(\Theta, \Phi, \delta_n) = (3D_a/16)[4 + 6R\cos 2\Phi + R\cos 2(\Theta - \Phi) - 4\cos 2\Theta + R\cos 2(\Phi + \Theta)]\sin^2\delta_n, \quad (16.7a)$$

$$C_2(\Theta, \Phi, \delta_n) = (-3D_a/2R)\cos\Theta\sin 2\Phi\sin^2\delta_n, \quad (16.7b)$$

$$C_3(\Theta, \Phi, \delta_n) = (3D_a/4)(R\cos 2\Phi - 2)\sin 2\Theta\sin 2\delta_n, \quad (16.7c)$$

$$C_4(\Theta, \Phi, \delta_n) = -6D_aR\sin\Theta\sin\Phi\cos\Phi\sin\delta_n\cos\delta_n, \quad (16.7d)$$

$$C_5(\Theta, \Phi, \delta_n) = (D_a/32)[4 + 6R\cos 2\Phi - 3R\cos(\Phi - \Theta) + 12\cos 2\Theta - 3R\cos 2(\Phi + \Theta)](3\cos 2\delta_n + 1). \quad (16.7e)$$

Equation (16.6) expresses D_{AB} explicitly in terms of the helical global orientation (Θ, Φ) and phase, ρ_0, of a structural element in the alignment tensor frame. Equation (16.6) has five unknown variables, D_a, R, Φ, Θ, and ρ_0, which can be fitted by the nonlinear least squares method (Walsh et al. 2004), which is implemented in the ORIENT program of the G2G toolkit (Wang et al. 2009).

The ORIENT program calculates the duplex axis using the base pair (bp) phosphate-to-phosphate vectors $\overrightarrow{p_ip_j}$, where i and j are the bp indices (Fig. 16.1f). The $\overrightarrow{p_ip_j}$ vector is from the 3′ to the 5′ strand (Wang et al. 2009). The vectors $\overrightarrow{\Delta P_i}(\overrightarrow{\Delta P_i} = \overrightarrow{p_{i+1}p_{j+1}} - \overrightarrow{p_ip_j})$ belong to the plane perpendicular to the duplex axis (Fig. 16.1g). The plane normal (which is also the duplex axis) is $n_i = \Delta P_i \otimes \Delta P_{i+1}$. The duplex axis is the average of the plane normal vectors, $\vec{D} = \sum_{i=1}^{k} n_i/k$, where $k + 2$ equals the number of bps in the duplex. The program allows a choice of fitting the RDC data either for an individual duplex or for all duplexes simultaneously, under a rigid-body assumption (Wang et al. 2009).

One of limitations in using RDCs to derive orientation information is the fourfold degeneracy that occurs if only one independent alignment medium is available (Fowler et al. 2000; Hus et al. 2001; Walsh et al. 2004). In an RNA consisting of three duplexes, there are 16 possible orientation combinations of the three duplexes (Wang et al. 2009). In theory, a second noncollinear alignment medium is required to resolve ambiguity in the relative orientation of the duplex. A shortcut solution to this problem can be obtained using SAXS, from which one can derive a global shape that is indicative of an approximate layout of the duplexes. Therefore, the shape can be utilized to identify the correct combination of the relative orientation, as we shall demonstrate in later sections (Wang et al. 2009).

16.2.2 Shape Derived from SAXS

Next, we briefly summarize the theories and protocols for using SAXS to derive global shapes of biomacromolecules in solution state, which have been described extensively in the literature (Svergun and Stuhrmann 1991; Chacon et al. 1998; Svergun 1999; Walther et al. 2000; Koch et al. 2003). For a molecule, which consists of N atoms and is randomly tumbling in solution, the scattering intensity, I, is given as a function of the momentum transfer, q, by Debye's formula (16.8) (Debye 1915):

$$I(q) = \sum_{i=1}^{N}\sum_{j=1}^{N} f_i(q) f_j(q) \frac{\sin(qr_{ij})}{qr_{ij}}, \tag{16.8}$$

where r_{ij} is the distance between pairs of atoms and $f_j(q)$ is the scattering factor of the individual atom, j. Beginning in the 1970s, Stuhrmann, Svergun, and coworkers developed an ab initio method to semianalytically reconstruct 3D shapes based on the expansion of q in spherical harmonics (Stuhrmann 1970; Svergun and Stuhrmann 1991). This approach is very powerful when applied to globular objects but limited for more complex shapes. In the late 1990s and early 2000s, Monte Carlo-based methods were also developed to reconstruct the low-resolution molecular envelopes/shapes, based on the idea that objects of arbitrary shape can be represented by collections of small beads (Chacon et al. 1998; Svergun 1999; Walther et al. 2000). Among the programs currently in widespread use, DALAI_GA uses the genetic algorithm (Chacon et al. 1998), the DAMMIN program uses a simulated annealing algorithm (Svergun 1999), and the Saxs3D program uses a "give-n-take" algorithm (Walther et al. 2000) to efficiently generate the bead molecular model that satisfies a given experimental SAXS profile. These programs also utilize fast algorithms to deconvolute the SAXS profiles to obtain bead models. For example, DAMMIN uses harmonic expansion, and GALAI_GA and Saxs3D use distance histogram methods. Bead model calculations using each of these efficient algorithms can be carried out on a PC in a few hours. To make the resulting bead models more physically meaningful, besides the goodness of fit of the SAXS profile, more regularizations have been imposed in the bead model calculations, such as extra penalties on bead model looseness and disconnectivity (Svergun 1999). In cases where the symmetry of the molecular shape is known (e.g., ellipsoidal, oblate, etc.), the programs provide options to restrain the shape of the bead models accordingly. Furthermore, due to the intrinsic degeneracy of SAXS, it is recommended that multiple bead model calculations be performed where the programs take a random seed to initiate calculations; the most probable model is usually the average of these calculated bead models.

16.3 Determining Global Structures of RNAs Using Global Restraints

16.3.1 A Simulated Case

Given the global orientations of the duplexes and the global shape of an RNA, could one, in principle, determine its global structure to useful resolution? To answer this question, we performed a simulation calculation. We used orientations of the three duplexes of the adenine riboswitch (riboA) from an X-ray crystal structure (Serganov et al. 2004) and the global dimension measurements of the shape to calculate an ensemble of global structures of this RNA. The shape of the RNA was derived from an experimental SAXS profile of a riboA RNA having a slightly different sequence (Wang et al. 2009). In addition to the orientation and the overall shape restraints, we applied generic distance restraints to restrain the three duplexes to the A-form conformation, enforce base-stacking throughout the RNA sequence, and position the adenine ligand correctly. The distance restraints for the ligand are readily calculated from the nuclear Overhauser effect (NOE) experiments (see below) (Wang et al. 2009). We used Xplor-NIH (Schwieters et al. 2003) and a hybrid rigid-body simulated annealing (SA) refinement protocol similar to that previously published (Zuo et al. 2008) for the calculation (Web page: https://ccrod.cancer.gov/confluence/display/public/SAXS). The regularization procedure available in Xplor-NIH connects linkers with duplex building blocks and removes any gross covalent geometry distortions before SA refinement. During the calculation, the orientations and phases of the three duplexes were fixed in space, but arbitrary linker motions were allowed using the internal variable module (IVM) facility (Schwieters et al. 2006) of the Xplor-NIH package (version 2.22 or newer). In addition to the loose distance and torsion-angle restraints that were applied to maintain the approximate A-form conformation of the duplexes, we used the following restraints in the calculation: (1) an explicit restraint on the radius of gyration (R_g), extracted from the SAXS data and (2) uniform distance restraints to maintain neighboring base stacking throughout the whole chain. The ensemble of structures using these simulated data is shown in Fig. 16.2. The backbone root-mean-square deviations (RMSDs) between the average structure of the ensemble and the X-ray crystal structure are 0.8 Å for residues in the three duplexes and 1.7 Å for all residues in the structure. This calculation suggests that it is possible, with accurate global shape information and duplex orientation restraints, to determine the global structure with a high degree of accuracy.

16.3.2 Experimental Case 1: riboA

The riboA RNA modulates the expression of associated genes in response to elevated concentrations of the cellular metabolite adenine (Mandal and Breaker

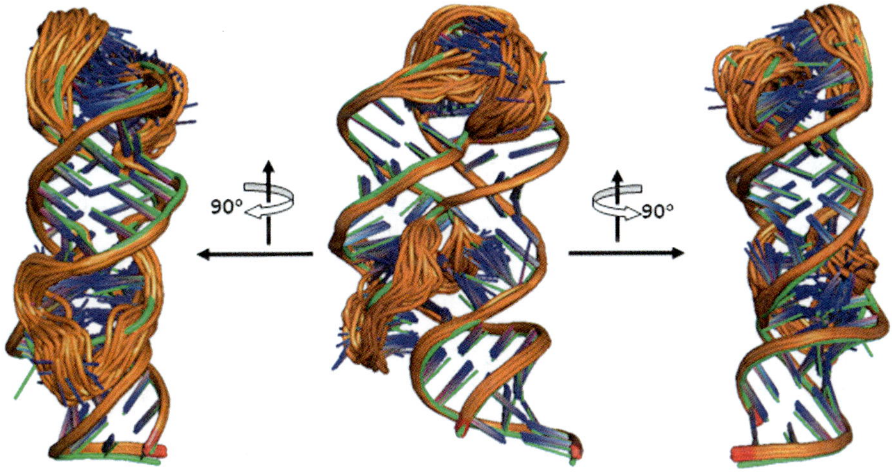

Fig. 16.2 Comparison of the ensemble of the G2G structures of riboA (*orange*) with the X-ray crystal structure (*green*, pdb access code: 1Y26) in three views. The G2G structures were calculated using the duplex orientations taken from the X-ray structure and the simulated SAXS profile of the X-ray crystal structure (Wang et al. 2009). Other generic restraints are described the text

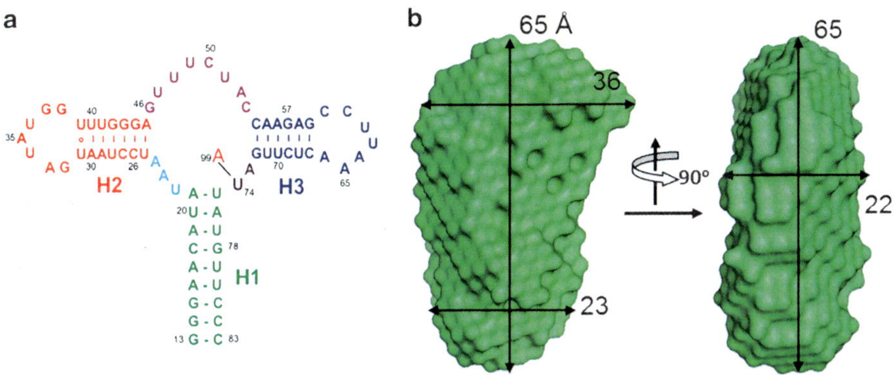

Fig. 16.3 The secondary structure (**a**) and a SAXS-derived molecular envelope (**b**) of the riboA RNA. In (**a**), A99 denotes an adenine ligand. In (**b**), the dimensional measures, which are indicative of the three duplexes aligned in parallel or antiparallel fashion, are depicted in the two views (Wang et al. 2009)

2004). The crystal structure of a 71-nucleotide (nt) riboA has been determined to a 2.6-Å resolution (PDB access code: 1Y26) (Serganov et al. 2004). In addition to its importance in regulating gene expression, this 71-nt riboA RNA was selected to test the method because of its relatively complex fold and large size with respect to the solution NMR method. Note that the sequence used in our test is slightly different from that of the crystal structure (Fig. 16.3) (Wang et al. 2009). The secondary structure of the riboA RNA consists of three duplexes, H1, H2, and H3, comprising 9, 7, and 6 bps, respectively. The three duplexes are joined by three short linkers, consisting of 3, 7, and 2 nt between H1 and H2, H2 and H3, and H3 and H1, respectively. For details, readers are referred to the original publication (Wang et al. 2009). The following is a brief summary of the experimental aspects of the method, which we refer to as G2G, or the "global restraints to global structures" method.

The G2G method requires assigning imino proton signals and identifying hydrogen bonds involved in canonical and noncanonical bps in duplexes. The assignments of imino signals of the riboA RNA were accomplished by an NOE walk of the 2D NOESY spectrum of the imino region, aided by the 2D $^{15}N...H–^{15}N$ HNN-COSY spectrum (Dingley et al. 2000; Wang et al. 2009). As we will show later, these 2D homonuclear NOESY and HNN-COSY spectra were also sufficient to assign imino signals of a 102-nt RNA as described in the next section (Zuo et al. 2010).

The experimental RDC data were measured from the in-phase/antiphase (IPAP) heteronuclear single-quantum coherence spectra (Ottiger et al. 1998), which were recorded using ^{15}N isotopic RNA samples in isotropic and anisotropic conditions. The anisotropic riboA sample was prepared by adding about 9.7 mg/ml of pf1 phage to weakly align the RNA molecule (ASLA Biotech, Burlington, NC), which produced a split of 9.8 Hz in the deuterium signal. RDC values ranged from 2.2 to 22.7 Hz for the imino $^{15}N–^{1}H$ of riboA. The IPAP spectra of the imino resonance splitting of the riboA in the alignment medium are shown in Fig. 16.4 (Wang et al. 2009).

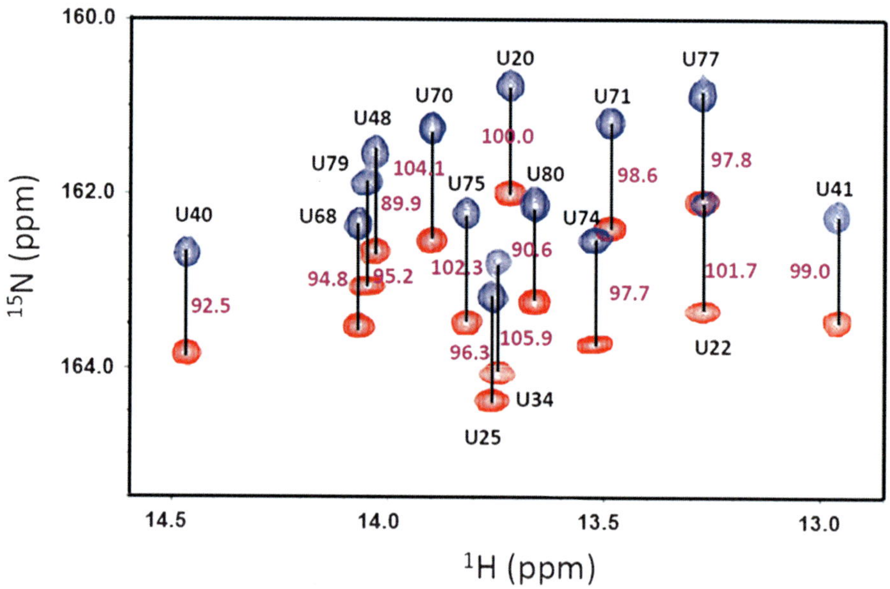

Fig. 16.4 A portion of the imino $^{15}N–^{1}H$ IPAP spectra of the ^{15}N-labeled riboA recorded in an anisotropic solution containing about 0.5 mM of the riboA RNA and 9.7 mg/ml of pf1 alignment medium (Wang et al. 2009)

The imino ^{15}N–^{1}H J-coupling was about 94 Hz, measured with a nonaligned isotropic sample. The RDC values were calculated with the difference in splitting recorded in the presence and absence of the alignment medium.

In principle, the unique orientation of each duplex can be determined unambiguously by utilizing a second independent alignment tensor (Losonczi et al. 1999). However, a second, truly independent, alignment medium for RNA is currently lacking (Latham et al. 2005). One way to get around this problem is to utilize the shape information that is derived from SAXS data to determine the unique orientation of each duplex in the molecular frame. In the case of riboA, the overall shape of the RNA suggests that the three duplexes are packed either parallel or antiparallel, consistent with the general packing pattern for RNA architectures (Leontis et al. 2006; Cruz and Westhof 2009). The quantitative determination of orientations and phases of the three duplexes in the riboA was then carried out using the ORIENT program in the G2G toolkit. In the calculation, the angles between the duplexes were restrained to either $0° \pm 30°$ or $180° \pm 30°$ for all three pairs of duplexes to allow for parallel or antiparallel arrangements. The best simultaneous fit yielded the axial component $D_a = -26.4$ Hz and the rhombic component $R = 0.35$ in (16.6), with orientations and phases (Θ, Φ, ρ_0) of (151°, 281°, 101°), (22°, 97°, 259°), and (40°, 101°, 63°) for duplexes H1, H2, and H3, respectively; these measurements represent the angles of 173° between H1 and H2, 169° between H1 and H3, and 18° between H2 and H3 (Wang et al. 2009). The RDC waves of the fits are shown in Fig. 16.5. The structural topology of the riboA RNA, based on the orientation and phase information derived from the above calculations, is depicted in Fig. 16.5. This topology illustrates the global arrangement of the three duplexes and makes it possible to build initial 3D coordinates using the BLOCK and PACK programs in the G2G toolkit. Before the rigid-body refinement, the regularization procedure is applied by using the IVM facility of the Xplor-NIH package to remove distortions of the covalent geometry in the linker regions, and especially the joints (Schwieters et al. 2006). The orientations and phases of the duplexes, but not the linker segments, were fixed, and only translational movements were allowed during the regularization and later the hybrid rigid-body SA refinement (Wang et al. 2009).

In addition to the restraints that are described in the simulated case in the last section, we applied experimental imino RDC restraints for residues in the nonduplex regions and approximate dimension restraints derived from the envelope of riboA in the form of approximate phosphorus–phosphorus distances. To avoid gross close contacts, we also added a minimum distance-repulsive restraint, 6.0 Å, which is the approximate sequential phosphorus distance in an A-form duplex and is generally the shortest possible distance separating any two phosphate groups in RNA structures (Wang et al. 2009). To speed up the calculation, SAXS data "sparsening" was carried out using a previously reported protocol (Grishaev et al. 2005). Specifically, an evenly spaced representative 20 points of the experimental SAXS data, ranging from momentum transfer $q = 0$–0.3 Å^{-1}, were used during the refinement. The pseudopotential force constants of various terms were empirically adjusted. Readers can find a sample script of the hybrid rigid-body refinement protocol on the Web (https://ccrod.cancer.gov/confluence/display/

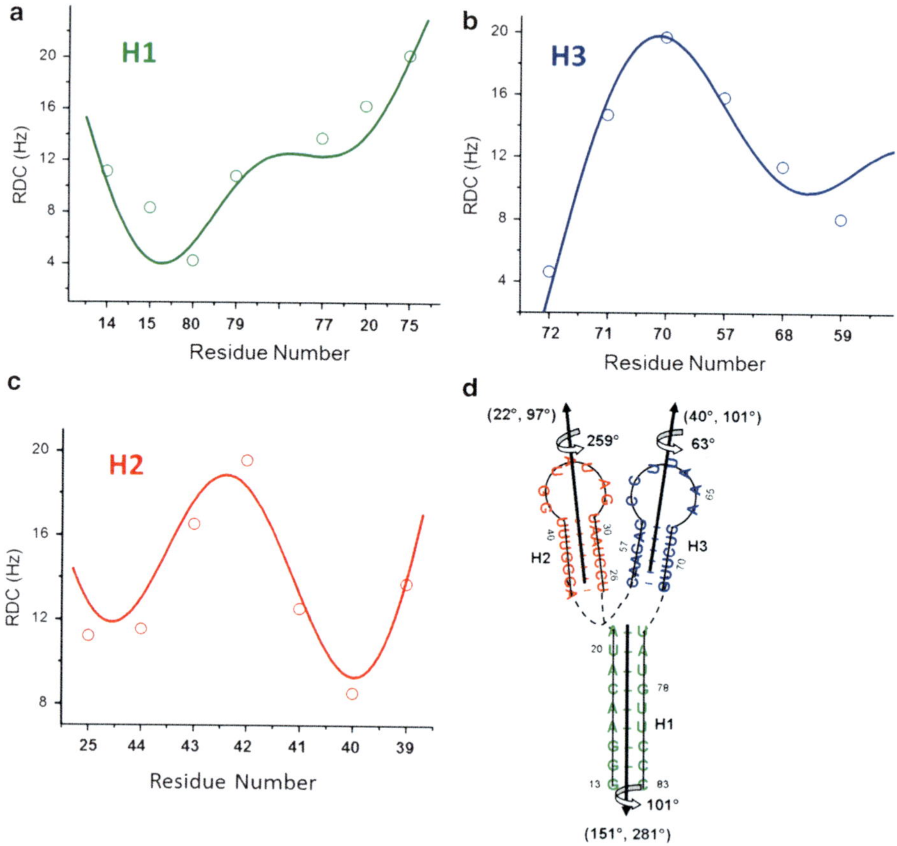

Fig. 16.5 The dipolar waves for H1, H2, and H3 in riboA, and the topological drawing for riboA based on the duplex orientations extracted from the RDC fits. In (**a–c**), the RDC data for these three duplexes are fitted simultaneously. In (**d**), orientations and phases, (Θ, Φ, ρ_0), depicted in the riboA topology were obtained from the best simultaneous fit. The broken lines represent linker residues in arbitrary conformations (Wang et al. 2009)

public/SAXS;jsessionid=7596149C96403A3E9073321879670359). The backbone RMSD of the average structure of the top 10% of the lowest energy structures for the overall structure, excluding the flexible loops from the rigid-body SA calculation, is about 3.3 Å, compared with the X-ray crystal structure (Fig. 16.6, top) (Wang et al. 2009). The ensemble of the structures also satisfies the experimental SAXS and RDC data (see the following sections). For a more detailed assessment of the structures, readers are referred to the original report (Wang et al. 2009).

The initial global structure, shown in Fig. 16.6, led to identification of close contacts, including hydrogen bonds, in the junction and loop regions (Wang et al. 2009). For example, this initial global structure puts the two loops in H1 and H2 facing each other in space (Fig. 16.5) and led to assignments of a number of sequential GC pairs, G38–C60 and G37–C61, because they immediately follow

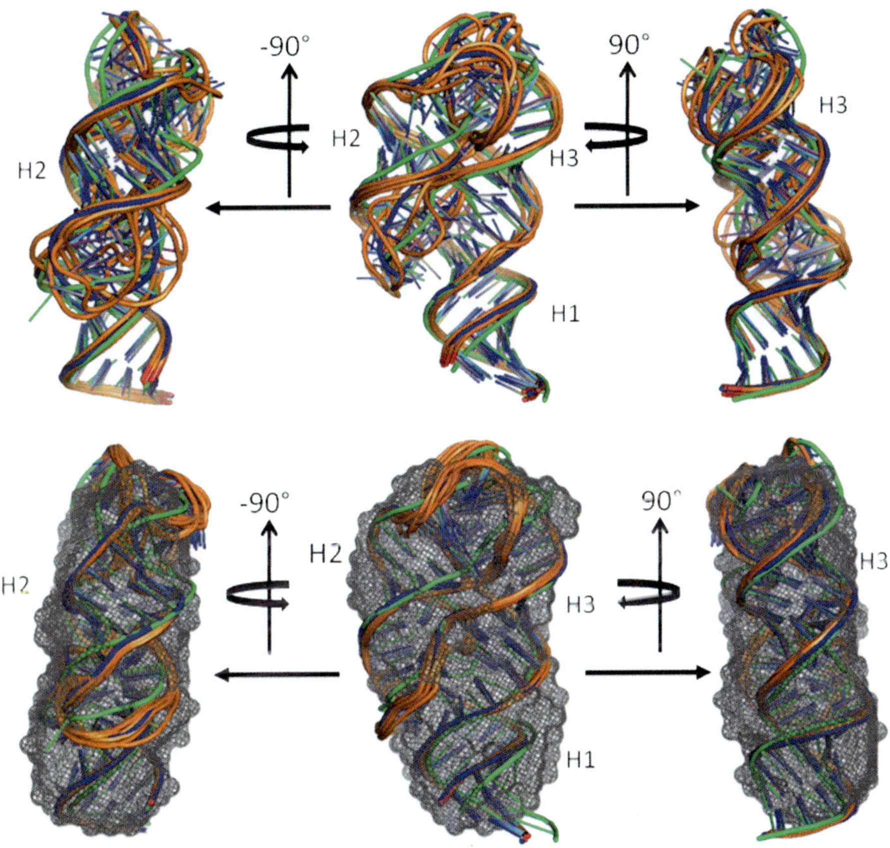

Fig. 16.6 *Top*: three views of the initial fold ensemble (*orange*) of the riboA structure that was calculated using the rigid-body SA refinement protocol and the Xplor-NIH. In the SA refinement, no tertiary interaction restraints were used (Wang et al. 2009). The average of the ensemble is shown in *blue*, and the X-ray crystal structure (1Y26) (Serganov et al. 2004) is shown in *green*. *Bottom*: the ensemble of the riboA structures that were calculated with experimental junction interaction restraints (details in text) that were readily assigned with the aid of the initial global fold (*top structures*). The average of the ensemble is shown in *blue* and the crystal structure in *green* (Wang et al. 2009). The *gray mesh* represents the SAXS-derived envelope of riboA

the G59–C67 pair in H3 in the imino NOE walk path. Wang et al. (2009) describe other close contacts in detail. These long-range distance restraints were then applied to further restrain the structure in the rigid-body SA calculation and improve structure accuracy. A comparison of the ensemble of the top 10% of the lowest energy structures and the X-ray crystal structure shows that the backbone RMSDs of the ensemble of the calculated global structures relative to the X-ray crystal structure are 3.0 ± 0.3 Å for the whole molecule and 2.5 ± 0.2 Å for the three orientation-and-phase-restrained duplexes (Fig. 16.6, bottom).

The agreement between the G2G structure and the experimental RDC and SAXS data was also examined. The correlation coefficient between the back-calculated RDCs, based on the ensemble of the top 10% of the lowest energy structures from the rigid-body calculation and the experimental data, is about 0.83 (Fig. 16.7, bottom). The correlation coefficient between the back-calculated RDCs, based on the regularized average structure of the ensemble and the experimental RDCs, is about 0.95 (Fig. 16.7). For comparison, the correlation coefficient between the RDCs calculated based on the X-ray crystal structure and the experimental data is about 0.77 (Fig. 16.7). The relatively low correlation coefficient between the experimental RDCs and the back-calculated ones from the X-ray crystal structure may be in part due to differences between the structures in solution and in crystalline states, as well as the sequence difference in duplex H1 that might result in direct or indirect changes in the structure nearby, or even in the entire structure (Wang et al. 2009). The quality of the G2G structure is also evaluated by the comparison of the back-calculated SAXS curves with the experimental ones, and the RMSD between the two, which is about 0.29 ± 0.04 (Fig. 16.8). The pair distance distribution function (PDDF) comparison is shown in Fig. 16.7.

It is possible to estimate the accuracy of the G2G structure using an empirical formula:

$$\text{RMSD} = \left[\alpha^2 P^{\text{duplex}} + \beta^2\left(1 - -P^{\text{duplex}}\right)\right]^{1/2}, \quad (16.9)$$

where α is the possible RMSD between the "true" and the database-derived duplex structures in the context of the structure, β is the possible RMSD between the "true" and the G2G structures of nonduplex regions, such as long linkers and underdetermined loops, and P^{duplex} is the %age of duplex residues in the RNA (Wang et al. 2009). For A-form-like duplexes, α of the individual duplex is well below 2.0 Å, based on RMSDs from the database (Wang et al. 2009). The value of β can vary significantly, depending on the length of the nonduplex regions, such as linkers and loops. In the case of riboA, duplexes make up more than 60% of the total residues and the linkers between H1 and H2 and H2 and H3 are relatively short; the overall RMSD between the G2G and the "true" structure is estimated to be about 3.3 Å or better, assuming α and β are about 2.5 and 4.0 Å for duplexes and the long linker/loops, respectively (Wang et al. 2009).

With knowledge of this global structure, the approximate position of the adenine ligand can also be determined (Wang et al. 2009). Moreover, the residues that are involved in the intricate interaction networks in this molecule, including the three-way helical junction, are all brought together in this global structure, and their distance restraints can also be extracted from NOE spectra, given this approximate G2G structure.

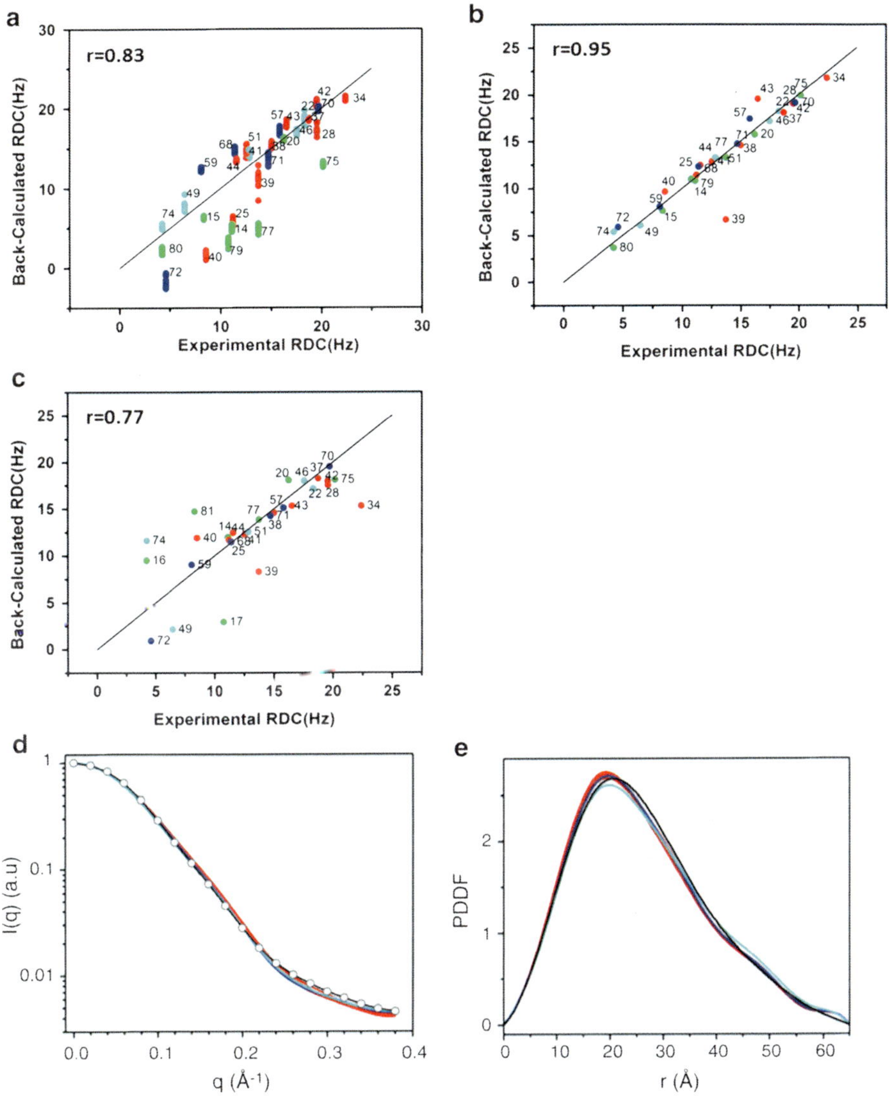

Fig. 16.7 (**a**) The correlation plots between the imino RDCs that were back-calculated from the ensemble of the top 10% lowest energy structures (Fig. 16.6, *bottom*) and the experimental RDCs. (**b**) The correlation plot between the imino RDCs that were back-calculated from the regularized average structure of the ensemble of the top 10% and the experimental RDCs. (**c**) The correlation plot between the imino RDCs that were back-calculated from the X-ray crystal structure (1Y26) (Serganov et al. 2004) and the experimental RDCs. (**d**) The comparison of experimental (*circle*), back-calculated SAXS curves based on the ensemble (*red*), average (*blue*), and the X-ray crystal structure (*cyan*); the RMSD between the first and the third is about 0.29 ± 0.04. (**e**) The comparison of experimental PDDF (*black*) and those for the ensemble (*red*), the average (*blue*), and the X-ray crystal structure (*cyan*) (Wang et al. 2009)

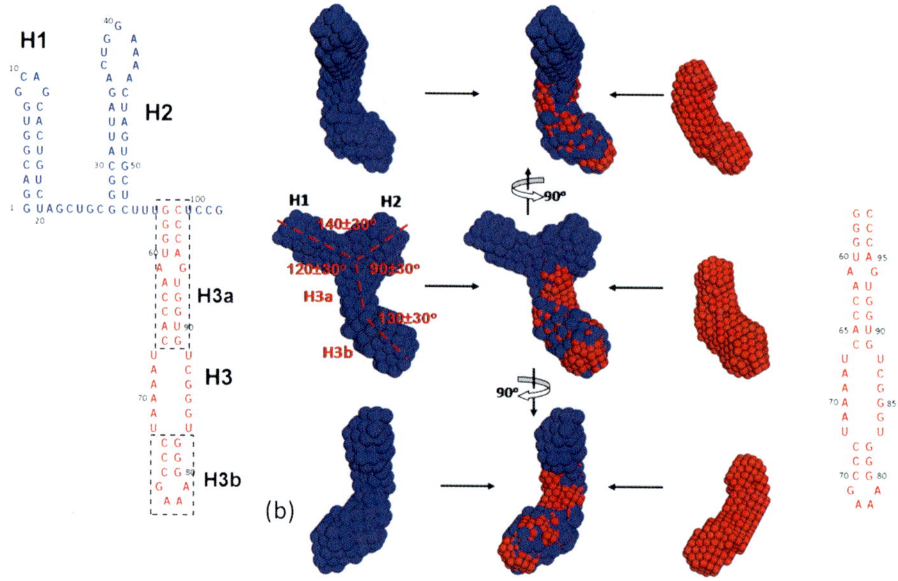

Fig. 16.8 The secondary structures of the TCV RBSE, subconstruct H3, and overlays of SAXS derived molecular envelopes of RBSE (*blue*) with H3 (*red*). The estimated angles among duplexes H1, H2, and H3a are depicted on the RBSE SAXS envelope (Zuo et al. 2010)

16.3.3 Experimental Case 2: The Global Structure of the 102-nt Ribosome-Binding Structural Element of the Turnip Crinkle Virus Genomic RNA

The 3′ untranslated region (3′ UTR) of turnip crinkle virus (TCV) genomic RNA contains a cap-independent translation element (CITE), which includes a ribosome-binding structural element (RBSE) (Fig. 16.8) that is involved in recruitment of the large ribosomal subunit (Stupina et al. 2008; Yuan et al. 2009). This RNA binds to 60S ribosomal subunits with an affinity of about 400 nM and competes with *N*-acetylated phe-tRNA for the P-site of the ribosome (Stupina et al. 2008; Zuo et al. 2010). Mutations that disrupt hairpin H1 repress ribosome binding (McCormack et al. 2008). The location of hairpin H1 is equivalent to that of the amino-acceptor arm of a tRNA structure. In addition, RNA-dependent RNA polymerase (RdRp) binding to the region causes a substantial conformational switch that disrupts the H1 region and likely promotes transcription of complementary strands while suppressing translation (McCormack et al. 2008; Stupina et al. 2008). A previous computational study suggested that this RNA folds into a structure that resembles a tRNA-like shape (McCormack et al. 2008) (see Chap. 7 by Shapiro). To understand the mechanism of 3′ UTR participation in translation and replication, it is important to determine the experimental global structure that outlines the spatial arrangements of the three hairpins, H1, H2, and H3. A 3D global structure of RBSE will also address, in part, the structural basis for the accessibility of the H3 for interaction with surrounding sequences.

The solution structure determination of mid- to large-size RNA molecules with complex folds is a daunting task. Solution RNA structures of similar-sized RNA molecules with relatively simpler folding took many years of effort to determine using the conventional bottom-up approach (Lukavsky et al. 2003; D'Souza et al. 2004). We therefore applied the G2G method to determine the solution global structure of this RNA.

We first verified the base pairing scheme by the conventional NOE walk method, aided by HNN-COSY spectra and the spectra of the mutants (Zuo et al. 2010). The molecular envelope of the RBSE (Fig. 16.8) was derived from the SAXS data using the DAMMIN program (Svergun 1999) and was found to form a twisted "r" shape, with approximate angles and dimensions depicted in Fig. 16.8. We identified the location of hairpin H3 by comparing the envelope shape of the RBSE to that of hairpin H3 and a number of constructs (Zuo et al. 2010). In particular, the long arm of the RBSE envelope matches remarkably well with that of the hairpin H3 construct (Zuo et al. 2010). We also assigned the locations of H1 and H2. The hairpin loop capping H2 contains residues complementary to those in the 3′ end of the RBSE, with which they form a pseudoknot (Stupina et al. 2008). The H2 hairpin is considerably larger than a simple hairpin. This example clearly illustrates that the SAXS data alone provide unique information about the global arrangement of the stem loops that is not attainable using any other method. Therefore, mapping out the shape using SAXS constitutes a considerable shortcut to the global structures of large RNAs in solution.

The next step is to derive the atomic coordinates of the global structure, which requires determining the orientations of the duplexes in the molecules. We determined the relative orientation and phase for each duplex using the ORIENT program as described for the riboA RNA in the previous section (Wang et al. 2009). The degenerate combinations of orientations that were not consistent with the molecular envelope were filtered out in the calculation by using the angle restrictions between the duplexes with a $\pm 30°$ error range. It is noteworthy that determining the duplex orientation proved relatively easier in TCV RBSE than in the riboA RNA, although the former is a larger RNA, as revealed by a clear outline of the duplexes. This shape-aided orientation determination greatly simplified the experimental design and the interpretation of RDC data and eliminated the need for a second independent alignment tensor. The dipolar waves of the RBSE duplexes are shown in Fig. 16.9, with the average orientations and phases (Θ, Φ, ρ_0) given in the topology drawing. The standard deviations in the angles are produced from the top fits with an RDC RMSD cutoff of 1.2 Hz.

The bending angle between H3a and the segment involving H3b in hairpin H3 was determined using a construct H3m, in which the A61–G94 mismatch was mutated to a C–G Watson–Crick pair in H3a, and H3b was extended by inserting a stretch of four bps between the triple CGs and the GAAA tetraloop (Fig. 16.10). This construct provides 7–9 imino RDCs for H3a/H3b, allowing for more accurate

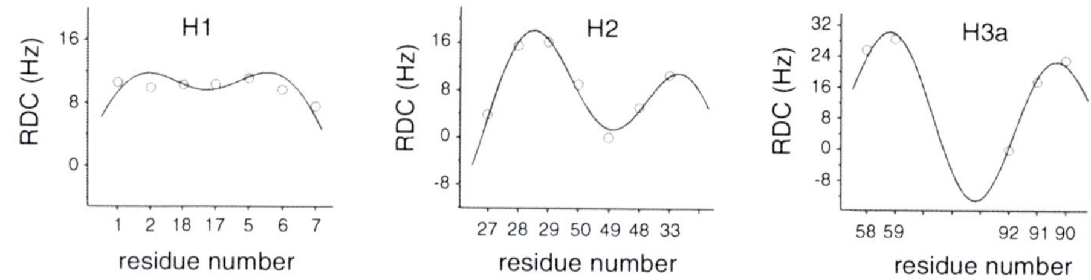

Fig. 16.9 The dipolar waves and simultaneous fits for H1, H2, and H3a in the TCV RBSE. The orientations and phases, (Θ, Φ, ρ_0), of these duplexes obtained from the best fit are displayed in Fig. 16.10 (Zuo et al. 2010)

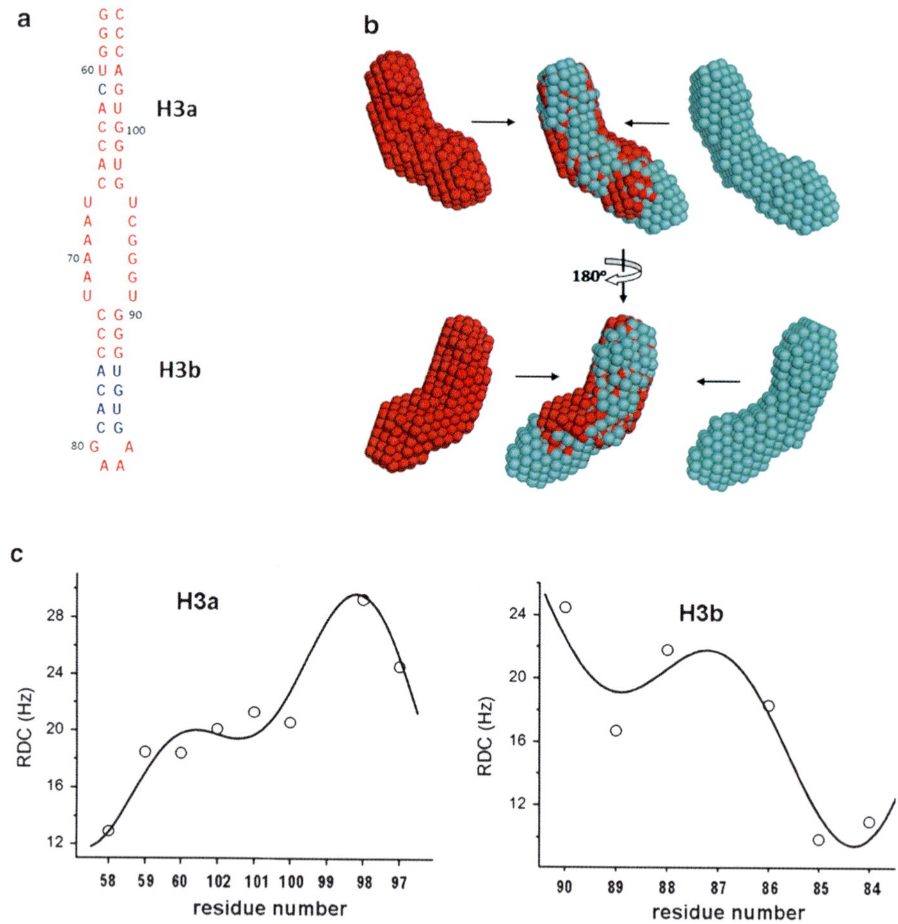

Fig. 16.10 (**a**) The secondary structure, (**b**) the SAXS molecular envelope, and (**c**) the dipolar wave fits for H3m, an extended construct of H3. In (**a**), four inserted bps and a mutation, A61–C61, compared to H3, are marked in *blue*. In (**b**), two views of the SAXS molecular envelope of H3m (*cyan*) and the overlays with that of H3 are displayed. In (**c**), the best simultaneous fit yields that the orientations and phases of H3a and H3b are (6°, 57°, 307°) and (16°, 256°, 180°), respectively, and an angle between H3a and H3b is 158° (Zuo et al. 2010)

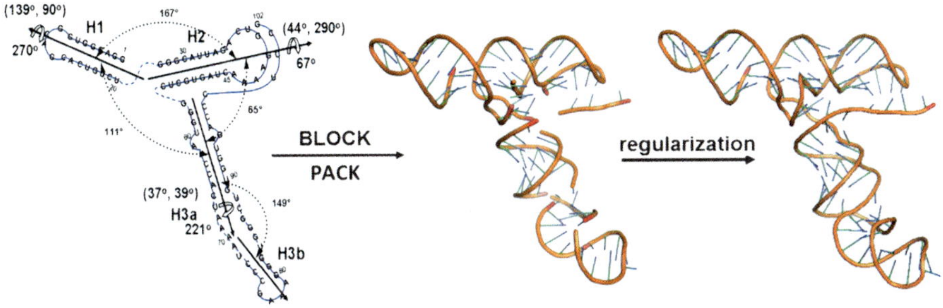

Fig. 16.11 A 2D topology drawing (*left*) of RBSE, the initial structure (*middle*) generated with the G2G toolkit, and the structure after regularization that fixes the bond breaks (*right*) using Xplor-NIH. The orientations and phases, (Θ, Φ, ρ_0), of H1, H2, and H3a, obtained from the best simultaneous fit, are given in the figure (*left*). The linker residues are represented with broken lines in the topology drawing (*left*), and residue numbers are drawn on the regularized structure (*right*) (Zuo et al. 2010)

orientation and phase determination. The single mutation in H3a and the insertion of a stretch of four bps after the triple CGs in H3b had little impact on the original angle between H3a and H3b, as seen in the low-resolution envelope (Fig. 16.10). The angle between H3a and H3b in the mutated and extended version of H3 is similar (140° ± 30°) to that in the intact RBSE or in the hairpin H3 construct, judging by the shapes. The top simultaneous RDC fits with an RMSD cutoff of 1.0 Hz produced an average angle between H3a and H3b in this RNA of 159° ± 2°; this angle was considered the approximate angle between H3a and H3b in H3m and was used to generate a starting structure.

The global shape and the orientations and phases of the duplexes led to elucidation of the topological arrangements of the hairpins and the initial 3D structure (Fig. 16.11). In addition, the pseudoknot formed between the residues in the terminal (hairpin) loop of H2 and those at the 3′ end of the RNA were restrained by Watson–Crick pairings and loosely restrained A-form duplex torsion angles. The linkers between the hairpins were set free without any restraint during the calculation; their possible structures were only indirectly restrained by the orientations, phases, and positions of duplexes and directly by the covalent linkages between the duplexes and linkers. This initial structure was regularized and was subjected to the hybrid rigid-body SA refinement (Zuo et al. 2010). The ensemble of the refined global structure of the RBSE is shown in Fig. 16.12a. The "goodness" of the global structure in terms of the global orientations of the duplexes and the overall shape is simultaneously benchmarked by the correlation coefficients of RDCs before (Fig. 16.12b) and after (Fig. 16.12c) the SA refinement and by the SAXS profiles (Fig. 16.12d) and the PDDF (Fig. 16.12e) curves. These correlation coefficients before and after the nonrigid-body SA refinement remain similar, suggesting the orientations of the duplexes are consistent with the SAXS data that restrain the overall shape of the molecule and indirectly restrain the duplex orientations. The comparison of the back-calculated SAXS curves based on the refined top 10% of

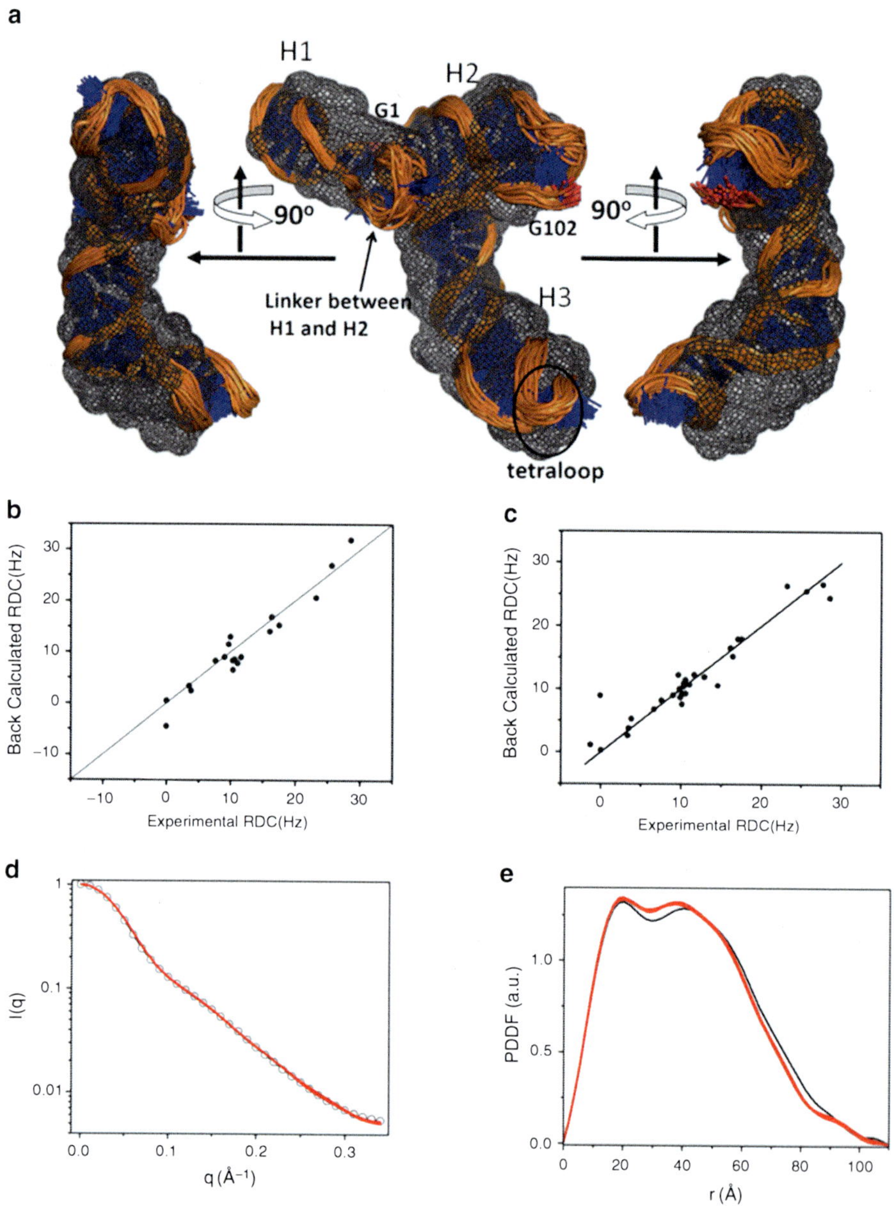

Fig. 16.12 The ensemble of global structures of the RBSE determined using the G2G "top-down" method and SAXS and PDDF curves comparison (Zuo et al. 2010). (**a**) The front (*middle*) and side (*left* and *right*) views of the RBSE structural ensemble (top 50% lowest energy) overlaid with the molecular envelope in *gray mesh*. (**b**) The correlation plot of the back-calculated RDCs based on the starting structure (Fig. 16.11, *right*), where the orientation of the three duplexes was determined using the RDC–structural periodicity correlation. Only RDCs in the duplex regions (the same experimental RDC data as shown in Fig. 16.9) were used in the correlation coefficient

the structures with the experimental SAXS data is displayed in Fig. 16.12d, and the RMSD between the two is about 0.20 ± 0.01. The comparison of PDDF curves of the corresponding SAXS profiles is shown in Fig. 16.12e.

As we have discussed in the introduction, due to lack of propensity correlation between an RNA sequence and its folded tertiary structure, it is not currently realistic to predict 3D RNA structures solely based on sequence, with the exception of simple RNA hairpins and duplexes. What we have demonstrated in this chapter is a novel way to delineate global structures of RNAs using global shape and orientation restraints. This global structure restrains the relative orientations and positions of structural elements within the structure and suggests possible tertiary contacts. In the case of the TCV RBSE RNA, the global structure alone is very intriguing. The TDV RBSE RNA consists of several structural moieties, shown in Fig. 16.13a. The global organization of these structural moieties shares a striking resemblance to that of a canonical tRNA (Fig. 16.13b), even though they differ in overall shape and sequence.

In summary, the G2G global structure of the TCV RBSE is consistent with global measurements in terms of overall shape, with the duplexes in their proper global orientations, and phases and positions that are consistent with the global measurements in solution. The accuracy of the backbone structure is estimated, using (16.9), as ~3.5 Å, comparable to that of the riboA structure (Wang et al. 2009).

16.4 Conclusions

The G2G method employs experimental measurements of global shape and orientations of helices to determine global structures of RNAs in solution. This experimental method differs in philosophy from previously reported approaches that are primarily computational (Martinez et al. 2008; Parisien and Major 2008; Jonikas et al. 2009b). While this method may open the possibility for determining high-resolution structures of relatively large RNAs in solution using NMR spectroscopy, it also poses an interesting challenge to computational RNA biochemists: Is it possible to compute atomic-resolution structures using motif libraries, given

Fig. 16.12 (continued) calculation. The correlation coefficient is approximately 0.97. (**c**) The correlation plot of the back-calculated RDCs based on the top 10% lowest G2G structures (see the text) vs. the experimental ones. The correlation is near unit (Zuo et al. 2010). (**d**) The comparison of experimental (*circle*) and back-calculated SAXS curves (*red*) based on the top 10% ensemble. The RMSD between experimental data and the back-calculated curves is 0.20 ± 0.01. RMSD is calculated based on the logarithm of the normalized SAXS intensities. (**e**) The comparison of PDDFs of the corresponding experimental SAXS (*black*) and back-calculated SAXS curves (*red*) in (**c**) (Zuo et al. 2010)

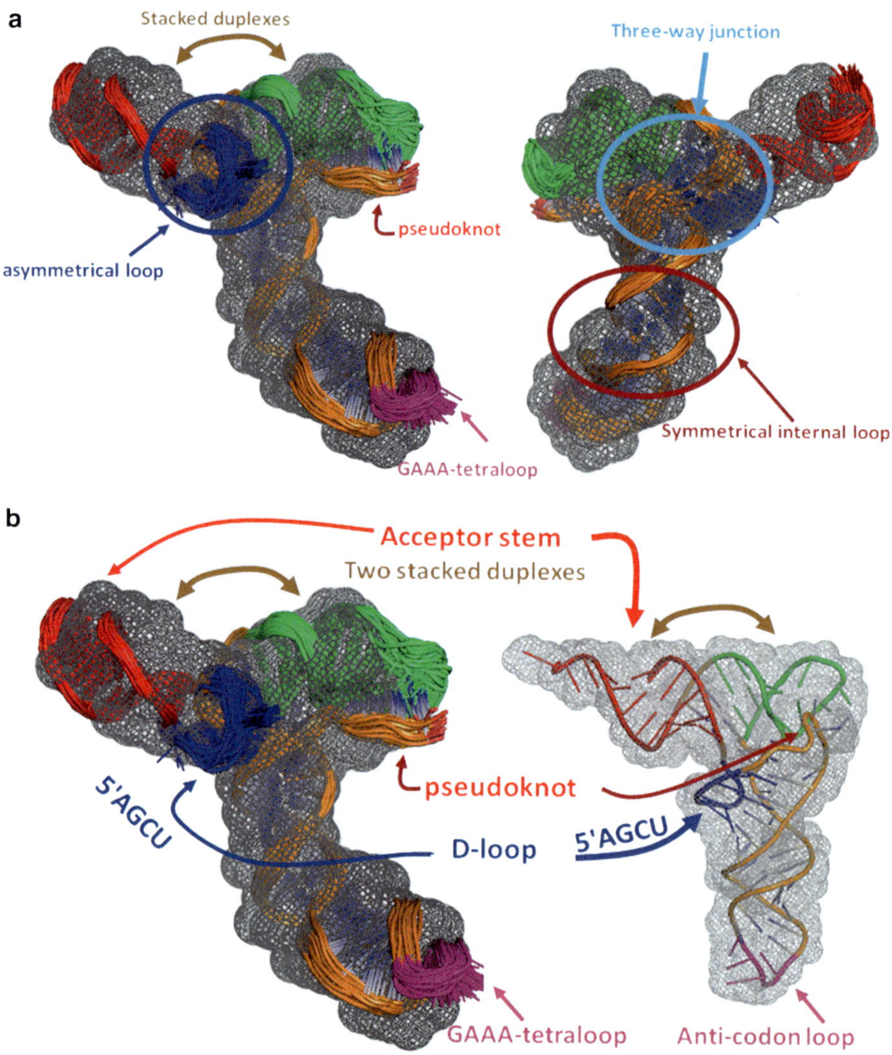

Fig. 16.13 The dissect of the structural elements in the TCV RBSE and the structural parallelism between the TCV RBSE and a $tRNA^{phe}$. (**a**) The TCV RBSE consists of a number of substructural elements and tertiary interactions, including stacked duplexes, a pseudoknot, an asymmetrical loop, a symmetrical internal loop, a three-way junction, and a GAAA-tetraloop; (**b**) the illustration of equivalency in arrangement of structural elements in the TCV RBSE and $tRNA^{phe}$. This comparison clearly demonstrates the parallelism between the two RNAs, even though two RNAs share little resemblance in sequence and overall molecular shape

the G2G global structures (Leontis and Westhof 2001, 2002, 2003; Leontis et al. 2006; Jonikas et al. 2009a)?

All calculations, programs, and scripts used for this chapter can be downloaded from the author's Web page: http://ccr.cancer.gov/staff/links.asp?profileid=5546.

Acknowledgment The research presented in this chapter is supported by the Intramural Research Program of the National Institutes of Health (NIH), National Cancer Institute, Center for Cancer Research to Y-X.W.

References

Blackburn EH (1992) Telomerases. Annu Rev Biochem 61:113–129

Bothner AA (ed) (1996) Magnetic field induced alignment of molecules. Encyclopedia of nuclear magnetic resonance. Wiley, Chichester

Chacon P, Moran F, Diaz JF, Pantos E, Andreu JM (1998) Low-resolution structures of proteins in solution retrieved from X-ray scattering with a genetic algorithm. Biophys J 74(6):2760–2775

Cruz JA, Westhof E (2009) The dynamic landscapes of RNA architecture. Cell 136(4):604–609

D'Souza V, Dey A, Habib D, Summers MF (2004) NMR structure of the 101-nucleotide core encapsidation signal of the Moloney murine leukemia virus. J Mol Biol 337(2):427–442

Debye P (1915) Zerstreuung von Roentgenstrahlen. Ann Phys 46:809–823

Dingley AJ, Masse JE, Feigon J, Grzesiek S (2000) Characterization of the hydrogen bond network in guanosine quartets by internucleotide 3hJ(NC)′ and 2hJ(NN) scalar couplings. J Biomol NMR 16(4):279–289

Ernst RR, Bodenhausen G, Wokaun A (1987) Principles of nuclear magnetic resonance in one and two dimensions. Oxford University Press, Oxford

Fire A, Xu S, Montgomery MK, Kostas SA, Driver SE, Mello CC (1998) Potent and specific genetic interference by double-stranded RNA in *Caenorhabditis elegans*. Nature 391 (6669):806–811

Fowler CA, Tian F, Al-Hashimi HM, Prestegard JH (2000) Rapid determination of protein folds using residual dipolar couplings. J Mol Biol 304(3):447–460

Gayathri C, Bothnerby AA, Vanzijl PCM, Maclean C (1982) Dipolar magnetic-field effects in NMR-spectra of liquids. Chem Phys Lett 87(2):192–196

Grishaev A, Wu J, Trewhella J, Bax A (2005) Refinement of multidomain protein structures by combination of solution small-angle X-ray scattering and NMR data. J Am Chem Soc 127 (47):16621–16628

Hus JC, Marion D, Blackledge M (2001) Determination of protein backbone structure using only residual dipolar couplings. J Am Chem Soc 123(7):1541–1542

Jonikas MA, Radmer RJ, Altman RB (2009a) Knowledge-based instantiation of full atomic detail into coarse-grain RNA 3D structural models. Bioinformatics 25(24):3259–3266

Jonikas MA, Radmer RJ, Laederach A, Das R, Pearlman S, Herschlag D, Altman RB (2009b) Coarse-grained modeling of large RNA molecules with knowledge-based potentials and structural filters. RNA 15(2):189–199

Koch MH, Vachette P, Svergun DI (2003) Small-angle scattering: a view on the properties, structures and structural changes of biological macromolecules in solution. Q Rev Biophys 36(2):147–227

Kruger K, Grabowski PJ, Zaug AJ, Sands J, Gottschling DE, Cech TR (1982) Self-splicing RNA: autoexcision and autocyclization of the ribosomal RNA intervening sequence of Tetrahymena. Cell 31(1):147–157

Kung HC, Wang KY, Goljer I, Bolton PH (1995) Magnetic alignment of duplex and quadruplex DNAs. J Magn Reson B 109(3):323–325

Latham MP, Brown DJ, McCallum SA, Pardi A (2005) NMR methods for studying the structure and dynamics of RNA. ChemBioChem 6(9):1492–1505

Leontis NB, Westhof E (2001) Geometric nomenclature and classification of RNA base pairs. RNA 7(4):499–512

Leontis NB, Westhof E (2002) The annotation of RNA motifs. Comp Funct Genomics 3 (6):518–524

Leontis NB, Westhof E (2003) Analysis of RNA motifs. Curr Opin Struct Biol 13(3):300–308
Leontis NB, Lescoute A, Westhof E (2006) The building blocks and motifs of RNA architecture. Curr Opin Struct Biol 16(3):279–287
Lescoute A, Westhof E (2006) The interaction networks of structured RNAs. Nucleic Acids Res 34 (22):6587–6604
Losonczi JA, Andrec M, Fischer MW, Prestegard JH (1999) Order matrix analysis of residual dipolar couplings using singular value decomposition. J Magn Reson 138(2):334–342
Lukavsky PJ, Kim I, Otto GA, Puglisi JD (2003) Structure of HCV IRES domain II determined by NMR. Nat Struct Biol 10(12):1033–1038
Mandal M, Breaker RR (2004) Adenine riboswitches and gene activation by disruption of a transcription terminator. Nat Struct Mol Biol 11(1):29–35
Martinez HM, Maizel JV Jr, Shapiro BA (2008) RNA2D3D: a program for generating, viewing, and comparing 3-dimensional models of RNA. J Biomol Struct Dyn 25(6):669–683
McCormack JC, Yuan X, Yingling YG, Kasprzak W, Zamora RE, Shapiro BA, Simon AE (2008) Structural domains within the 3′ untranslated region of Turnip crinkle virus. J Virol 82 (17):8706–8720
Mesleh MF, Veglia G, DeSilva TM, Marassi FM, Opella SJ (2002) Dipolar waves as NMR maps of protein structure. J Am Chem Soc 124(16):4206–4207
Moore PB, Steitz TA (2002) The involvement of RNA in ribosome function. Nature 418 (6894):229–235
Ottiger M, Delaglio F, Bax A (1998) Measurement of J and dipolar couplings from simplified two-dimensional NMR spectra. J Magn Reson B 131(2):373–378
Parisien M, Major F (2008) The MC-Fold and MC-Sym pipeline infers RNA structure from sequence data. Nature 452(7183):51–55
Schwieters CD, Kuszewski JJ, Tjandra N, Clore GM (2003) The Xplor-NIH NMR molecular structure determination package. J Magn Reson B 160(1):65–73
Schwieters CD, Kuszewski JJ, Clore GM (2006) Using Xplor-NIH for NMR molecular structure determination. Prog Nucl Magn Reson Spectrosc 48(1):47–62
Serganov A, Yuan YR, Pikovskaya O, Polonskaia A, Malinina L, Phan AT, Hobartner C, Micura R, Breaker RR, Patel DJ (2004) Structural basis for discriminative regulation of gene expression by adenine- and guanine-sensing mRNAs. Chem Biol 11(12):1729–1741
Stuhrmann HB (1970) Interpretation of small-angle scattering functions of dilute solutions and gases. A representation of the structures related to a one-particle-scattering function. Acta Crystallogr A26:297–306
Stupina VA, Meskauskas A, McCormack JC, Yingling YG, Shapiro BA, Dinman JD, Simon AE (2008) The 3′ proximal translational enhancer of Turnip crinkle virus binds to 60S ribosomal subunits. RNA 14(11):2379–2393
Svergun DI (1999) Restoring low resolution structure of biological macromolecules from solution scattering using simulated annealing. Biophys J 77(6):2879
Svergun DI, Stuhrmann HB (1991) New developments in direct shape determination from small-angle scattering. 1. Theory and model-calculations. Acta Cryst A 47:736–744
Tolman JR, Flanagan JM, Kennedy MA, Prestegard JH (1995) Nuclear magnetic dipole interactions in field-oriented proteins – information for structure determination in solution. Proc Natl Acad Sci USA 92(20):9279–9283
Tucker BJ, Breaker RR (2005) Riboswitches as versatile gene control elements. Curr Opin Struct Biol 15(3):342–348
Walsh JD, Wang YX (2005) Periodicity, planarity, residual dipolar coupling, and structures. J Magn Reson B 174(1):152–162
Walsh JD, Cabello-Villegas J, Wang YX (2004) Periodicity in residual dipolar couplings and nucleic acid structures. J Am Chem Soc 126(7):1938–1939
Walther D, Cohen FE, Doniach S (2000) Reconstruction of low-resolution three-dimensional density maps from one-dimensional small-angle X-ray solution scattering data for biomolecules. J Appl Crystallogr 33:350–363

Wang J, Walsh JD, Kuszewski J, Wang YX (2007) Periodicity, planarity, and pixel (3P): a program using the intrinsic residual dipolar coupling periodicity-to-peptide plane correlation and phi/psi angles to derive protein backbone structures. J Magn Reson 189(1):90–103

Wang J, Zuo X, Yu P, Xu H, Starich MR, Tiede DM, Shapiro BA, Schwieters CD, Wang YX (2009) A method for helical RNA global structure determination in solution using small-angle X-ray scattering and NMR measurements. J Mol Biol 393(3):717–734

Yuan X, Shi K, Meskauskas A, Simon AE (2009) The 3′ end of Turnip crinkle virus contains a highly interactive structure including a translational enhancer that is disrupted by binding to the RNA-dependent RNA polymerase. RNA 15(10):1849–1864

Zhang Q, Sun X, Watt ED, Al-Hashimi HM (2006) Resolving the motional modes that code for RNA adaptation. Science 311(5761):653–656

Zhang Q, Stelzer AC, Fisher CK, Al-Hashimi HM (2007) Visualizing spatially correlated dynamics that directs RNA conformational transitions. Nature 450(7173):1263–1267

Zuo XB, Wang JB, Foster TR, Schwieters CD, Tiede DM, Butcher SE, Wang YX (2008) Global molecular structure and interfaces: refining an RNA: RNA complex structure using solution X-ray scattering data. J Am Chem Soc 130(11):3292–3293

Zuo X, Wang J, Yu P, Eyler D, Xu H, Starich MR, Tiede DM, Simon AE, Kasprzak W, Schwieters CD, Shapiro BA, Wang YX (2010) Solution structure of the cap-independent translational enhancer and ribosome-binding element in the 3′ UTR of turnip crinkle virus. Proc Natl Acad Sci USA 107(4):1385–1390

Chapter 17
RNA Structure Determination by Structural Probing and Mass Spectrometry: MS3D

A.E. Hawkins and D. Fabris

Abstract Recent advances of detection strategies based on mass spectrometry (MS) have reawakened the interest in chemical methods for RNA structural elucidation by enabling experimental protocols that minimize their typical pitfalls. At the same time, the development of ever more sophisticated modeling techniques has helped close the resolution gap by providing atomic-level details that were previously beyond reach. Here, we describe the integration of MS-assisted structural probing with appropriate computational techniques, which has been termed MS3D, and illustrate its application to the elucidation of RNA substrates of biological significance. We address typical concerns faced by probing applications and possible solutions supported by the MS platform. We describe strategies for translating sparse spatial constraints afforded by footprinting and cross-linking reagents into testable all-atom structures. We also discuss future advances that would take further advantage of the synergy between experimental and computational approaches to increase the accuracy of chemical methods and to expand their scope to progressively larger and more complex targets.

17.1 Introduction

A series of events in the last decade has keenly heightened the interest in the development of alternative experimental approaches for the investigation of the structure–function relationships in RNA and its functional assemblies. The completion of the Human Genome Project (Lander et al. 2001; Venter et al. 2001;

A.E. Hawkins
University of Maryland Baltimore County, Catonsville, MD, USA

D. Fabris (✉)
The RNA Institute, University at Albany, Life Sciences Research Building, Room 1109, 1400 Washington Avenue, Albany, NY 12222, USA
e-mail: fabris@albany.edu

N. Leontis and E. Westhof (eds.), *RNA 3D Structure Analysis and Prediction*, Nucleic Acids and Molecular Biology 27, DOI 10.1007/978-3-642-25740-7_17,

Consortium 2004) has prompted the observation that, although more than 70% of the entire genome is transcribed into RNA, less than 1.5% may be coding for actual proteins, according to conservative estimates (Birney et al. 2007; Pheasant and Mattick 2007). Extensive reevaluation of sequences that were once dismissed as "selfish" or "junk" (Doolittle and Sapienza 1980; Orgel and Crick 1980) has led to the discovery of numerous new classes of non-coding elements (ncRNA) (Mattick and Makunin 2006; Volff 2006; Zuckerkandl and Cavalli 2007), which taken together still cannot account for the entire transcribed pool (Claverie 2005). The elucidation of the mechanism of riboswitch-mediated gene regulation (Winkler et al. 2002; Nudler and Mironov 2004) has further reinforced the conclusion that sequence information alone is not sufficient to understand the functions of many ncRNAs. In fact, riboswitches regulate the expression of downstream genes by binding specific metabolites that induce conformational changes in the mRNA of which they are a part and modulate its transcription or translation (Nahvi et al. 2002; Edwards et al. 2007). For this reason, novel approaches are needed to link knowledge of RNA structure with the identity of cognate ligands, the nature of their interactions, and the effects of binding on conformational dynamics. Mass spectrometry (MS)-based approaches have the ability of providing this type of information and, therefore, are poised to play a significant role in the structure–function investigation of ncRNA (Fabris 2010). In this context, we are exploring MS technologies, collectively known as MS3D (Young et al. 2000; Yu and Fabris 2003, 2004), which aim at improving the use of chemical probing for RNA structure determination.

Assessing the susceptibility of functional groups to specific chemical reagents constitutes a very versatile strategy for obtaining insights into their solvent accessibility and, by extension, into their proximity to the substrate surface and local structural context. In the case of nucleic acid structures, reagents targeting the H-bonding edge of the nucleobases can reveal their participation in base-pairing interactions (Peattie and Gilbert 1980; Ehresmann et al. 1987; Brunel and Romby 2000). Monitoring the ability of bifunctional reagents to establish conjugates that bridge across susceptible groups affords the possibility of determining their mutual distance in the substrate fold, which can reveal the position of long-range tertiary interactions between domains brought into contact by the 3D fold (Kenny et al. 1979; Stiege et al. 1983; Yu et al. 2008a). For these reasons, chemical probing represents a very appealing complement to established techniques for structural determination, which can be successfully utilized to study species of limited availability, or present in complex sample mixtures, such as those encountered during in vivo investigations. Furthermore, the concerted application of advanced strategies for molecular modeling has demonstrated the possibility of compensating for the intrinsically low resolution of the sparse spatial constraints afforded by structural probing. Indeed, the utilization of experimental data to guide model building operations can lead to all-atom models that constitute accurate representations of the substrate structure in solution and enable the formulation of testable hypotheses that otherwise would be unwarranted (Yu et al. 2005, 2008a).

Owing to their unique mass signatures, any adduct produced by chemical probing can be readily characterized by mass mapping and sequencing strategies (Yu and Fabris 2003, 2004; Kellersberger et al. 2004). Unlike methods that rely on polyacrylamide gel electrophoresis (PAGE) to identify probed nucleotides, MS technologies do not require termination of primer elongation or probe-specific chemistry to induce strand cleavage at the modification site. This favorable feature encourages the addition of new effective reagents to the available toolkit, which otherwise would not be considered viable (Zhang et al. 2006). Furthermore, the applicability of this analytical platform to virtually any type of biopolymer enables full characterization of protein–RNA as well as RNA–RNA cross-links, thus supporting the elucidation of ribonucleoprotein assemblies (Jensen et al. 1996; Urlaub et al. 1997; Golden et al. 1999). Finally, the possibility of implementing nuclease (and protease) digestion after the probing reaction is complete, which is aimed at obtaining hydrolytic products that are more readily amenable to mass mapping and sequencing, extends the accessible size of potential targets far beyond the practical limits allowed, for example, by NMR. In this chapter, we describe methods for MS-aided structural probing and provide suggestions on effective experimental design. We discuss critical issues that need to be considered for successful probing and provide examples of MS3D structure elucidation.

17.2 Selecting the Proper Approach: Mono- vs. Bifunctional Probing

Secondary and tertiary structures of nucleic acids are largely defined by base-pairing interactions between complementary regions that may be located anywhere in their sequence. Essential elements of secondary structure, such as stem–loop hairpins, are stabilized by the pairing of strand segments that are adjacent in the sequence (Fig. 17.1). The annealing of distal sections establishes long-range tertiary interactions that determine the RNA global fold. Base-pairing between complementary segments of distinct strands is a very common way for creating intermolecular contacts that define the quaternary structure of multisubunit assemblies. Therefore, identifying the positions of nucleotides involved in base-pairing interactions constitutes a very fundamental step toward elucidating RNA structure. This objective can be achieved by monofunctional probes that are capable of assessing directly or indirectly the pairing status of susceptible nucleotides.

17.2.1 Footprinting Information from Monofunctional Reagents

A direct reading of pairing status is provided by solvent-accessibility reagents that only modify the H-bonding edge of a certain nucleotide when that edge is not

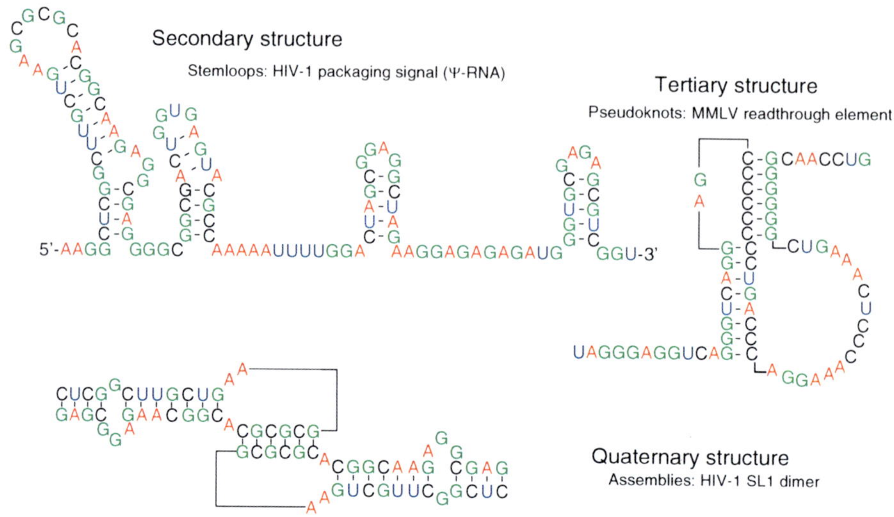

Fig. 17.1 Levels of RNA structure

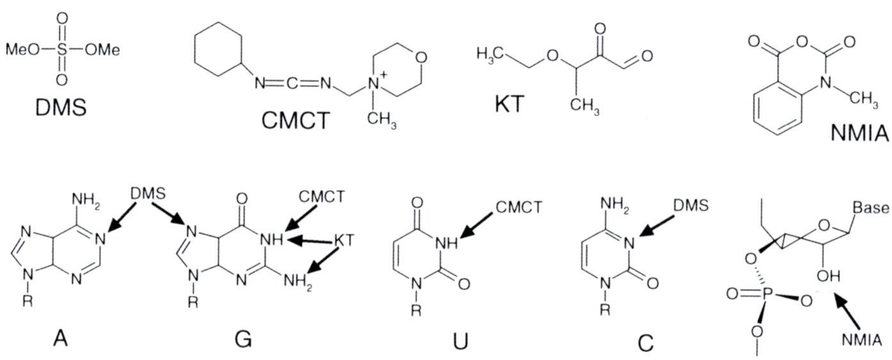

Fig. 17.2 Selected monofunctional reagents for footprinting applications: dimethyl sulfate (DMS), 1-cyclohexyl-3-(2-morpholinoethyl) carbodiimide metho-*p*-toluenesulfonate (CMCT), kethoxal (KT), and *N*-methylisatoic anhydride (NMIA). The positions modified by each probe are indicated

employed in a base-pairing interaction. In the case of the Watson–Crick edge, the entire nucleotide spectrum can be readily covered by the combination of dimethyl sulfate (DMS), kethoxal (KT), and 1-cyclohexyl-3-(2-morpholinoethyl) carbodiimide metho-*p*-toluenesulfonate (CMCT) (Fig. 17.2). DMS induces irreversible methylation of the exposed Watson–Crick edges of adenine (at the N1 position) and, to a lesser extent, of cytosine (at the N3 position) (Lawley 1957; Brookes and Lawley 1961; Lawley and Brookes 1963), which results in the formal addition (incremental mass, ΔM) of 14.01 Da to the initial mass of the target. KT modifies the Watson–Crick edge of unpaired guanine at the N1 and N2 positions (Shapiro and Hachmann 1966; Shapiro et al. 1969), forming a 1,2-diol with a ΔM of 130.06 Da. The initial adduct is further stabilized by the formation of a boronate ester (ΔM = 138.05 Da) in environments containing boric acid buffer (Akinsiku et al. 2005). CMCT is active toward N3-uridine and N1-guanine with formation of a 251.21 Da adduct that tends to incorporate an additional water molecule (ΔM = 269.21 Da) upon protracted reaction (Metz and Brown 1969). We have successfully employed these reagents in individual (Yu and Fabris 2003) and multiplexed applications (Yu and Fabris 2004) to take advantage of their different mass shifts. These types of probes have shown to be exquisitely sensitive to subtle structural motifs, as exemplified by the ability of DMS to modify a critical adenine involved in an unusual triple-base platform in the stem of HIV-1 stem–loop 3 (SL3) (Yu and Fabris 2003).

Indirect base-pairing information can be inferred by monitoring the reactivity of DMS with the N7 position on the Hoogsteen edge of guanine. Reaction at this position is inhibited by the tightly stacked arrangement assumed by contiguous pairs in stable double-stranded regions. Nucleotides present in helical structures may still be susceptible to methylation when they occupy terminal positions that leave one side of the aromatic system exposed to the solvent or when they are involved in "wobble" GU pairs that distort the regular stacking pattern of the double helix (Yu and Fabris 2003). As observed for other footprinting probes, the possibility of producing modification of nucleotides present in helical structures is increased by dynamic effects, such as the transient opening and closing of the H-bonds between nucleotides known as base-pair breathing. In similar indirect fashion, participation in base-pairing interactions can be detected also by monitoring the susceptibility of the corresponding ribose 2′-hydroxyl to acylation by *N*-methylisatoic anhydride (NMIA, Fig. 17.2), which is hindered by the 3′-phosphate group in the puckering conformation imposed by the participation of the respective nucleotide to a double-helical structure. The position of modified bases is usually inferred from the ability of NMIA adducts to inhibit strand elongation by reverse transcriptase, which constitutes the basis for an approach termed selective 2′-hydroxyl acylation analyzed by primer extension (SHAPE) (Merino et al. 2005; Wilkinson et al. 2006, 2008). Automated platforms based on capillary electrophoresis are subsequently employed to differentiate products terminating at the probed positions, which enables high-throughput analysis of RNA secondary structure. Typical challenges faced by this approach are represented by relatively small targets that may not allow for the stable annealing of short primers and by the possible inhibition of primer extension by templates with particularly stable secondary structures. In these cases, the implementation of MS analysis in what has been termed selective 2′-hydroxyl acylation analyzed by mass spectrometry (SHAMS) (Fabris et al. 2009; Turner et al. 2009) allows one to dispense with the primer extension process and to directly characterize the NMIA adducts that exhibit a characteristic ΔM of 133.05 Da.

17.2.2 Spatial Contiguity from Bifunctional Probes

The use of monofunctional probes is subject to ambiguous interpretations because they provide a "negative" snapshot of the actual effects of base pairing, i.e., they only report reactivity for nucleotides that *are not* involved in base pairing. In principle, any steric effects that may reduce the accessibility of susceptible groups, including the binding of proteins and other ligands, can be expected to produce results that are indistinguishable from those produced by base pairing. Indeed, these reagents are broadly classified as footprinting probes because of their ability to define the interfaces between bound components. If presence of putative ligand(s) in solution cannot be assessed, careful controls should be performed by repeating experiments under different conditions, for example, by varying the pH, ionic strength, or temperature, so as to disrupt weak binding interactions or affect the initial substrate conformation. Even when these experiments confirm that the protection effects observed for a certain nucleotide are caused by its participation in a base pair, solvent-accessibility probes are still incapable of unambiguously revealing the identity of the base with which such nucleotide may be paired. Because footprinting data cannot indicate which nucleotides are mutually interacting, their interpretation relies heavily on structure prediction algorithms to find the best match between experimental results and possible pairing patterns that are calculated from thermodynamics principles or statistical knowledge of RNA structure (Zuker 1989; Major et al. 1991; Rivas and Eddy 1999; Mears et al. 2002; Mathews and Turner 2006). In contrast, cross-linking approaches lock pairs of susceptible nucleotides in conjugated products that reveal their spatial proximity in the native structure. Photo-activated techniques that produce zero-length cross-linking (Stiege et al. 1983; Atmadja et al. 1985; Doring et al. 1991) and bifunctional reagents that bridge across groups within cross-linking reach (Oste and Brimacombe 1979; Stiege et al. 1982; Zhang et al. 2006) enable the unambiguous identification of bases that are placed in direct contact or close mutual proximity by the 3D fold. In the context of large RNA structures, cross-linking of nucleotides that are distal in the strand sequence can identify tertiary interactions between discrete domains. Bridging of nucleotides across annealed strands determines the exact pairing register between them. This particular feature can help solve possible ambiguities arising from the ability of large sequences to form alternative pairing patterns, which may exhibit very similar reactivity profiles when treated with monofunctional probes.

In our work, we have explored bifunctional reagents that target well-defined nucleophilic groups present on substrates of both nucleic acid and protein nature. For example, *bis*(2-chloroethyl)-methylamine (nitrogen mustard, NM) (Fig. 17.3) includes two alkylating functions that react with the N7 position of guanine and, to a lesser extent, N3 of adenine in nucleic acids, as well as with the thiol of cysteine residues in proteins (Byrne et al. 1996; Zaia et al. 1996). The desired bifunctional product adds 83.07 Da to the combined masses of the conjugated species. Incompletely reacting monofunctional adducts in which the second 2-chloroethyl function

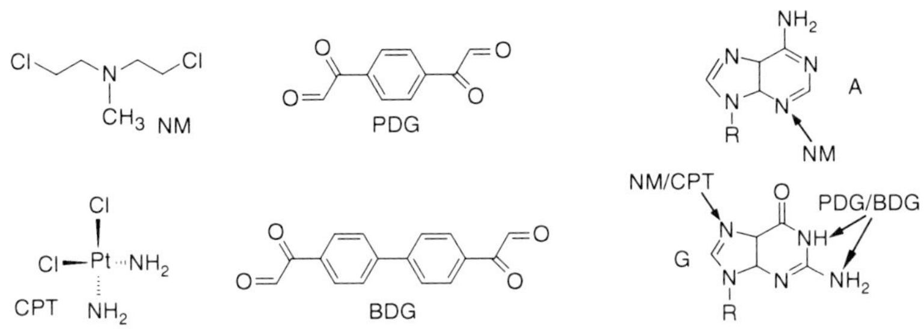

Fig. 17.3 Selected bifunctional reagents for cross-linking applications: *bis*(2-chloroethyl)-methylamine (nitrogen mustard, NM), 1,4-phenyl-diglyoxal (PDG), *cis*-diamminedichloro-platinum (II) (cisplatin, CPT), and 4,4′-biphenyl-diglyoxal (BDG). The positions modified by each probe are indicated

fails to react or undergoes hydrolysis to 2-hydroxyethyl exhibit characteristic mass shifts of 120.02 and 102.09, respectively (Zhang et al. 2006). The conformational freedom afforded by spacing arms that consist of flexible ethylene units allows for effective bridging across structures placed within 9.5 ± 1.5 Å of each other. With the objectives of reducing the allowable cross-linked range and increasing the precision of the corresponding constraint, we have investigated the properties of reagents based on more rigid aromatic scaffolds, such as 1,4-phenyl-diglyoxal (PDG) and 4,4′-biphenyl-diglyoxal (BDG) (Fig. 17.3) (Zhang et al. 2008). In analogy with the monofunctional probe KT, the 1,2-dicarbonyl functionalities can attack the N1 and N2 positions on the Watson–Crick edge of guanine and the side chain of arginine residues to produce 1,2-diol adducts with characteristic mass shifts of 190.03 and 266.06 Da for PDG and BDG bridges, respectively. With effective cross-linking spans of 6.14 ± 0.64 Å and 10.44 ± 0.80 Å, their aromatic spacers present decidedly narrower ranges that allow for more precise distance determinations than those afforded by popular reagents with aliphatic arms. The utilization of modular cross-linkers with repeating spacer units illustrates the possibility of developing a nested series of probes capable of bridging across different spans, which could be employed in multiplexed fashion by taking advantage of their unique mass signatures (Zhang et al. 2008). Finally, we have explored the utilization of *cis*-diamminedichloroplatinum (II) (cisplatin, CPT) (Fig. 17.3) as an example of chemical probe devoid of actual spacer structure (Zhang et al. 2006). CPT produces stable adducts with purines at their N7 position and, to a lesser extent, cytosine at their N3 position. The product exhibits a mass shift of 225.99 Da and a very characteristic isotopic signature associated with the presence of coordinated Pt(II). Adducts are formed between contiguous nucleotides in single- as well as double-stranded regions, with the latter taking place where base stacking is disrupted either by non-Watson–Crick pairing or helix breathing. In this case, a cross-link distance of 2.7 Å corresponds exclusively to the span between coordinated ligands because coordination bonds exhibit significantly fewer degrees of freedom than aliphatic spacers and their intrinsic dynamics are accounted for during the energy minimization process.

17.2.3 Targeted Probing

The choice of probes should be dictated by the overall goal of the investigation. In general, experimental elucidation of unknown (i.e., unsolved) targets cannot be accomplished by relying only on a single probe but requires instead the concerted application of series of mono- and bifunctional reagents to obtain complementary information. The accuracy of resulting models correlates closely with the number and diversity of the experimental constraints available for 3D modeling, which are used to establish the relative positions of nucleotides in 3D space. The level of confidence in a model increases when independent complementary data corroborate the spatial arrangement of the different components. In this direction, the application of MS-based approaches that overcome the limitations associated with traditional detection platforms is poised to greatly expand the assortment of chemicals that can be employed as structural probes (Zhang et al. 2006). The use of a single probe can still meet the challenge when initial structural information is already available from previous rounds of probing, other experimental techniques, or structure prediction algorithms. Similarly, when the goal is to test structural hypotheses, to confirm the presence of specific structural features, or to assess the dynamics of a known structure, smaller sets of probes selected for their ability to hit specific targets will readily provide the data necessary to complete the desired tasks.

17.3 The Intrinsic Hazards of Chemical Probing

The dynamic nature of substrates in solution and the fundamental principles of chemical probing, which require susceptible groups to be accessible on the substrate surface, or placed within mutual striking distance, expose these types of approaches to intrinsic hazards. Footprinting methods tend to be sensitive to transient conformational changes that may make nucleotides temporarily accessible to the probe. For known structures, the ability to determine the yield of modification at each position enables one to identify regions subjected to conformational effects and even to assess their dynamics in quantitative fashion. For the determination of unknown structures, however, this characteristic may represent a source of ambiguities brought about by the possibility that initial chemical modification may itself alter the substrate folding and make additional sites accessible, which are inaccessible in the folded structure. In the case of bifunctional cross-linkers, unfavorable matching of substrate dynamics and reaction kinetics may lead to the permanent bridging of functional groups that are placed only temporarily within the reagent span, thus artificially amplifying the detection of sparsely populated conformations. Potential probe-induced distortion and kinetic traps can lead to constraints that misrepresent the equilibrium structure in solution. The possibility of producing these types of artifacts cannot be completely eliminated, but their incidence can be minimized, or at least recognized, through careful experimental design.

17.3.1 Fine Tuning the Conditions of Probe Application

The identification of regions involved in prominent dynamics is a very important facet of structure characterization and a critical stage in the assessment of possible probe-induced conformational effects. Environmental factors, including pH, ionic strength, temperature, and ligand concentration, influence substrate dynamics and, consequently, the outcome of probing experiments. MS detection, unlike PAGE-based analysis, allows one to accurately determine the number of modifications per substrate molecule. This capability can be exploited to monitor the overall number of modifications as environmental factors are systematically varied (Kellersberger et al. 2004). Sudden increases of the observed number of modifications can reveal the threshold at which transitions may be triggered by the chemical modification, thus defining the limits of probe application beyond which structural integrity is compromised. Additionally, the MS analytical platform enables the implementation of alternative strategies based on nested bifunctional cross-linkers that include the same reactive groups at the ends of modular spacers of increasing span (Zhang et al. 2008). The detection of nucleotides bridged by reagents with widely different spans signals the ability of these bases to assume widely different positions in 3D space, which could be attributed to the possible occurrence of alternative folds or kinetic traps.

The outcome of probing reactions is also influenced by the amount of probe employed relative to the available substrate, which can be expressed as probe to substrate ratio (*P*/*S*). It is generally accepted that ideal *P*/*S* values should lead on average to only one modification per three substrate molecules, which would statistically rule out the potential that the same molecule might be hit twice and might possibly exhibit secondary modifications prompted by structure distortion. Unfortunately, a priori calculations of "single-hit" concentrations are complicated by factors that influence general chemical reactivity, in addition to those affecting substrate dynamics. Viable *P*/*S* values can be approximated by considering the total number of susceptible bases in the sequence regardless of their possible steric situation (Yu and Fabris 2004; Zhang et al. 2006), the typical reactivity of the selected probe (Yu and Fabris 2004; Zhang et al. 2006), and the presence of buffers that may interfere with the reaction (Richter et al. 2004). Experimentally, baseline conditions for tackling the target substrate could be initially obtained by employing a known analog with very similar characteristics, which would facilitate the evaluation of individual factors. Titration schemes should be devised in which the *P*/*S* is progressively increased and the number of modifications is determined directly by MS analysis. In this way, probe-induced disruption of the native RNA structure would be signaled by readily recognizable jumps in the total number of adducts. These types of strategies can be employed to safely increase the attainable yields of modified products beyond the limits prescribed by single-hit statistics. Boosting adduct production is expected to facilitate the application of MS approaches to samples of limited availability, overcoming the absence of the advantageous amplification effects that are characteristic of primer extension platforms.

17.3.2 Weighing the Validity of Probing Information

The possible occurrence of probe-induced artifacts can be also evaluated during the processes of structure refinement and validation, which typically take place after at least one round of modeling has been completed. This type of evaluation is accomplished, for example, by weighing the reproducibility of each experimental constraints and its consistency with other constraints and the overall 3D model. If a cross-link between two domains is not observed under all conditions and conflicts with other constraints, this is an indication that the structure is subject to prominent conformational effects. The consistent detection of cross-linked conjugates across discrete domains supports the possible presence of a long-range tertiary interaction that stabilizes their placement in mutual proximity. In this case, mutating key nucleotides so as to abolish the putative interaction can serve to validate the corresponding constraints. Confidence that the observed cross-link reflects a real contact in the wild-type structure increases when the cross-link is abolished in the mutant sequence.

Although these types of experimental controls test actual structural features, their outcome clearly reports on the validity of the corresponding spatial constraints that lead to such structure. In general, information associated with cross-linking products that are invariably detected under a broad range of experimental conditions carry higher confidence than products observed only under more narrowly defined conditions, which may not represent the most populated fold. Variability associated with possible substrate dynamics should still be noted as valuable information about the stability of the putative structure. Sensitivity to environmental conditions, reproducibility at different probe concentrations, and mutagenesis confirmation could represent the basis for scoring algorithms designed to rank the experimental constraints and to guide their selection for subsequent modeling operations (Yu et al. 2008a). Combined with other statistics pertaining to the modeling operations, these scores could be employed to express the confidence level in the final model and to compare the quality of MS3D structures.

17.4 Combining Chemical Probing with MS-Based Technologies

The intrinsic character of chemical probing, substantiated by the introduction of stable covalent modifications with unique mass signatures, presents an excellent fit for MS-based strategies. Indeed, treating target substrates with these types of reagents resembles exposing photographic film to light. The process leaves a permanent "impression" of the substrate structure in the form of a modification pattern specific to the 3D fold, which can be subsequently "developed" by any suitable means that do not necessarily have to preserve the original fold. This favorable feature has significant consequences on the applicability of probing approaches by allowing for the utilization of the widest possible range of analytical procedures for product characterization, without considering any associated denaturing effect. For example, the possibility of submitting probed material to a wide choice of hydrolytic operations, which are typically performed to obtain the smaller products necessary to map the modified positions, makes species of virtually any size accessible to structural elucidation. The possibility of performing denaturing separation procedures offers the ability of completing probing reactions in the presence of virtually any type of ligand, salt, or additive that may be required to ensure the correct folding of the target species, without jeopardizing subsequent MS characterization. The possibility of devising specific extraction protocols based on affinity capture and similar strategies, which enables one to isolate and concentrate RNA molecules of interest from very heterogeneous samples, could support in vivo applications calling for the analysis of whole cell lysates or other complex biological extracts.

17.4.1 MS Analysis of Probing Products

The strategies employed to identify modifications produced by chemical probes rely on either matrix-assisted laser desorption ionization (MALDI) (Karas et al. 1987; Tanaka et al. 1987) or electrospray ionization (ESI) (Aleksandrov et al. 1984; Yamashita and Fenn 1984), both of which have become the standard techniques for the characterization of biomolecules. Numerous reviews have been dedicated over the years to the MS analysis of nucleic acids, to which we refer the interested reader (Crain 1990; Limbach 1996; Murray 1996; Nordhoff et al. 1996; Hofstadler et al. 2005; Fabris 2010). It is important, however, to highlight the fact that MALDI applications are made possible by the utilization of suitable matrices that minimize possible fragmentation during the desorption process. These matrices are different from those employed for protein analysis, and their selection depends on the type of laser available. When UV lasers are utilized, typical matrices are 3′-hydroxypicolinic acid (Wu et al. 1993), picolinic acid (Tang et al. 1994), 2′, 4′, 6′-trihydroxyacetophenone (Pieles et al. 1993), and 6-aza-2-thiothymine (Lecchi et al. 1995). Instead, succinic acid and urea are typically employed with IR lasers (Nordhoff et al. 1992). The presence of phosphate groups in the biopolymer backbone, which can afford net negative charges, makes oligonucleotides better suited for analysis in negative-ion mode, but positive-ion spectra can be also obtained with lower sensitivity. Owing to the limited charging involved with the MALDI process, the size of analytes accessible by this ionization technique is greatly influenced by the available mass analyzer. For example, taking advantage of the virtually unlimited mass range afforded by time-of-flight (TOF) (Cotter 1997) analyzers, species comprising up to 2,180 nucleotides (~673 kDa nominal mass) were determined with better than 1% accuracy and low femtomole sample consumption (Berkenkamp et al. 1998).

Comparable sensitivity levels can be reached by ESI-MS analysis, especially when flow rates in the nanoL/min range are employed in the so-called nanospray mode (Wilm and Mann 1996). Further, the multiply charged character of electrosprayed ions makes very large nucleic acids readily amenable to analyzers of limited mass range. This is due to the fact that a certain ion with mass m and charge z will be detected with progressively lower m/z ratios as charge increases. Taking advantage of this characteristic, ions produced by the coliphage T4 DNA (~340,000 nucleotides and ~1.1×10^8 Da nominal mass) were successfully observed in the 2,700–3,700 m/z mass range (Chen et al. 1995) by Fourier transform ion cyclotron resonance mass spectrometry (FTICR-MS) (Comisarow and Marshall 1974; Hendrickson et al. 1999). An additional favorable feature of ESI is the ability to handle samples directly in solution, which enables direct online interfacing with separation techniques, such as high-performance liquid chromatography (HPLC) (Pomerantz and McCloskey 1990; Apffel et al. 1997). This characteristic is particularly advantageous for the analysis of complex sample mixtures.

17.4.2 Selecting Proper Sample Handling Procedures

A typical challenge faced by the MS analysis of nucleic acids is posed by the presence in solution of metal counterions, such as Na^+, K^+, and Mg^{2+}. Upon transfer to the gas phase, these cations tend to form stable adducts of unpredictable stoichiometry, which may significantly reduce the observed resolution and signal-to-noise ratio (S/N) to the point where the signal may be completely suppressed (Nordhoff et al. 1996). The irreversible nature of the bonds formed by structural probes favors the implementation of a wide variety of strategies for overcoming the potential hurdle. A common remedy consists of replacing metal cations with the more volatile ammonium ion (Stults et al. 1991; Pieles et al. 1993), which dissociates in the gas phase into NH_3 and H^+ and results in the formal neutralization of a negatively charged phosphate group (Amad et al. 2000). Adequate ammonium replacement/desalting is achievable by performing ethanol precipitation (Stults et al. 1991), adding sequestrating agents (Limbach et al. 1995; Muddiman et al. 1996; Turner et al. 2008), or using ion-exchange resins (Nordhoff et al. 1992), reversed phase HPLC (Little et al. 1995), and ultrafiltration or microdialysis (Liu et al. 1996; Xu et al. 1998; Hannis and Muddiman 1999). The utilization of salt-free solvents, materials, and plasticware reduces the risk of reintroducing unwanted counter-ions during downstream sample handling. Furthermore, these procedures are not only capable of minimizing the adverse effects of metal cations on analytical performance but can also achieve effective reaction quenching by eliminating unreacted reagent in solution. While some of these procedures may clearly lead to structure denaturation, none is typically expected to induce degradation of probe adducts and consequent loss of information.

The desalting levels achieved by the different methods are comparable across the board; however, sample recovery may vary widely. The fact that HPLC can be directly coupled with MS instruments offers the additional advantage of minimizing possible sample losses and nuclease contamination caused by handling operations. LC-MS analysis of nucleic acids is typically performed by ion-pairing chromatography, which uses tetraethylammonium (TEA) as a pairing agent and hexafluoroisopropanol (HFIP) as an organic additive for increased spray stability (Apffel et al. 1997; Azarani and Hecker 2001). Recently, the specific interactions between phosphate groups and stationary phases containing titanium dioxide (TiO_2) have been exploited to separate nucleic acid analytes and to achieve selective enrichment of peptide–RNA conjugates obtained from cross-linked ribonucleoproteins (Richter et al. 2009). In general, the absence of a front-end separation step increases the demands on sample preparation and on the resolving power of the MS platform. In this case, desalting should be performed as early as possible in the experimental workflow, while minimizing the utilization of any non-volatile components. Although MS-friendly divalent substitutes are not currently available for Mg^{2+}, higher concentrations of ammonium have proven capable of preserving the stability and binding properties of nucleic acid substrates (Hagan and Fabris 2003; Fisher et al. 2006). The ability to work with samples containing up to 2.5 M ammonium acetate (Gapeev et al. 2009), which far exceed the ionic strength observed in the vast majority of physiological environments, offers great flexibility for the successful MS analysis of oligonucleotides.

17.4.3 Strategies for Identifying Probed Nucleotides

The position of modified nucleotides is typically obtained through bottom-up approaches, in which the material is digested with specific nucleases to provide smaller products that are more suitable for mass mapping and sequencing (Fig. 17.4) (Fabris 2010). Alternatively, top-down approaches that dispense with specific hydrolysis can be followed to obtain direct sequence information by submitting the probed material to tandem mass spectrometry (MS/MS) (McLafferty 1981). The inclusion of a hydrolytic step, however, significantly facilitates the structural analysis of the larger targets, whereas direct MS/MS analysis can be hampered by the reduced efficiency of gas-phase fragmentation manifested by substrates with progressively greater numbers of bonds. Mapping experiments that provide the accurate mass of digestion products are frequently sufficient to infer the position of probed nucleotides, knowing the base specificity of probing and hydrolysis reagents. For example, if a G-specific probe is employed and the observed product contains only one G, then the position of a single adduct is all but certain. More frequently, however, multiple susceptible bases are present in the same hydrolytic product and, therefore, the adduct position cannot be identified with certainty solely from mapping data. In this case, MS/MS must be applied to obtain the sequence information necessary to solve the ambiguity.

In a typical MS/MS experiment, the precursor ion of interest is isolated in the mass analyzer from all the other ions present in the mixture, in what is called

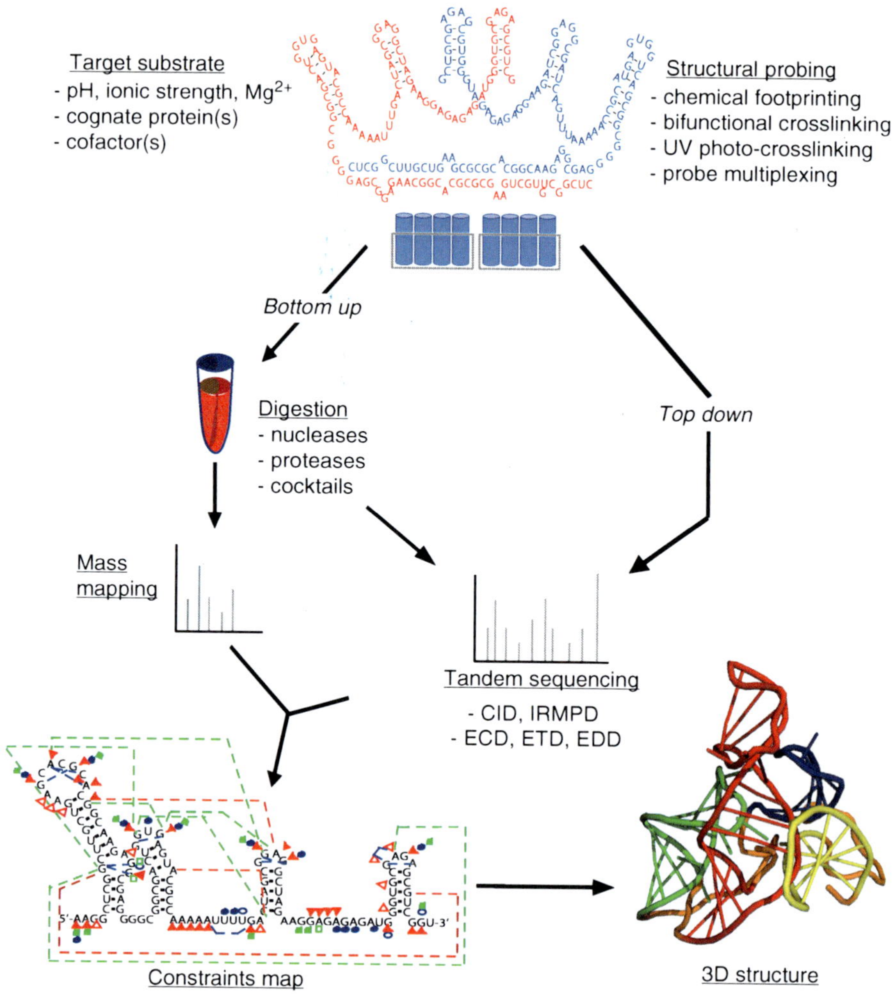

Fig. 17.4 General workflow for 3D structure determination of nucleic acids based on structural probing and MS analysis (MS3D). The substrate is probed under ideal conditions preserving its native fold. Characterization of ensuing covalent adducts can be performed under denaturing conditions, following either bottom-up or top-down approaches. The positions of probed nucleotides provide spatial constraints that are summarized on 2D maps, from which a complete, all-atom 3D structure can be generated through established molecular modeling protocols. Adapted with permission from Fabris (2010)

selection step (McLafferty 1981). Its gas-phase fragmentation is then activated by imparting energy in the form of collisions with inert gas, irradiation with infrared photons, or interactions with electrons or other ions (Woodin et al. 1978; Gauthier et al. 1991; Zubarev et al. 1998). The ensuing dissociation processes produce characteristic fragment ladders, or ion series, which provide the biopolymer sequence (Hunt et al. 1986; Biemann and Scoble 1987; McLuckey et al. 1991).

(i.e., 251.21 Da, as indicated by the asterisk), whereas d_l-H_2O does not. Proceeding from the other end of the sequence, y_1, y_2, and y_3 do not contain the mass shift, which is present in y_5*. As summarized in the inset using a compact notation, the CMCT adduct can be clearly located onto G2 (arrow). The characterization of cross-linked products is performed in similar fashion, but the interpretation of the relative data may be significantly complicated by the presence of conjugated strands that can independently generate overlapping ion series. While software tools, such as Mongo Oligo Calculator (Rozenski 1999), SOS (Rozenski and McCloskey 2002), and others have been developed to aid the interpretation of spectra obtained from unmodified nucleic acids and their monofunctional adducts, the programs Links and MS2Links (Kellersberger et al. 2004; Yu et al. 2008b) have been designed to support the characterization of conjugates produced by bifunctional cross-linkers. These algorithms can handle not only RNA–RNA products but also DNA–DNA, protein–protein, and any of their heteroconjugate combinations. Links and MS2Links programs are freely available to the public through the ms3d.org portal, which was established to support a growing community of investigators interested in alternative approaches to structural determination (Yu et al. 2008b).

17.5 Translating Probing Information into Full-Fledged 3D Structures

The strategy followed to generate 3D models is dictated by the types of experimental data and prior structural information that may be available. A range of software tools can be employed to support the different operations, but no individual package currently integrates data interpretation, constraint assessment, and model generation. Although steps in this direction are underway (Yu et al. 2008b), there is still no replacement for direct experience with the overall process.

17.5.1 Rationalizing Probing Information

Monofunctional probes discriminate nucleotides exposed to the solvent from those that may be protected by pairing, ligand binding, or other steric effects. At first sight, the value of this type of information may seem limited, but careful examination of unreacted/protected stretches can reveal mutually complementary sequences capable of defining stable elements of secondary and tertiary structure. The process can be aided by computational tools, such as MFold (Zuker 2003), MC-Fold (Parisien and Major 2008), PKNOTS (Rivas and Eddy 1999), ILM (Ruan et al. 2004), and many others, which were developed for assessing the folding of predefined sequences according to rigorous thermodynamics considerations and statistical information. In contrast, bifunctional cross-linkers identify bases that are

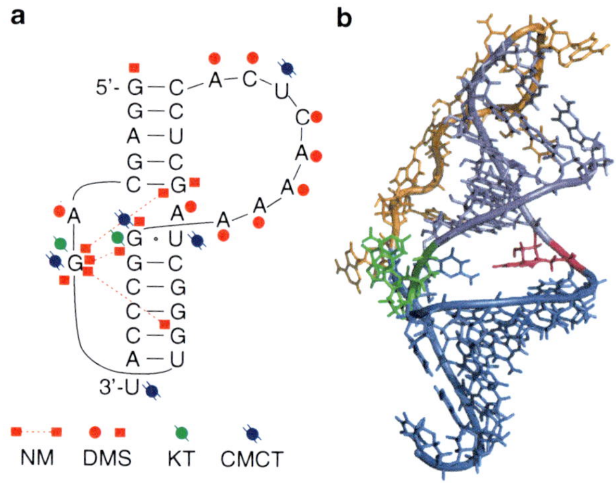

Fig. 17.7 (**a**) Modifications map and (**b**) MS3D structure of mouse mammary tumor virus (MMTV) ribosome frameshifting pseudoknot. *Circle* and *square* symbols indicate modifications of the Watson–Crick or Hoogsteen edge, respectively. The coordinates of atoms in the double-stranded regions provided an average of ~3 Å root mean square deviation (rmsd) from the coordinates of the corresponding atoms in the NMR structure. Adapted with permission from Yu et al. (2005)

placed within a certain distance from one another, which is determined by the probe span. Considering that this information can be utilized directly in modeling operations, its value can be comparable to that of distance determinations provided by nuclear Overhauser effect spectroscopy (NOESY) or Förster resonance energy transfer (FRET) techniques. Additionally, this information can be employed indirectly to corroborate the interpretation of results afforded by monofunctional probes. This feature is particularly important for larger RNA constructs in which partially complementary sequences may anneal off-register or produce alternative pairing patterns, which may lead to ambiguous footprinting data. By conjugating complementary strand segments, bifunctional cross-links delimit the placement of the conjugated bases to within a determined distance that can only match a well-defined annealing register. In this way, the concerted application of mono- and bifunctional probes can provide experimental confirmation of base-pairing patterns predicted by folding algorithms.

A first step for rationalizing probing data involves drawing graphic representations of the spatial relationships between nucleotides in the target sample. Detailed 2D maps can be obtained to summarize the inferred base pairings and observed cross-linked positions in the context of putative secondary structures. For example, Fig. 17.7a summarizes the probing results for the mouse mammary tumor virus (MMTV) ribosome frameshifting pseudoknot (Yu et al. 2005). This graphic representation was completed by placing segments that did not exhibit detectable reactivity in double-stranded structures, in keeping with the notion that bases involved in putative Watson–Crick pairs are generally unreactive to chemical probes targeting the Watson–Crick edges. The only exceptions are bases at the ends of helices or crossover sites. Conversely, nucleotides that showed susceptibility to chemical probes were placed in single-stranded loops and hinge regions, which are readily accessible by the selected reagents. Additional reactivity observed at the end of double-stranded stems is consistent with typical breathing dynamics. The functional groups that were bridged by bifunctional probes are connected by dashed lines to imply that those points are situated within a certain distance in 3D space. At the end, the graphic representation obtained from this exercise in rationalization matched very closely the putative pseudoknot structures calculated by the folding algorithms PKNOTS (Rivas and Eddy 1999) and pknotsRG (Reeder et al. 2007), which contributed additional confidence to the proposed data interpretation.

17.5.2 Generating Full-Fledged 3D Structures

This type of modification map constitutes an excellent starting point for generating full-fledged 3D structures. Base-pairing relationships and distance information can be processed directly by constraint satisfaction algorithms to generate all-atom structures, or employed as filter for selecting the best possible model from a pool of plausible structures, or decoys, produced by computational methods. In the case of the MMTV frameshifting pseudoknot, we employed the algorithm included in the Macromolecular Conformations by SYMbolic programming (MC-SYM) suite (Major et al. 1991, 1993, Parisien and Major 2008) to obtain initial models for subsequent refinement. The program is capable of building polynucleotide structures from fundamental RNA information, such as chemical structure, bond distances, torsion and dihedral angles, etc., which was extracted from a large number of high-resolution structures obtained by established techniques. From an input consisting of the sequence of the MMTV construct, 11 base-pairing relationships confirmed by structural probing, and 3 additional distance constraints highlighted in Fig. 17.7a, MC-SYM produced a small set of possible models. Their examination revealed that the double-stranded stems were indistinguishable from one another, whereas the single-stranded loops presented rather large variability. The initial models were submitted to rounds of simulated annealing and Cartesian molecular dynamics (MD) by using the modules *anneal.inp* and *minimize.inp* of the crystallography and NMR system (CNS) suite (Brünger et al. 1998). These calculations were restrained by standard backbone torsion angles, planarity, and hydrogen-bonding information included in the modified CharmM force field employed by CNS. The minimized models were averaged together to provide the final structure shown in Fig. 17.7b, which includes an unpaired purine at the stems hinge and exhibits all the characteristic features of a classic type-1 pseudoknot (Yu et al. 2005).

Generating complete models is greatly facilitated by the availability of partial structural information from other sources. This concept is exemplified by the elucidation of the HIV-1 packaging signal (Ψ-RNA), an ~120-nt region of viral RNA involved in genome recognition, dimerization, and packaging (Yu et al. 2008a). While size and flexibility considerations have thus far precluded its comprehensive determination by NMR or crystallography, high-resolution structures have been separately obtained for four putative stem–loops folded by contiguous stretches of the Ψ-RNA sequence. Initial probing experiments revealed a global morphology characterized by well-defined domains that matched the previously described stem–loops, thus ruling out the possible formation of different secondary structures by alternative folding of the full-length construct. This observation allowed us to utilize the high-resolution coordinates of the individual stem–loops, which are available in the Protein Data Bank (PDB) (Berman et al. 2000), as the initial building blocks for assembling a full-length model. Single-stranded regions connecting contiguous domains, for which no high-resolution information was available, were modeled de novo by MC-SYM, as described. The assembly operation was accomplished by using the *merge_structure.inp* module of CNS. Spatial constraints provided by bifunctional probes were utilized to triangulate the proper placement of the separate sections in 3D space. In particular, a total of 17 interdomain and 29 intradomain cross-links were used to restrain rounds of simulated annealing performed in CNS. The distance information provided by the cross-linking experiments was input to *anneal.inp* by using the restraint set file that is typically employed for NOESY constraints. The restraints were enforced throughout the energy minimization process but were lifted in final rounds to relieve modeling strain and to test whether the structure would maintain its overall morphology. The resulting model possessed the atomic-level detail afforded by the initial high-resolution structures and the global morphology determined by the cross-linking data (Fig. 17.8) (Yu et al. 2008a).

17.5.3 Model Evaluation

Once a structure has been obtained, evaluation and validation can be accomplished according to standard methods. In the case of substrates for which high-resolution structures are available, optimal superimposition of the respective coordinates enables direct comparisons by calculating their average root mean square deviation (rmsd). In the example of the MMTV frameshifting pseudoknot (Fig. 17.7b), the coordinates of all the atoms present in double-stranded regions provided an average rmsd of ~3 Å from the coordinates of the corresponding atoms in the NMR structure deposited in the PDB (Shen and Tinoco 1995). This value of rmsd falls within the range matched by similar comparisons between high-resolution structures obtained for the same substrate by different instrumentation and data processing methods. It is important to note that the coordinates of flexible single-stranded loops were omitted from the rmsd calculation because, in the original NMR report, these regions had yielded an insufficient number of constraints to support confident assignment. The omission helped minimize an intrinsic drawback of average rmsd as a measure of fitness, which is represented by its tendency of spreading the error over the entire molecule while failing to identify the actual sources of discrepancy between structures. For this reason, new metrics have been recently introduced to better account for local deviations, intradomain deformations, and interdomain discrepancies (Parisien et al. 2009), which should not only enable more

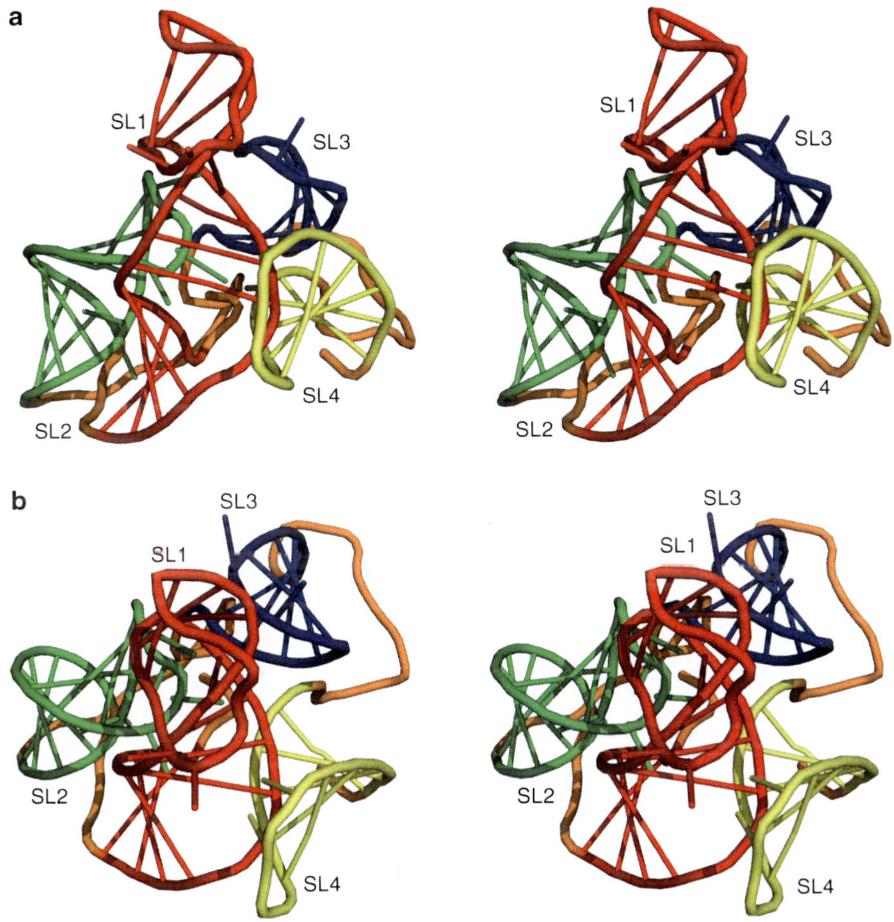

Fig. 17.8 Stereoview diagrams of full-length Ψ-RNA viewed from the side (**a**) and top (**b**). *Red*, SL1; *green*, SL2; *blue*, SL3; *yellow*, SL4. Linker sequences between adjoining stem–loops are *orange*. The relatively compact cloverleaf conformation is stabilized by a long-range tertiary interaction between the apical loop of SL4 and the upper stem region of SL1. Reproduced with permission from Yu et al. (2008a)

meaningful comparisons but could also help the refinement process by guiding the design of additional probing experiments.

Direct comparisons with high-resolution structures are certainly helpful for assessing the quality of structures obtained by MS-based approaches and, by extension, for evaluating the performance of such approaches. However, this option is not feasible for rating structures of species that have not been solved by established techniques, i.e., the expected targets of alternative technologies. In this case, structures must be evaluated on their own merits according to objective criteria, including their conformity to the general knowledge of RNA structure extracted by surveying the growing collection of available high-resolution data.

Aberrant conformations and geometric violations of normal RNA structure decrease the confidence level in the final results. A second set of criteria regards internal consistency of the experimental data employed to generate it. In this case, the confidence level will be increased by the absence of conspicuous steric clashes and specific structural features that may contradict the actual probing data. Models that place cross-linked nucleotides out of mutual range, or insert hindering structures in the space between conjugated bases, do not faithfully reflect the initial experimental input and raise doubts about the validity of the model or the corresponding spatial constraints. When planning this type of analysis, it is helpful to exclude a subset of experimental constraints from model generation, reserving them only for verification purposes. In the case of Ψ-RNA, an additional set of 16 inter and intradomain constraints were reserved for checking the final structure for possible inconsistencies. The majority of these cross-links were found to be consistent with the structural context, but a few bridged bases located within range only when the intrinsic flexibilities of both probe and RNA structure were taken in account and only one bridged positions that were completely out of range (Yu et al. 2008a). While it is evident that the confidence level can be greatly increased by these types of qualitative observations, only the development of an actual algorithm could provide the means for obtaining unambiguous quantitative assessments. In combination with typical statistical assessments associated with modeling operations, this tool could provide the basis for a comprehensive evaluation system, which should account also for the number, type, and quality of experimental data to reach a more complete evaluation of structures obtained from chemical probing approaches.

17.6 Connecting the Dots

While RNA secondary structure is predominantly defined by the Watson–Crick pairing of complementary bases, non-Watson–Crick base pairs involving also the Hoogsteen and sugar edges contribute to interactions determining tertiary and quaternary structure. Widely recurring structural motifs, such as base triples/quadruples, platforms, ribose zippers, loop–receptor interactions, and others, have been recognized as modular building blocks that shape RNA architecture through long-range contacts (Batey et al. 1999; Hermann and Patel 1999; Leontis et al. 2006). Ideally, the development of monofunctional reagents targeting the functional groups that are not involved in Watson–Crick pairing could enable direct correlations between distinctive footprinting patterns and corresponding structural elements, thus leading to their direct determination by structural probing. Development and characterization of probes that react preferentially with functional groups on the Hoogsteen or sugar edges of each nucleotide would be most useful in this regard. Future exploration of this type of approach will count on the increasing availability of suitable reagents made possible by the implementation of MS-based detection.

A step toward recognizing the presence of recurring long-range contacts is afforded by bifunctional cross-linkers, which are capable of identifying their constitutive components by bridging structures in mutual proximity. In the case of Ψ-RNA, for example, numerous cross-links placed the single-stranded loop of the stem–loop 4 (SL4) domain alongside the stem of stem–loop 1 (SL1) (Yu et al. 2008a). The fact that SL4 assumes a typical GNRA tetraloop structure (Amarasinghe et al. 2001; Kerwood et al. 2001) suggested that its contact with SL1 may consist of a GNRA loop–receptor interaction (Cate et al. 1996; Costa and Michel 1997). Based on the loop structure and the effects of Mg^{2+} on cross-linking yields, we hypothesized that SL1 may fit the mold of the broad-spectrum class II receptors indentified by in vitro selection methods (Geary et al. 2008). For these reasons, we employed one of these receptors as a template for homology modeling the putative Ψ-RNA interaction (Fig. 17.9). The resulting structure was tested experimentally by replacing A345 with C, in such a way as to eliminate specific hydrogen bonding and stacking interactions that contributed to the stability of the long-range contact. The mutation induced the loss of the specific interdomain cross-links observed for the wild-type construct, thus providing strong experimental support to the proposed interaction (Yu et al. 2008a).

The structural elucidation of Ψ-RNA exemplifies the possible synergies between chemical probing and computational approaches. In fact, cross-linking information has the potential to facilitate the stitching together of building blocks provided by high-resolution techniques or prediction algorithms to build viable 3D models in modular fashion. Structural elements indentified as putative interacting partners could be used to search a database of recurring interactions compiled from the ever

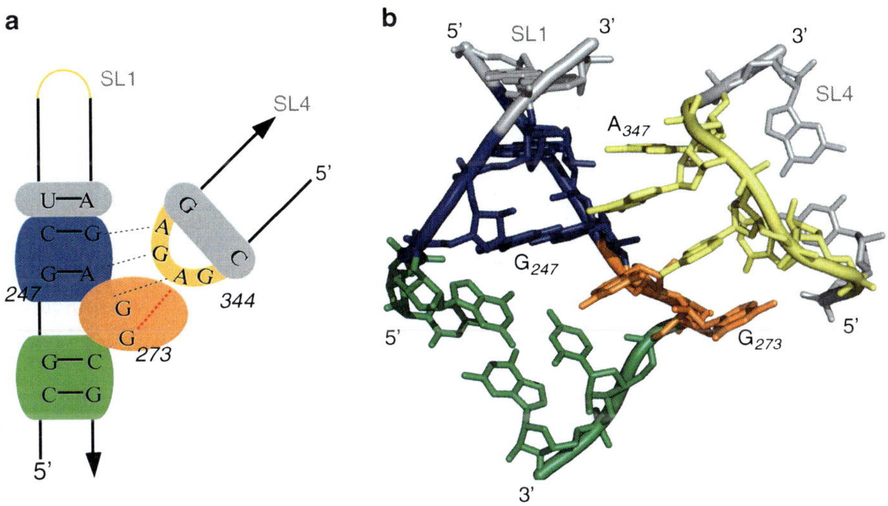

Fig. 17.9 Loop–receptor interaction in Ψ-RNAs. (**a**) Diagram showing the placement of the SL1 and SL4 domains in the proposed interaction. (**b**) All-atom model detailing the specific molecular contacts. Reproduced with permission from Yu et al. (2008a)

expanding knowledge of RNA structure. Recognized motifs could then constitute convenient templates for homology modeling or become the starting modules for assembling the structure of interest. Alternatively, probing data could serve to rank decoys generated from sequence information by prediction algorithms, course-grained modeling, or other strategies. Selecting the decoy that best fits the experimental information would be made more stringent by the inclusion of pairwise constraints afforded by bifunctional reagents. In turn, computational methods could help fill the resolution gap by providing atomic-level details that are typically beyond the reach of chemical probing approaches. At the end, these synergistic strategies would be expected to produce full-fledged models that comply with rigorous thermodynamic/statistical principles of RNA structure and, at the same time, maintain a firm grounding in direct empirical observations. Only a sound confluence of both theoretical and experimental knowledge can provide final structures possessing the confidence level required by the elucidation of previously unsolved substrates.

17.7 Conclusions

The examples discussed here offer a glimpse of the great potential afforded by MS3D for structural elucidation. The effectiveness of chemical probing approaches will be expected to grow with the continued development of MS technologies and computational methods. From the experimental point of view, the broader implementation of MS detection will lead to greater utilization of bifunctional probes and to the introduction of new reagents, which are not usable with traditional technologies. Owing to the suitability of this analytical platform for samples of heterogeneous nature, this approach will find widespread application to the structural elucidation of ribonucleoprotein assemblies inaccessible by established techniques. The capacity of MS-based technologies to handle very complex sample mixtures will spur an expansion toward in vivo applications. As the ability of supporting experimental strategies to minimize or eliminate the pitfalls of chemical approaches will improve, the reliability of the information afforded by structural probes will also improve. Improvements in our abilities to accurately determine diverse spatial constraints will stimulate the development of new computational strategies for their effective utilization. Improved methods are also needed to assess the quality of these constraints and to take this information into account when evaluating the final structure. We envision that these advances will help establish MS3D as the approach of choice for the structural elucidation of progressively larger substrates immersed in their natural cellular environments.

Acknowledgments This work was supported by National Institutes of Health Grant GM643208 and National Science Foundation Grant CHE-0439067.

Wilkinson KA et al (2006) Selective 2′-hydroxyl acylation analyzed by primer extension (SHAPE): quantitative RNA structure analysis at single nucleotide resolution. Nat Protoc 1(3):1610–1616

Wilkinson KA et al (2008) High-throughput SHAPE analysis reveals structures in HIV-1 genomic RNA strongly conserved across distinct biological states. PLoS Biol 6(4):e96

Wilm M, Mann M (1996) Analytical properties of the nanoelectrospray ion source. Anal Chem 68:1–8

Winkler W et al (2002) Thiamine derivatives bind messenger RNAs directly to regulate bacterial gene expression. Nature 419(6910):952–956

Woodin RL et al (1978) Multiphoton dissociation of molecules with low power continuous wave infrared laser radiation. J Am Chem Soc 100:3248–3250

Wu KJ et al (1993) Matrix-assisted laser desorption time-of-flight mass spectrometry of oligonucleotides using 3-hydroxypicolinic acid as an ultraviolet-sensitive matrix. Rapid Commun Mass Spectrom 7(2):142–146

Xu N et al (1998) A microfabricated dialysis device for sample cleanup in electrospray ionization mass spectrometry. Anal Chem 70(17):3553–3556

Yamashita M, Fenn JB (1984) Electrospray ion source. Another variation on the free-jet theme. J Phys Chem 88:4671–4675

Young MM et al (2000) High throughput protein fold identification by using experimental constraints derived from intramolecular cross-links and mass spectrometry. Proc Natl Acad Sci USA 97(11):5802–5806

Yu E, Fabris D (2003) Direct probing of RNA structures and RNA protein interactions in the HIV-1 packaging signal by chemical modification and electrospray ionization Fourier transform mass spectrometry. J Mol Biol 330(2):211–223

Yu ET, Fabris D (2004) Toward multiplexing the application of solvent accessibility probes for the investigation of RNA three-dimensional structures by electrospray ionization – Fourier transform mass spectrometry. Anal Biochem 344:356–366

Yu ET et al (2005) Untying the FIV frameshifting pseudoknot structure by MS3D. J Mol Biol 345:69–80

Yu ET et al (2008a) MS3D structural elucidation of the HIV-1 packaging signal. Proc Natl Acad Sci USA 105:12248–12253

Yu ET et al (2008b) The collaboratory for MS3D: a new cyberinfrastructure for the structural elucidation of biological macromolecules and their assemblies using mass spectrometry-based approaches. J Proteome Res 7(11):4848–4857

Zaia J et al (1996) A binding site for chlorambucil on metallothionein. Biochemistry 35(9):2830–2835

Zhang Q et al (2006) Toward building a database of bifunctional probes for the MS3D investigation of nucleic acids structures. J Am Soc Mass Spectrom 17:1570–1581

Zhang Q et al (2008) Nested Arg-specific bifunctional crosslinkers for MS-based structural analysis of proteins and protein assemblies. Anal Chim Acta 627:117–128

Zubarev RA et al (1998) Electron capture dissociation of multiply charged protein cations: a nonergodic process. J Am Chem Soc 120:3265–3266

Zuckerkandl E, Cavalli G (2007) Combinatorial epigenetics, "junk DNA", and the evolution of complex organisms. Gene 390(1–2):232–242

Zuker M (1989) Computer prediction of RNA structure. Methods Enzymol 180:262–288

Zuker M (2003) Mfold web server for nucleic acid folding and hybridization prediction. Nucleic Acids Res 31(13):3406–3415

Appendix

The bases of RNA, like DNA, can interact edge-to-edge, when arrays of hydrogen-bond donors and acceptors on the interacting bases align appropriately. The resulting, planar arrangements are called base pairs. Analysis of RNA structures shows that each unmodified RNA base presents three edges for H-bond mediated pairing, the Watson–Crick (W), Hoogsteen (H), and Sugar (S) edges (see Fig. A1 panel a). RNA base pairs can therefore be conveniently classified according to the interacting edges of the paired bases. For each pair of interacting edges, two relative orientations of the glycosidic bonds (*cis* or *trans*) are possible (see Fig. A1, panel b), giving rise to 12 geometric base pairs. These are shown schematically in Fig. A1, panel c, using triangles to represent each nucleobase. Each base pair family is named and classified according to the glycosidic bond orientation (*cis* or *trans*) and the interacting edges, as previously described (Leontis and Westhof 2001; Leontis et al. 2002). For example, pairing between the Watson–Crick edge of one base and the Hoogsteen edge of a second base with the glycosidic bonds in *trans* produces a base pair belonging to the *trans* Watson–Crick/Hoogsteen or "tWH" base pair family. The common (AU and GC) Watson–Crick base pairs, as well as the "wobble" (GU or AC) base pairs, belong to the *cis* Watson–Crick/Watson–Crick family, abbreviated as "cWW." Symbols have also been proposed for annotating base pairs in 2D diagrams of RNA structure, using circles, squares, and triangles to represent the interacting Watson–Crick, Hoogsteen, and Sugar edges, respectively (Leontis and Westhof 2001). *cis* base pairs are indicated by using filled symbols and *trans* pairs with open symbols. The geometric base pair classification has proven useful in annotating and analyzing RNA 3D structures and understanding RNA sequence variation and evolution (Leontis et al. 2006; Brown et al. 2009; Stombaugh et al. 2009; Hoehndorf et al. 2011). Given that base triples are sets of three nucleotides interacting by hydrogen bonding, this approach can be applied to systematically group and name base triples (see Fig. A1, panel d) and higher order nucleo-base clusters (Abu Almakarem et al. 2011).

N. Leontis and E. Westhof (eds.), *RNA 3D Structure Analysis and Prediction*, Nucleic Acids and Molecular Biology 27, DOI 10.1007/978-3-642-25740-7,

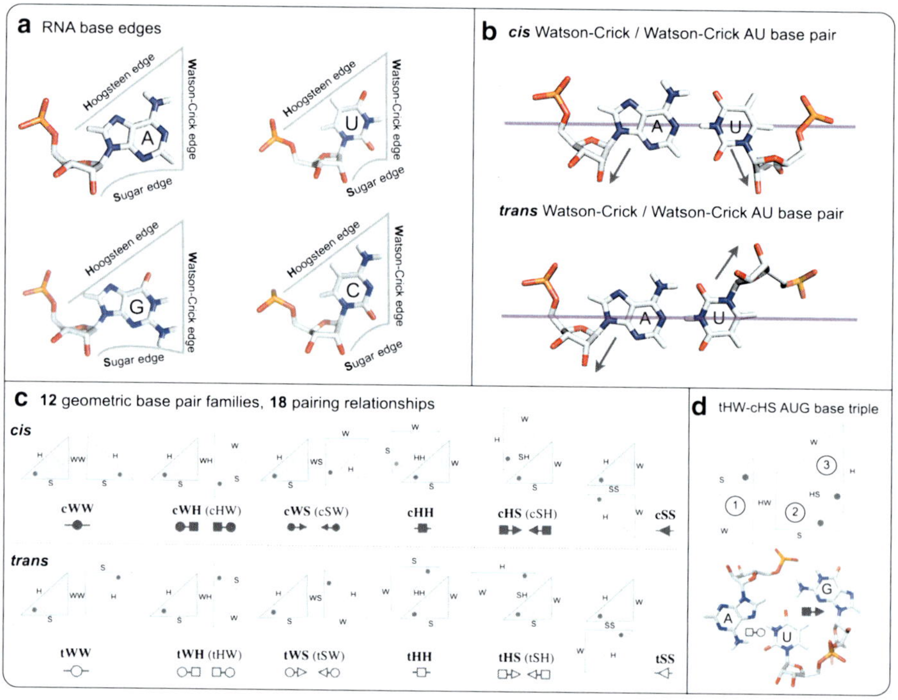

Fig. A1 Summary of Leontis/Westhof base pairing classification. (A) Each unmodified RNA nucleotide presents three edges for base pairing interactions, the Hoogsteen (H), Watson–Crick (W) and Sugar (S) edges. Therefore, nucleobases can be conveniently represented by triangles as shown. Note that the sugar edges include the 2′-OH group of the riboses. (B) For each pair of edges, nucleotides can pair in two distinct ways, designated cis and trans, and related by 180° rotation of one nucleotide about the magenta axis that bifurcates the nucleobases perpendicular to the interacting edges. The glycosidic bonds of the nucleotides are on the same side of this axis in the cis configuration, and on opposite sides in the trans configuration (indicated by arrows). (C) Schematic representations of each of the 12 basic base pair families, using triangles to represent each base. Symbols for annotating secondary structures of RNA with non-Watson–Crick base pairs are also provided. The symbols associate circles with W edges, squares with H edges and triangles with S edges. Filled in symbols represent *cis* base pairs and open symbols, *trans* base pairs. Note that the 12 base pair families result in 18 base pairing relations due to the asymmetry of some base pairs. (D) Schematic showing a representative regular base triple, AUG tHW/cHS. The central base (U), numbered base '2', pairs with each of the other two bases of the triple using a distinct base edge. A is base 1 and G is base 3. The triple is named according to the base pairs formed by bases 1 and 2 (tHW in this case) and by bases 2 and 3 (cHS in this case)

FTICR-MS. *See* Fourier transform ion cyclotron resonance mass spectrometry (FTICR-MS)
Full-atom reconstruction, 84
Funnel, 81–83

G
GA. *See* Genetic algorithm (GA)
GAAA tetraloop, 351
Gag and gag-pol, 203
Gap statistic, 326
Gaussian chain approximation, 191
Gel electrophoresis, 325
Generic distance restraints, 342
Genetic algorithm (GA), 122, 123
Geometric discrepancy, 285
G2G
  accuracy, 348
  global structure, 355
  structure, 348
  toolkit, 340
GNRA tetraloop, 132
Go potential, 232
Graphical processing unit (GPU), 143, 163

H
Hairpin, 45
High-performance liquid chromatography (HPLC), 372
HIV-1 packaging signal, Psi-RNA, 379
Homologous RNA, 287
Homologs, 147
Homology modeling, 7, 10, 67, 70, 71, 92, 94–96, 100, 101, 105, 106, 108, 113
HPLC. *See* High-performance liquid chromatography (HPLC)
H-type pseudoknots, 124, 191
Hybrid rigid-body simulated annealing (SA) refinement protocol, 342
Hydrogen bonds, 52
Hydroxyl radical footprinting, 325
3′-Hydroxypicolinic acid, 371

I
ILM, 377
Incremental mass, 365
Indels, 70, 73, 76
Infernal, 74
Interaction energies, 248, 250, 261–265, 267, 269–271
Intermediate folds, 124
Internal coordinate mechanics, 147
In vitro evolution, 61

K
Kethoxal (KT), 364
Kinetic data, 325
Kinetic model, 323
Kinetic traps, 368
KinFold, 331
Kissing loop complexes, 193
Knowledge-based, 47, 51
KT. *See* Kethoxal (KT)

L
Larger tertiary folds, 206
Lennard-Jones potential, 83
Levinthal, 55
Ligands, 286
Limitations, 52–54
Links, 377
Livebench, 85, 86
Local structural alignment, 294
LocaRNA, 74
Loop entropy, 174–175, 190
Low-resolution, 195
L-21 *Tetrahymena thermophila*, 319

M
Macromolecular Conformations by SYMbolic programming (MC-SYM), 379
MALDI. *See* Matrix-assisted laser desorption ionization (MALDI)
Mass-balance equation, 327
Mass mapping, 373
MathWorks' Matlab, 320
Matrix-assisted laser desorption ionization (MALDI), 371
MC-Fold, 377
MC-Fold/MC-Sym, 51, 72
MC-SYM. *See* Macromolecular Conformations by SYMbolic programming (MC-SYM)
Metal counterions, 372
Metrics for model evaluation, 13–14
mfe structure. *See* Minimum free energy (mfe) structure
Mfold, 125, 126, 377
Minimal salt models, 300
Minimum free energy (mfe) structure, 22

ModeRNA, 67, 72–78, 86
Modifications, 67, 72, 75, 76
Modified nucleosides, 220
Modular cross-linkers, 369
Molecular dynamics, 51, 129, 133, 135, 143
Molecular envelope, 341
Molecular interactions, 241, 245, 251, 253, 254, 257, 269, 272
Molecular mechanics, 133
Molecular signatures, 215
Molecular simulations, 78, 259
Mongo Oligo Calculator, 377
Monte Carlo, 49, 68, 78, 80, 81, 83
Motif library, 101, 103–105, 112, 113
MPGAfold, 119, 120, 122–127, 139
MRM. *See* Multi-resolution modeling (MRM)
ms3d org portal, 377
MS2Links, 377
MS/MS. *See* Tandem mass spectrometry (MS/MS)
Multiplexed, 367
Multiplexed applications, 365
Multi-resolution modeling (MRM), 147
Multiscale approach, 168, 195
Mutation, 123

N

Nanospray, 372. *See also* Electrospray
NAST. *See* Nucleic acid simulation tool (NAST)
Network analysis, 229
Nitrogen mustard (NM), *bis* (2-chloroethyl)-methylamine, 367
N-methylisatoic anhydride (NMIA), 365
NMIA. *See* N-methylisatoic anhydride (NMIA)
NOESY. *See* Nuclear Overhauser effect spectroscopy (NOESY)
Non-redundant, 282
Non-redundant lists, 290
Nuclear Overhauser effect spectroscopy (NOESY), 378
Nucleases, 373
Nucleic acid simulation tool (NAST), 51, 143
Nucleotide modifications, 75
11-Nucleotide motif, 296–297
Nucleotide-to-nucleotide alignment, 294

O

OH footprinting data, 326
OH radicals, 325
OpenMM, 164
ORIENT, 340
Over-packing, 170–172

P

Pair distance distribution function (PDDF), 348
Parallel cluster computer, 123
Partition function, 21
PDDF. *See* Pair distance distribution function (PDDF)
1,4-Phenyl-diglyoxal (PDG), 367
Physics-based, 51
Picolinic acid, 371
PKNOTS, 377, 379
pknotsRG, 379
PMF. *See* Potential of mean force (PMF)
Potential of mean force (PMF), 177
Precision *vs.* accuracy, 11–12
Predicting RNA 3D structure, 202
Probe-induced distortion, 368
Probe to substrate ratio (P/S), 369
-1 programmed ribosomal frameshift, 203
Programs
  DotKnot, 31
  HotKnots, 31
  ILM, 30
  KnotSeeker, 31
  MaxExpect, 25
  mfold, 19, 24
  PKNOTS, 29
  pknotsRG, 29–30
  RNAcast, 32
  RNAfold, 25
  RNAshapes, 24
  UNAfold, 19, 23
Progress curves, 325
Protein folding, 93, 99
Provisional redundancy, 288
P/S. *See* Probe to substrate ratio (P/S)
Pseudoknot, 190, 378, 380
  H-type, 26
  structure and stability, 197–198
Pulse-chase mass spectrometry, 332
PyMOL, 225
Python, 72

Q

Quadruplex, 57
Quantum-chemical computations, 258
Quaternary structure, 95
Query, 95, 105, 106, 112

Printed by Publishers' Graphics LLC
LSI20121220.19.34.340